AIAA AEROSPACE DESIGN ENGINEERS GUIDE

Sixth Edition

This book belongs to

____

AIAA AEROSPACE DESIGN ENGINEERS GUIDE

Sixth Edition

LIBRARY OF FLIGHT

Ned Allen, Editor-in-Chief
Lockheed Martin Corporation
Bethesda, Maryland

American Institute of Aeronautics and Astronautics, Inc.
1801 Alexander Bell Drive
Reston, Virginia 20191-4344

March 2017

American Institute of Aeronautics and Astronautics, Inc., Reston, Virginia

Library of Congress Cataloging-in-Publication Data

AIAA aerospace design engineers guide.—6th ed.
p. cm.—(Library of flight)
Includes bibliographical references and index.
ISBN 978-1-60086-912-9 (alk. paper)
1. Airplanes—Design and construction—Handbooks, manuals, etc. 2. Space vehicles—Design and construction—Handbooks, manuals, etc. I. American Institute of Aeronautics and Astronautics.
TL551.A32 2012
629.1'2—dc23

2012020127

ISBN 978-1-60086-912-9

Foreword

This sixth edition of the *AIAA Aerospace Design Engineers Guide* (ADEG) is a revised version of the previous edition. It has been prepared under the charter of the AIAA Design Engineering Technical Committee to assist the design engineer in creating and defining practical aerospace products. The purpose of this guide is to provide a condensed collection of commonly used aerospace engineering reference data and to also function as a general reference guide for disciplines related specifically to aerospace design.

The previous editions were published in 1983, 1987, 1993, 1998, and 2003, and the guide will be updated whenever the committee has accumulated sufficient material to warrant a new edition. Materials included in the guide were compiled principally from design manuals and handbooks. The Design Engineering Technical Committee is indebted to many people and companies for their voluntary cooperation.

In addition to revisions of previous material, the sixth edition has fully adopted the AIAA policy of using SI (International System of Units) as the primary system of units for its technical publications. Where practical, all reference material, data, formulas, and graphs now use SI units or contain SI units in addition to the units used in previous editions of this guide.

The information in this guide cannot be guaranteed, and it should not be referenced as a final authority for certification of designs.

Comments and suggestions for improvement should be sent to the Design Engineering Technical Committee at DesignGuide@aiaa.org for their consideration. Receipt of all errors and corrections will be acknowledged.

ADEG Subcommittee
AIAA Design Engineering Technical Committee
1801 Alexander Bell Drive, Suite 500
Reston, VA 20191-4344
email: DesignGuide@aiaa.org

Table of Contents

PREFACE

USING THIS GUIDE

The *AIAA Aerospace Design Engineers Guide* has been compiled to assist aerospace design engineers during design inception and development. The purpose of this guide is to serve as a specialized-purpose, in-field handbook used by design engineers to perform back-of-the-envelope/rough-order-of-magnitude estimates and calculations for early preliminary and conceptual aerospace design. The guide is not intended to be a comprehensive handbook for producing highly detailed production designs, although some of the design data may be suitable for the designer's objectives. Other specialized handbooks and detailed company handbooks/manuals and specifications are generally available to the designer to support comprehensive detail and production design efforts.

This guide is divided into 11 major sections, as shown in the Table of Contents, each with a summary topic list at the beginning of the section to help the user quickly and easily find information. The sections provide the following categories of aerospace design engineering information: Sections 1, 2, and 3 provide mathematical definition, conversion factors, and general materials and specifications information. Sections 4, 5, and 6 provide detailed information on section properties, structures, and mechanical design. Sections 7 and 8 provide universal product definition nomenclature in the form of "geometric dimensioning and tolerancing" and electrical, electronic, and electromagnetic design information. Sections 9, 10, and 11 provide aircraft and helicopter, air breathing propulsion, and spacecraft and launch vehicle design information.

The 11 sections of the Design Guide are arranged to maximize useability of design data, which appear in the form of visual and written explanations, tabular data, formulas/equations, graphics, figures, maps, glossaries, standards, specifications, references, and design rules of thumb.

SYSTÈME INTERNATIONAL D'UNITÉS

This modernized version of the metric system uses the international abbreviation SI. Throughout this guide, the primary units of measure will be SI units. Where warranted, Imperial or English units will also be provided. The SI system contains seven base units as in the table below. These base units are further used to develop derived units such as velocity, acceleration, force, pressure, power, and many others required in engineering and science. There are also standards associated with SI units on the proper use of symbols and how they should be displayed in text, formulas, and equations. The reader is encouraged to become familiar with these standards for their proper application.

Base Quantity or "Dimension"	SI Unit	SI Unit Symbol ("Abbreviation"), Use Roman (Upright) Type
length	meter	m
mass	kilogram	kg
time	second	s
electric current	ampere	A
thermodynamic temperature	kelvin	K
amount of substance	mole	mol
luminous intensity	candela	cd

Introduction

Design engineering is fundamental to every aerospace project. The role of the design engineer is the creation, synthesis, iteration, optimization, and presentation of design solutions. It is the primary discipline that creates and transforms ideas into a product definition that satisfies customer as well as business requirements.

This Design Guide provides the design engineer with reference data that can be used during the execution of systems analyses, trade studies, and detailed designs. The integrated product development (IPD) concept leverages the multidisciplined teaming environment to create successful products. The IPD team brings sales, marketing, engineering, manufacturing, quality assurance, and customer service disciplines together, focusing on the value of the product to the customer. Whatever the designer's role in the design process, an understanding of some simple basic systems engineering concepts will ensure more successful application of the information in this Design Guide to create better products.

Systems Engineering Concepts

Understand Customer's Need

Whether the customer is an internal product team or an external client company, it is important to understand its needs. Every product starts with a need. Sometimes product requirements are written that do not reflect the true needs of the customer; if so, they should be revised to reflect those needs. In the end, the value of the design is always measured against the customer's needs; therefore those needs must be accurately captured in the product requirements specification.

Develop Concept of Operations

Most products have multiple users in their lifetimes. Products are tested, deployed, operated, maintained, and eventually retired from service. The requirements of all operational phases must be considered for a design to be successful. Complex systems typically have a formal concept of operations that accompanies the product requirements. If a concept of operations is not included with the product requirements, then one should be developed with the customer. Scenarios for how the product will be used throughout its life cycle should be developed, and the concept of operations and the product requirements should be used as the basis for the design.

Review Product Requirements for Completeness

Product requirements that the design must satisfy likely will be furnished. Use a checklist to verify that the product requirements are complete. Typical requirement categories include 1) performance, 2) lifetime/duty cycle, 3) affordability, 4) reliability, 5) human factors, 6) field support/logistics, and 7) deployment/disposal. If the product requirements are not complete, appropriate assumptions should be made and then validated with the customer. Just because the product requirements within a category are incomplete does not mean that they are not important.

Use Trade Study Methods to Develop the Product Design

After all of the important up-front work necessary for good product design is completed, there are further systems engineering techniques that can help complete the product design. Trade study methodology is an effective way to choose among design alternatives and to develop the product design. Essential elements of a trade study include the following.

Identify design alternatives. If only one design concept is developed, review it with the future users of the product. This may result in additional design alternatives. The level of detail to which the design alternatives are defined varies with the program phase. Early on, high-level concepts are sufficient whereas during full-scale development, detailed designs and operating concepts are required.

Develop evaluation criteria and weighting factors. Top-level evaluation criteria usually include performance, safety, reliability, cost, risk, flexibility, and growth. Not all of these will apply in every case. Detailed criteria should be tailored to a particular situation. For example, under the top-level criterion of performance, detailed criteria might include weight, volume, and/or power consumption. If cost is an evaluation criterion, be sure to consider the total life-cycle cost. Finally, weighting factors should be developed based on the relative importance of the evaluation criteria. Work closely with the customer to develop the evaluation criteria and appropriate weighting factors.

Analyze design alternatives and select final design concept. Each design alternative should be analyzed and scored against the evaluation criteria, applying the weighting factors and calculating numerical scores for each alternative. Risk mitigation approaches should be developed for the key risk areas. Sensitivity studies should be performed for the highest ranked alternatives to ensure that small changes in one design parameter do not cause major changes in the other parameters. The design alternative with the highest score and acceptable sensitivities is the preferred solution. For preferred solutions that involve high risk, developing a second design alternative in parallel with the first should be pursued until the feasibility of the preferred design is demonstrated.

Benefits of a systems engineering approach. A systems engineering approach is recommended for any aerospace design project. The systems engineering philosophy integrates and links requirements, schedule, cost, design solutions, alternative concepts, risk considerations, and product verification and validation in a manner that enables the design team to have adequate visibility into and accountability of the complete development effort.

In summary, design engineering is the creative process by which ideas from one or many contributors are converted to documents that define a product that can be profitably manufactured and that meets the design, performance, and functional specifications required. Design engineering seeks an optimal whole, rather than attempting to perfect each individual part within a system, thus obtaining a balanced, well-designed product that fulfills the requirements and satisfies customer and business needs.

AIAA Systems Engineering Technical Committee
1801 Alexander Bell Drive, Suite 500
Reston, VA 20191-4344

Section 1

Mathematics

GREEK ALPHABET

Capital	Lowercase	Letter	Capital	Lowercase	Letter
Α	α	Alpha	Ν	ν	Nu
Β	β	Beta	Ξ	ξ	Xi
Γ	γ	Gamma	Ο	o	Omicron
Δ	δ	Delta	Π	π	Pi
Ε	ε	Epsilon	Ρ	ρ	Rho
Ζ	ζ	Zeta	Σ	σ	Sigma
Η	η	Eta	Τ	τ	Tau
Θ	θ	Theta	Υ	υ	Upsilon
Ι	ι	Iota	Φ	ϕ	Phi
Κ	κ	Kappa	Χ	χ	Chi
Λ	λ	Lambda	Ψ	ψ	Psi
Μ	μ	Mu	Ω	ω	Omega

SI PREFIXES

Multiplication Factor	Prefix	Symbol
1 000 000 000 000 000 000 000 000 = 10^{24}	yotta	Y
1 000 000 000 000 000 000 000 = 10^{21}	zetta	Z
1 000 000 000 000 000 000 = 10^{18}	exa	E
1 000 000 000 000 000 = 10^{15}	peta	P
1 000 000 000 000 = 10^{12}	teta	T
1 000 000 000 = 10^{9}	giga	G
1 000 000 = 10^{6}	mega	M
1 000 = 10^{3}	kilo	k
100 = 10^{2}	hecto[a]	h
10 = 10^{1}	deka[a]	da
0.1 = 10^{-1}	deci[a]	d
0.01 = 10^{-2}	centi[a]	c
0.001 = 10^{-3}	milli	m
0.000 001 = 10^{-6}	micro	μ
0.000 000 001 = 10^{-9}	nano	n
0.000 000 000 001 = 10^{-12}	pico	p
0.000 000 000 000 001 = 10^{-15}	femto	f
0.000 000 000 000 000 001 = 10^{-18}	atto	a
0.000 000 000 000 000 000 001 = 10^{-21}	zepto	z
0.000 000 000 000 000 000 000 001 = 10^{-24}	yocto	y

[a]To be avoided where possible.

ALGEBRA

POWERS AND ROOTS

$$a^n = a \cdot a \cdot a \ldots \text{to } n \text{ factors} \quad a^{-n} = \frac{1}{a^n}$$

$$a^m \cdot a^n = a^{m+n} \quad \frac{a^m}{a^n} = a^{m-n}$$

$$(ab)^n = a^n b^n \quad \left(\frac{a}{b}\right)^n = \frac{a^n}{b^n}$$

$$(a^m)^n = (a^n)^m = a^{mn} \quad (\sqrt[n]{a})^n = a$$

$$a^{1/n} = \sqrt[n]{a} \quad a^{m/n} = \sqrt[n]{a^m}$$

$$\sqrt[n]{ab} = \sqrt[n]{a}\sqrt[n]{b} \quad \sqrt[n]{\frac{a}{b}} = \frac{\sqrt[n]{a}}{\sqrt[n]{b}}$$

$$\sqrt[n]{\sqrt[m]{a}} = \sqrt[mn]{a}$$

ZERO AND INFINITY OPERATIONS

$a \cdot 0 = 0$	$a \cdot \infty = \infty$	$0 \cdot \infty$	indeterminate
$\frac{0}{a} = 0$	$\frac{a}{0} = \infty$	$\frac{0}{0}$	indeterminate
$\frac{\infty}{a} = \infty$	$\frac{a}{\infty} = 0$	$\frac{\infty}{\infty}$	indeterminate
$a^0 = 1$	$0^a = 0$	0^0	indeterminate
$\infty^a = \infty$		∞^0	indeterminate
$a - a = 0$	$\infty - a = \infty$	$\infty - \infty$	indeterminate

$$a^{\infty} = \infty \quad \text{if } a^2 > 1 \qquad a^{\infty} = 0 \quad \text{if } a^2 < 1 \qquad a^{\infty} = 1 \quad \text{if } a^2 = 1$$

$$a^{-\infty} = 0 \quad \text{if } a^2 > 1 \qquad a^{-\infty} = \infty \quad \text{if } a^2 < 1 \qquad a^{-\infty} = 1 \quad \text{if } a^2 = 1$$

BINOMIAL EXPANSIONS

$$(a \pm b)^2 = a^2 \pm 2ab + b^2$$

$$(a \pm b)^3 = a^3 \pm 3a^2b + 3ab^2 \pm b^3$$

$$(a \pm b)^4 = a^4 \pm 4a^3b + 6a^2b^2 \pm 4ab^3 + b^4$$

$$(a \pm b)^n = a^n \pm \frac{n}{1}a^{n-1}b + \frac{n(n-1)}{1 \cdot 2}a^{n-2}b^2$$

$$\pm \frac{n(n-1)(n-2)}{1 \cdot 2 \cdot 3}a^{n-3}b^3 + \cdots$$

Note: n may be positive or negative, integral or fractional. If n is a positive integer, the series has $(n + 1)$ terms; otherwise, the number of terms is infinite.

ALGEBRA, CONTINUED

LOGARITHMS

$$\log_b b = 1, \log_b 1 = 0, \log_b 0 = \begin{cases} +\infty & \text{when } b \text{ lies between 0 and 1} \\ -\infty & \text{when } b \text{ lies between 1 and } \infty \end{cases}$$

$$\log_b M \cdot N = \log_b M + \log_b N \qquad \log_b \frac{M}{N} = \log_b M - \log_b N$$

$$\log_b N^p = p \log_b N \qquad \log_b \sqrt[r]{N^p} = \frac{p}{r} \log_b N$$

$$\log_b N = \frac{\log_a N}{\log_a b} \qquad \log_b b^N = N \quad b^{\log_b N} = N$$

THE QUADRATIC EQUATION

If

$$ax^2 + bx + c = 0$$

then

$$x = \frac{-b \pm \sqrt{b^2 - 4ac}}{2a} = \frac{2c}{-b \mp \sqrt{b^2 - 4ac}}$$

The second equation serves best when the two values of x are nearly equal.

$$\text{If } b^2 - 4ac \begin{matrix} > \\ = 0 \\ < \end{matrix} \begin{cases} \text{the roots are real and unequal} \\ \text{the roots are real and equal} \\ \text{the roots are imaginary} \end{cases}$$

THE CUBIC EQUATIONS

Any cubic equation $y^3 + py^2 + qy + r = 0$ may be reduced to the form $x^3 + ax + b = 0$ by substituting for y the value $[x - (p/3)]$. Here $a = 1/3(3q - p^2)$, $b = 1/27(2p^3 - 9pq - 27r)$.

ALGEBRAIC SOLUTION OF $x^3 + ax + b = 0$

$$A = \sqrt[3]{-\frac{b}{2} + \sqrt{\frac{b^2}{4} + \frac{a^3}{27}}} \qquad B = \sqrt[3]{-\frac{b}{2} - \sqrt{\frac{b^2}{4} + \frac{a^3}{27}}}$$

$$x_1 = A + B \quad x_2 = -\frac{A+B}{2} + \frac{A-B}{2}\sqrt{-3} \quad x_3 = -\frac{A+B}{2} - \frac{A-B}{2}\sqrt{-3}$$

If

$$\frac{b^2}{4} + \frac{a^3}{27} \begin{matrix} > \\ = 0 \\ < \end{matrix} \begin{cases} \text{1 real root, 2 conjugate imaginary roots} \\ \text{3 real roots, at least 2 equal} \\ \text{3 real and unequal roots} \end{cases}$$

Algebra, Continued

The Cubic Equations, Continued

Trigonometric Solution for $x^3 + ax + b = 0$

For the case when $(b^2/4) + (a^3/27) < 0$, the formulas for the algebraic solution of cubic equations give the roots in impractical form for numerical computation. (In this case, a is negative.) Compute the value of angle ϕ derived from

$$\cos\phi = \sqrt{\frac{b^2}{4} \div \left(-\frac{a^3}{27}\right)}$$

Then

$$x_1 = \mp 2\sqrt{-\frac{a}{3}}\cos\frac{\phi}{3} \quad x_2 = \mp 2\sqrt{-\frac{a}{3}}\cos\left(\frac{\phi}{3} + 120°\right)$$

$$x_3 = \mp 2\sqrt{-\frac{a}{3}}\cos\left(\frac{\phi}{3} + 240°\right)$$

where the upper or lower signs describe b as positive or negative.

Where $(b^2/4 + (a^3/27) >= 0$, compute the values of the angles ψ and ϕ from $\cot 2\psi = [(b^2/4) \div (a^3/27)]^{1/2}$, $\tan\phi = (\tan\psi)^{1/3}$. The real root of the equation then becomes

$$x = \pm 2\sqrt{\frac{a}{3}}\cot 2\phi$$

where the upper or lower sign describes b as positive or negative.

When $(b^2/4) + (a^3/27) = 0$, the roots become

$$x = \mp 2\sqrt{-\frac{a}{3}} \quad \pm\sqrt{-\frac{a}{3}} \quad \pm\sqrt{-\frac{a}{3}}$$

where the upper or lower signs describe b as positive or negative.

The Binomial Equation

When $x^n = a$, the n roots of this equation become

a positive:

$$x = \sqrt[n]{a}\left(\cos\frac{2k\pi}{n} + \sqrt{-1}\sin\frac{2k\pi}{n}\right)$$

a negative:

$$x = \sqrt[n]{-a}\left[\cos\frac{(2k+1)\pi}{n} + \sqrt{-1}\sin\frac{(2k+1)\pi}{n}\right]$$

where k takes in succession values of 0, 1, 2, 3, …, $n - 1$.

TRIGONOMETRY

TRIGONOMETRIC FUNCTIONS OF AN ANGLE

$$\text{sine (sin)}\alpha = \frac{y}{r}$$

$$\text{cosine (cos)}\alpha = \frac{x}{r}$$

$$\text{tangent (tan)}\alpha = \frac{y}{x}$$

$$\text{cotangent (cot)}\alpha = \frac{x}{y}$$

$$\text{secant (sec)}\alpha = \frac{r}{x}$$

$$\text{cosecant (csc)}\alpha = \frac{r}{y}$$

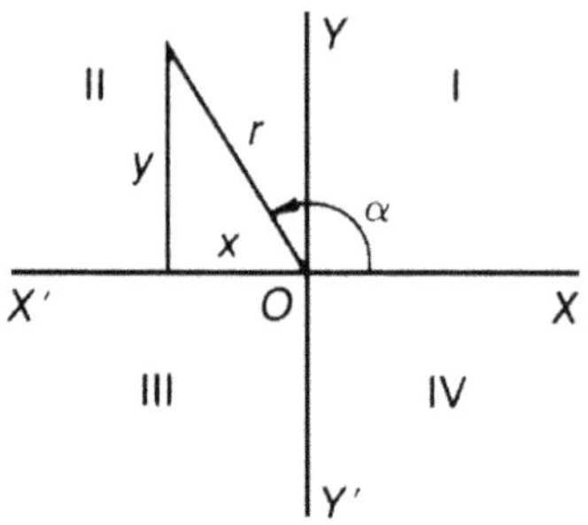

SIGNS OF FUNCTIONS

Quadrant	sin	cos	tan	cot	sec	csc
I	+	+	+	+	+	+
II	+	−	−	−	−	+
III	−	−	+	+	−	−
IV	−	+	−	−	+	−

FUNCTIONS OF 0°, 30°, 45°, 60°, 90°, 180°, 270°, 360°

	0°	30°	45°	60°	90°	180°	270°	360°
sin	0	1/2	$\sqrt{2}/2$	$\sqrt{3}/2$	1	0	−1	0
cos	1	$\sqrt{3}/2$	$\sqrt{2}/2$	1/2	0	−1	0	1
tan	0	$\sqrt{3}/3$	1	$\sqrt{3}$	∞	0	∞	0
cot	∞	$\sqrt{3}$	1	$\sqrt{3}/3$	0	∞	0	∞
sec	1	$2\sqrt{3}/3$	$\sqrt{2}$	2	∞	−1	∞	1
csc	∞	2	$\sqrt{2}$	$2\sqrt{3}/3$	1	∞	−1	∞

FUNDAMENTAL FUNCTION RELATIONS

$\sin\alpha = 1/\csc\alpha$ $\quad\cos\alpha = 1/\sec\alpha$ $\quad\tan\alpha = 1/\cot\alpha = \sin\alpha/\cos\alpha$

$\csc\alpha = 1/\sin\alpha$ $\quad\sec\alpha = 1/\cos\alpha$ $\quad\cot\alpha = 1/\tan\alpha = \cos\alpha/\sin\alpha$

$\sin^2\alpha + \cos^2\alpha = 1$ $\quad\sec^2\alpha - \tan^2\alpha = 1$ $\quad\csc^2\alpha - \cot^2\alpha = 1$

TRIGONOMETRY, CONTINUED

FUNCTIONS OF SEVERAL ANGLES

$$\sin 2\alpha = 2\sin\alpha\cos\alpha$$

$$\cos 2\alpha = 2\cos^2\alpha - 1 = 1 - 2\sin^2 a = \cos^2\alpha - \sin^2\alpha$$

$$\sin 3\alpha = 3\sin\alpha - 4\sin^3\alpha$$

$$\cos 3\alpha = 4\cos^3\alpha - 3\cos\alpha$$

$$\sin 4\alpha = 4\sin\alpha\cos\alpha - 8\sin^3\alpha\cos\alpha$$

$$\cos 4\alpha = 8\cos^4\alpha - 8\cos^2\alpha + 1$$

$$\sin n\alpha = 2\sin(n-1)\alpha\cos\alpha - \sin(n-2)\alpha$$

$$\cos n\alpha = 2\cos(n-1)\alpha\cos\alpha - \cos(n-2)\alpha$$

HALF-ANGLE FUNCTIONS

$$\sin\frac{\alpha}{2} = \sqrt{\frac{1-\cos\alpha}{2}} \qquad \cos\frac{\alpha}{2} = \sqrt{\frac{1+\cos\alpha}{2}}$$

$$\tan\frac{\alpha}{2} = \frac{1-\cos\alpha}{\sin\alpha} = \frac{\sin\alpha}{1+\cos\alpha} = \sqrt{\frac{-\cos\alpha}{1+\cos\alpha}}$$

POWERS OF FUNCTIONS

$$\sin^2\alpha = \frac{1}{2}(1-\cos 2\alpha) \qquad \cos^2\alpha = \frac{1}{2}(1+\cos 2\alpha)$$

$$\sin^3\alpha = \frac{1}{4}(3\sin\alpha - \sin 3\alpha) \qquad \cos^3\alpha = \frac{1}{4}(\cos 3\alpha + 3\cos\alpha)$$

$$\sin^4\alpha = \frac{1}{8}(\cos 4\alpha - 4\cos 2\alpha + 3) \qquad \cos^4\alpha = \frac{1}{8}(\cos 4\alpha + 4\cos 2\alpha + 3)$$

FUNCTIONS: SUM OR DIFFERENCE OF TWO ANGLES

$$\sin(\alpha \pm \beta) = \sin\alpha\cos\beta \pm \cos\alpha\sin\beta$$

$$\cos(\alpha \pm \beta) = \cos\alpha\cos\beta \mp \sin\alpha\sin\beta$$

$$\tan(\alpha \pm \beta) = \frac{\tan\alpha \pm \tan\beta}{1 \mp \tan\alpha\tan\beta}$$

TRIGONOMETRY, CONTINUED

SUMS, DIFFERENCES, AND PRODUCTS OF TWO FUNCTIONS

$$\sin\alpha \pm \sin\beta = 2\sin\frac{1}{2}(\alpha \pm \beta)\cos\frac{1}{2}(a \mp \beta)$$

$$\cos\alpha + \cos\beta = 2\cos\frac{1}{2}(\alpha + \beta)\cos\frac{1}{2}(a - \beta)$$

$$\cos\alpha - \cos\beta = -2\sin\frac{1}{2}(\alpha + \beta)\sin\frac{1}{2}(a - \beta)$$

$$\tan\alpha \pm \tan\beta = \frac{\sin(\alpha \pm \beta)}{\cos\alpha\cos\beta}$$

$$\sin^2\alpha - \sin^2\beta = \sin(\alpha + \beta)\sin(\alpha - \beta)$$

$$\cos^2\alpha - \cos^2\beta = -\sin(\alpha + \beta)\sin(\alpha - \beta)$$

$$\cos^2\alpha - \sin^2\beta = \cos(\alpha + \beta)\cos(\alpha - \beta)$$

$$\sin\alpha\sin\beta = \frac{1}{2}\cos(\alpha - \beta) - \frac{1}{2}\cos(\alpha + \beta)$$

$$\cos\alpha\cos\beta = \frac{1}{2}\cos(\alpha - \beta) + \frac{1}{2}\cos(\alpha + \beta)$$

$$\sin\alpha\cos\beta = \frac{1}{2}\sin(\alpha + \beta) + \frac{1}{2}\sin(\alpha - \beta)$$

RIGHT TRIANGLE SOLUTION

Given any two sides, or one side and any acute angle α, find the remaining parts:

$$\sin\alpha = \frac{a}{c} \quad \cos\alpha = \frac{b}{c} \quad \tan\alpha = \frac{a}{b} \quad \beta = 90^\circ - \alpha$$

$$a = \sqrt{(c+b)(c-b)} = c\sin\alpha = b\tan\alpha$$

$$b = \sqrt{(c+a)(c-a)} = c\cos\alpha = \frac{a}{\tan\alpha}$$

$$c = \frac{a}{\sin\alpha} = \frac{b}{\cos\alpha} = \sqrt{a^2 + b^2}$$

$$A = \frac{1}{2}ab = \frac{a^2}{2\tan\alpha} = \frac{b^2\tan\alpha}{2} = \frac{c^2\sin 2\alpha}{4}$$

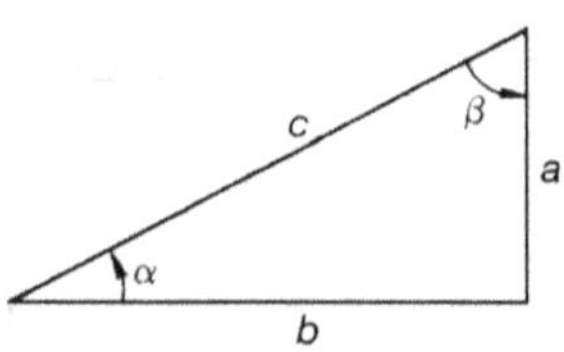

MENSURATION

Note: A = area, V = volume

OBLIQUE TRIANGLE SOLUTION

RIGHT TRIANGLE

$$A = \frac{1}{2}ab \qquad c = \sqrt{a^2 + b^2}$$

$$a = \sqrt{c^2 - b^2} \qquad b = \sqrt{c^2 - a^2}$$

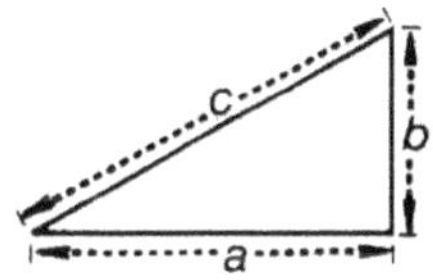

OBLIQUE TRIANGLE

$$A = \frac{1}{2}bh$$

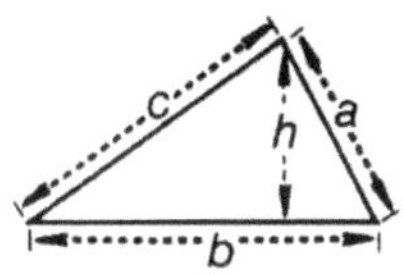

EQUILATERAL TRIANGLE

$$A = \frac{1}{2}ah = \frac{1}{4}a^2\sqrt{3}$$

$$h = \frac{1}{2}a\sqrt{3}$$

$$r_1 = \frac{a}{2\sqrt{3}}$$

$$r_2 = \frac{a}{\sqrt{3}}$$

PARALLELOGRAM

$$A = ah = ab\sin\alpha$$

$$d_1 = \sqrt{a^2 + b^2 - 2ab\cos\alpha}$$

$$d_2 = \sqrt{a^2 + b^2 + 2ab\cos\alpha}$$

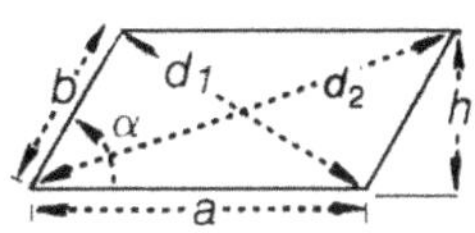

TRAPEZOID

$$A = \frac{1}{2}h(a + b)$$

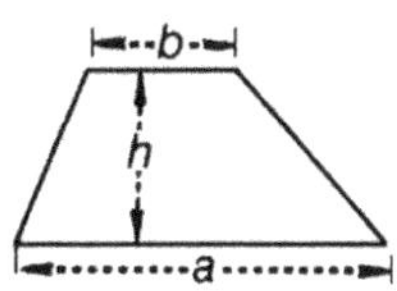

MENSURATION, CONTINUED

CIRCLE

$C =$ circumference

$\alpha =$ central angle in radians

$$C = \pi D = 2\pi R$$

$$c = R\alpha = \frac{1}{2}D\alpha = D\cos^{-1}\frac{d}{R} = D\tan^{-1}\frac{l}{2d}$$

$$l = 2\sqrt{R^2 - d^2} = 2R\sin\frac{\alpha}{2} = 2d\tan\frac{\alpha}{2} = 2d\tan\frac{c}{D}$$

$$d = \frac{1}{2}\sqrt{4R^2 - l^2} = \frac{1}{2}\sqrt{D^2 - l^2} = R\cos\frac{\alpha}{2}$$

$$= \frac{1}{2}l\cot\frac{\alpha}{2} = \frac{1}{2}\cot\frac{c}{D}$$

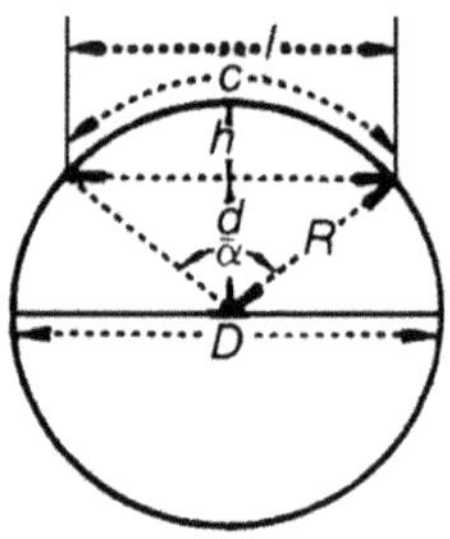

$$h = R - d$$

$$\alpha = \frac{c}{R} = \frac{2c}{D} = 2\cos^{-1}\frac{d}{R}$$

$$= 2\tan^{-1}\frac{l}{2d} = 2\sin^{-1}\frac{l}{D}$$

$$A = \pi R^2 = \frac{1}{4}\pi D^2 = \frac{1}{2}RC = \frac{1}{4}DC$$

$$A_{\text{sector}} = \frac{1}{2}Rc = \frac{1}{2}R^2\alpha = \frac{1}{8}D^2\alpha$$

$$A_{\text{segment}} = A_{\text{sector}} - A_{\text{triangle}} = \frac{1}{2}R^2(\alpha - \sin\alpha)$$

$$= \frac{1}{2}R\left(c - R\sin\frac{c}{R}\right)$$

$$= R^2\sin^{-1}\frac{l}{2R} - \frac{1}{4}l\sqrt{4R^2 - l^2}$$

$$= R^2\cos^{-1}\frac{d}{R} - d\sqrt{R^2 - d^2}$$

$$= R^2\cos^{-1}\frac{R-h}{R} - (R-h)\sqrt{2Rh - h^2}$$

MENSURATION, CONTINUED

CIRCLE, CONTINUED

AREA BY APPROXIMATION

If $y_0, y_1, y_2, \ldots, y_n$ are the lengths of a series of equally spaced parallel chords and if h is their distance apart, the area enclosed by boundary is given approximately by

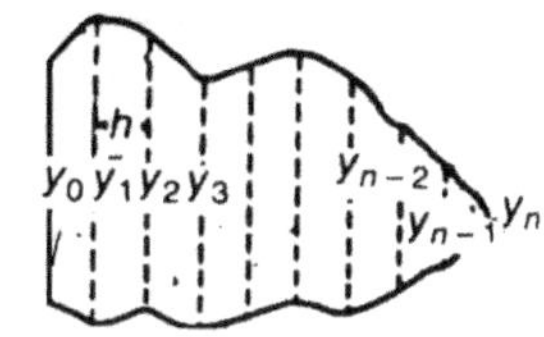

$$A_s = \frac{1}{3}h[(y_0 + y_n) + 4(y_1 + y_3 + \cdots + y_{n-1}) + 2(y_2 + y_4 + \cdots + y_{n-1})]$$

where n is even (Simpson's Rule).

ELLIPSE

A = $\pi ab/4$

Perimeter s

$$= \frac{\pi(a+b)}{2}\left[1 + \frac{1}{4}\left(\frac{a-b}{a+b}\right)^2 + \frac{1}{64}\left(\frac{a-b}{a+b}\right)^4 + \frac{1}{256}\left(\frac{a-b}{a+b}\right)^6 + \cdots\right]$$

$$\approx \pi\frac{(a+b)}{8}\left[3(1+\lambda) + \frac{1}{1-\lambda}\right]$$

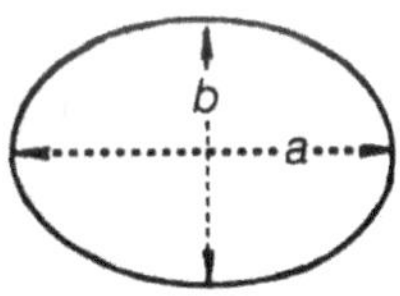

where $\lambda = \left[\frac{a-b}{2(a+b)}\right]^2$

PARABOLA

A = $(2/3)ld$

Length of arc s

$$= \frac{1}{2}\sqrt{16d^2 + l^2} + \frac{l^2}{8d}\ln\left(\frac{4d + \sqrt{16d^2 + l^2}}{l}\right)$$

$$= l\left[1 + \frac{2}{3}\left(\frac{2d}{l}\right)^2 - \frac{2}{5}\left(\frac{2d}{l}\right)^4 + \cdots\right]$$

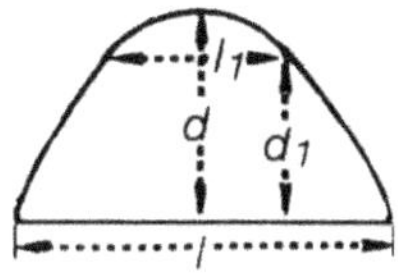

Height of segment $d_1 = \frac{d}{l^2}(l^2 - l_1^2)$

Width of segment $l_1 = l\sqrt{\frac{d - d_1}{d}}$

MENSURATION, CONTINUED

CATENARY

Length of arc s

$$= l\left[1+\frac{2}{3}\left(\frac{2d}{l}\right)^2\right]$$

approximately, if d is small in comparison with l.

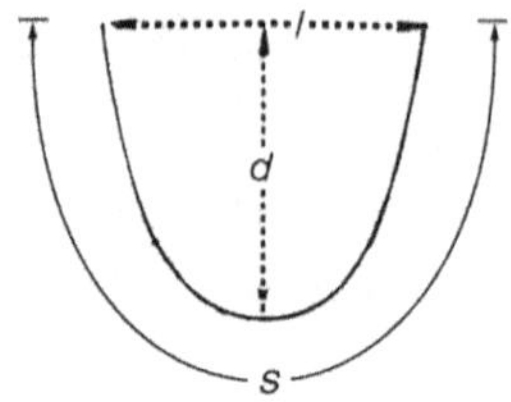

SPHERE

$$A = 4\pi R^2 = \pi D^2$$

$$A_{zone} = 2\pi Rh = \pi Dh$$

$$V_{sphere} = 4/3\pi R^3 = 1/6\pi D^3$$

$$V_{spherical\ sector} = 72/3\pi R^2 h = 1/6\pi D^2 h$$

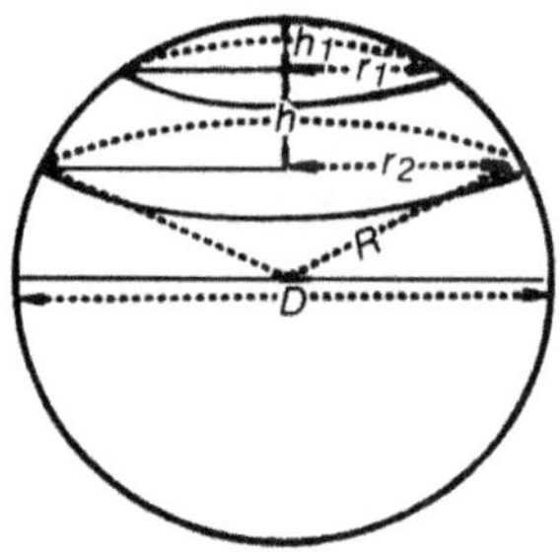

V for spherical segment of one base:

$$1/6\pi h\left(3r_1^2 + 3r_2^2 + h^2\right)$$

ELLIPSOID

$$V = 4/3\pi\, abc$$

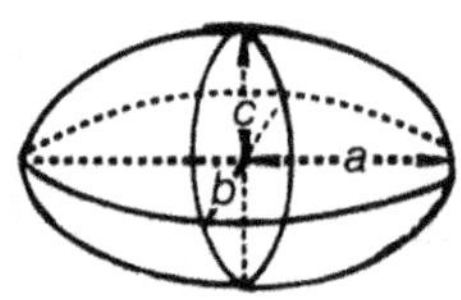

PARABOLOID SEGMENT

$$V(\text{segment of one base}) = \frac{1}{2}\pi r_1^2 h$$

$$V(\text{segment of two bases}) = \frac{1}{2}\pi d\left(r_1^2 + r_2^2\right)$$

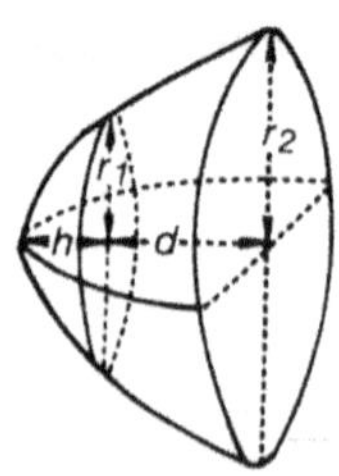

TORUS

$$V = 2\,\pi^2\,Rr^2$$

$$\text{Surface area} = 4\pi^2\,Rr$$

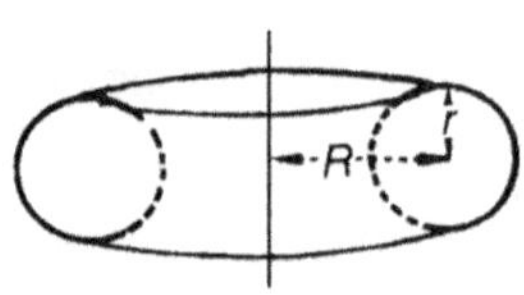

ANALYTIC GEOMETRY

RECTANGULAR AND POLAR COORDINATE RELATIONS

$$x = r\cos\theta \quad y = r\sin\theta$$

$$r = \sqrt{x^2 + y^2} \quad \theta = \tan^{-1}\frac{y}{x} \quad \sin\theta = \frac{y}{\sqrt{x^2 + y^2}}$$

$$\cos\theta = \frac{x}{\sqrt{x^2 + y^2}} \quad \tan\theta = \frac{y}{x}$$

POINTS AND SLOPES

For any two points P_1 (x_1, y_1) and P_2 (x_2, y_2) and the angle α_1 from OX to P_1 P_2 measured counterclockwise, the distance, d, between these two points,

$$P_1P_2 = d = \sqrt{(x_2 - x_1)^2 + (y_2 - y_1)^2}$$

$$P_1P_2 \text{ midpoint} = \frac{x_1 + x_2}{2}, \frac{y_1 + y_2}{2}$$

For the point that divides P_1P_2 in the ratio $m_1 : m_2$ going from P_1 to P_2,

$$\left(\frac{m_1x_2 + m_2x_1}{m_1 + m_2}, \frac{m_1y_2 + m_2y_1}{m_1 + m_2}\right)$$

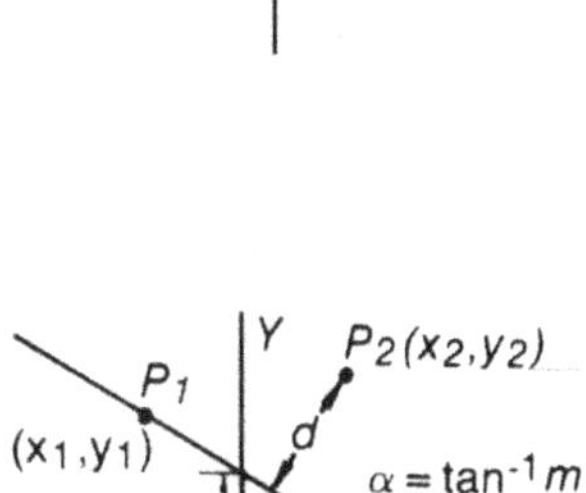

$$\text{Slope of } P_1P_2 = \tan\alpha = m = \frac{y_2 - y_1}{x_2 - x_1}$$

For angle β between two lines of slopes m_1 and m_2,

$$\beta = \tan^{-1}\frac{m_2 - m_1}{1 + m_1m_2}$$

STRAIGHT LINE

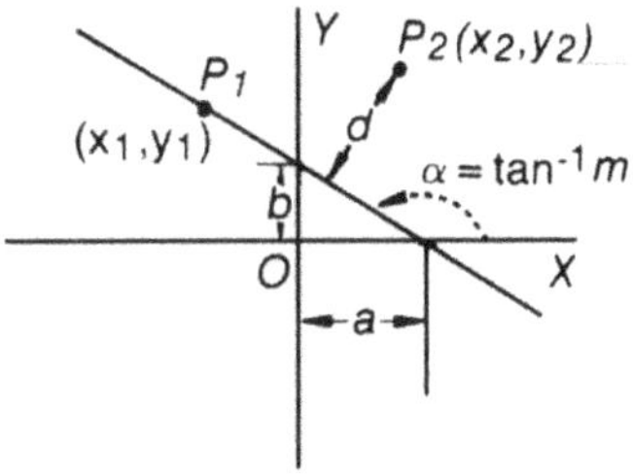

$$Ax + By + C = 0$$

$$-A \div B = \text{slope}$$

$$y = mx + b$$

(where m = slope, b = intercept on OY)

$$y - y_1 = m(x - x_1)$$

[m = slope, where $P_1(x_1, y_1)$ is a known point on line]

$$d = \frac{Ax_2 + By_2 + C}{\pm\sqrt{A^2 + B^2}}$$

[where d equals distance from a point $P_2(x_2, y_2)$ to the line $Ax + By + C = 0$]

ANALYTIC GEOMETRY, CONTINUED

CIRCLE

Locus of points at a constant distance (radius) from a fixed point C (center).

CIRCLE A

$$(x-h)^2+(y-k)^2=a^2$$
$$r^2+b^2-2br\cos(\theta-\beta)=a^2$$

$C(h, k)$, radius $= a$ $\quad C(b, \beta)$, radius $= a$

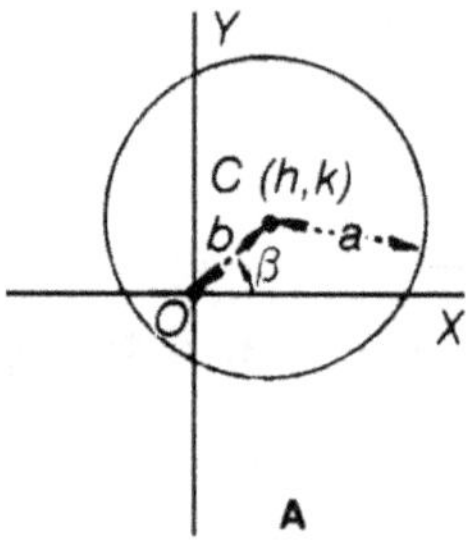

A

CIRCLE B_{top}

$$x^2+y^2=2ay$$
$$r=2a\sin\theta$$

$C(0, a)$, radius $= a$ $\quad C\left(a, \frac{\pi}{2}\right)$, radius $= a$

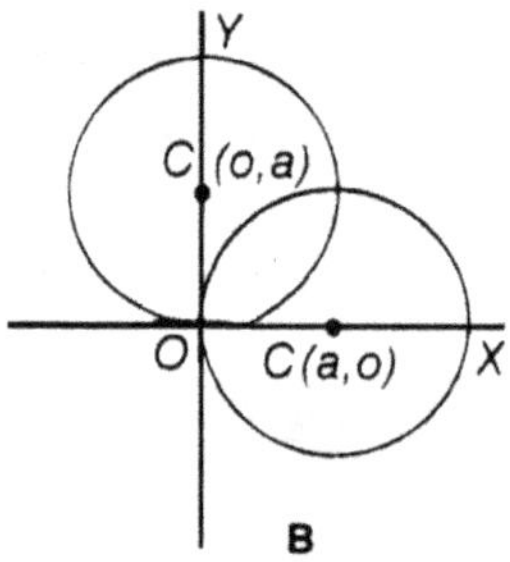

B

CIRCLE B_{bottom}

$$x^2+y^2=2ax$$
$$r=2a\cos\theta$$

$C(a, 0)$, radius $= a$ $\quad C(a, 0)$, radius $= a$

CIRCLE C

$$x^2+y^2=a^2$$
$$r=a$$

$$x=a\cos\phi \quad y=a\sin\phi$$

$C(0,0)$, radius $= a$ $\quad \phi =$ angle from OX to radius

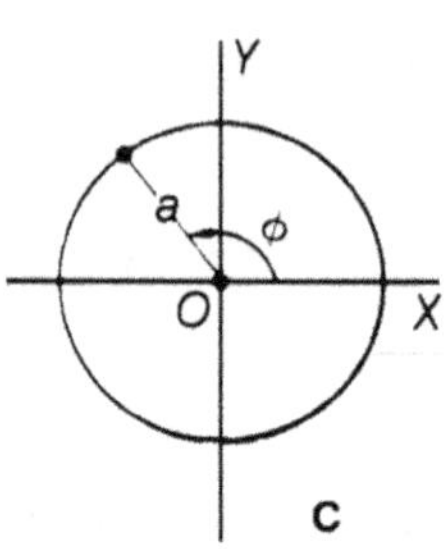

C

ANALYTIC GEOMETRY, CONTINUED

PARABOLA

Conic where $e = 1$; latus rectum, $a = LR$.

PARABOLA A

$(y-k)^2 = a(x-h)$	V, vertex (h, k), axis $\parallel OX$
$y^2 = ax$	Vertex (0, 0), axis along OX

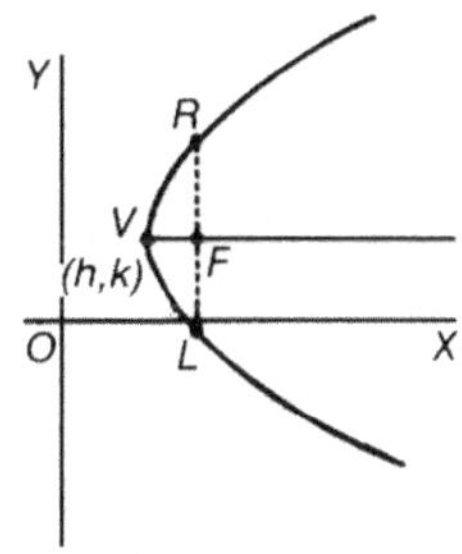

A

PARABOLA B

$(x-h)^2 = a(y-k)$	Vertex (h, k), axis $\parallel OY$
$x^2 = ay$	Vertex (0, 0), axis along OY

Distance from vertex V to focus F: $\frac{1}{4}a$

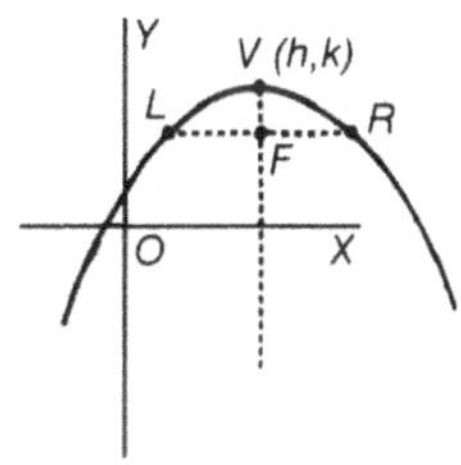

B

SINE WAVE

$$y = a\sin(bx + c)$$

$$y = a\cos(bx + c') = a\sin(bx + c) \quad (\text{where } c = c' + \pi/2)$$

$$y = m\sin bx + n\cos bx = a\sin(bx + c)$$

$$(\text{where } a = \sqrt{m^2 + n^2},\ c = \tan^{-1} n/m)$$

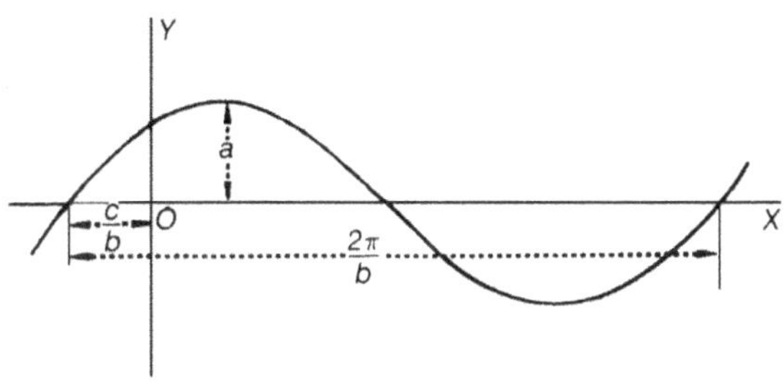

$$\text{wavelength, } \lambda = \frac{2\pi}{b}$$

ANALYTIC GEOMETRY, CONTINUED

EXPONENTIAL OR LOGARITHMIC CURVES

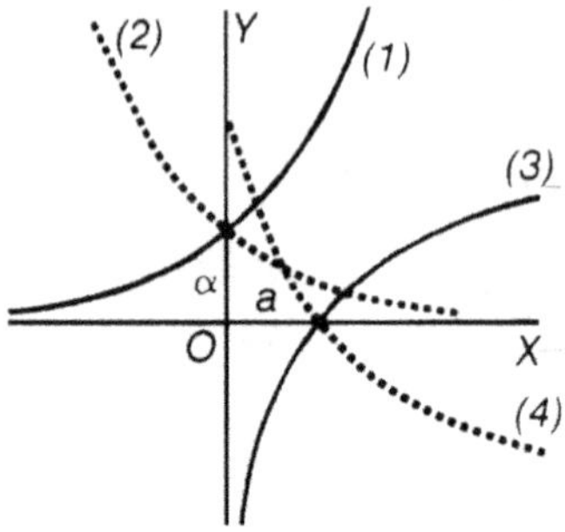

1) $y = ab^x$ or $x = \log_b \frac{y}{a}$

2) $y = ab^{-x}$ or $x = -\log_b \frac{y}{a}$

3) $x = ab^y$ or $y = \log_b \frac{x}{a}$

4) $x = ab^{-y}$ or $y = -\log_b \frac{x}{a}$

The equations $y = ae^{\pm nx}$ and $x = ae^{\pm ny}$ are special cases.

CATENARY

Line of uniform weight suspended freely between two points at the same level:

$$y = a\left[\cosh\left(\frac{x}{a}\right) - 1\right] = \frac{a}{2}\left(e^{x/a} + e^{-x/a} - 2a\right)$$

(where y is measured from the lowest point of the line, w is the weight per unit length of the line, T_h is the horizontal component of the tension in the line, and a is the parameter of the catenary; $a = T_h / w$)

HELIX

Curve generated by a point moving on a cylinder with the distance it transverses parallel to the axis of the cylinder being proportional to the angle of rotation about the axis:

$$x = a\cos\theta$$
$$y = a\sin\theta$$
$$z = k\theta$$

(where a = radius of cylinder, $2\pi k$ = pitch)

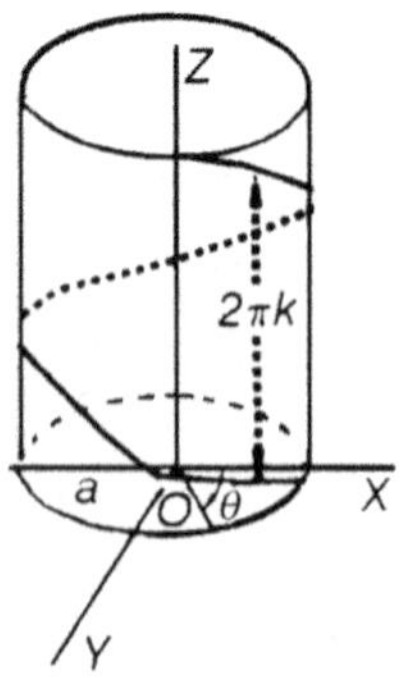

ANALYTIC GEOMETRY, CONTINUED

POINTS, LINES, AND PLANES

Distance between points $P_1(x_1, y_1, z_1)$ and $P_2(x_2, y_2, z_2)$:

$$d = \sqrt{(x_2 - x_1)^2 + (y_2 - y_1)^2 + (z_2 - z_1)^2}$$

Direction cosine of a line (cosines of the angles α, β, γ which the line or any parallel line makes with the coordinate axes) is related by

$$\cos^2\alpha + \cos^2\beta + \cos^2\gamma = 1$$

If $\cos\alpha : \cos\beta : \cos\gamma = a : b : c$, then

$$\cos\alpha = \frac{a}{\sqrt{a^2 + b^2 + c^2}} \qquad \cos\beta = \frac{b}{\sqrt{a^2 + b^2 + c^2}} \qquad \cos\gamma = \frac{c}{\sqrt{a^2 + b^2 + c^2}}$$

Direction cosine of the line joining $P_1(x_1, y_1, z_1)$ and $P_2(x_2, y_2, z_2)$:

$$\cos\alpha : \cos\beta : \cos\gamma = x_2 - x_1 : y_2 - y_1 : z_2 - z_1$$

Angle θ between two lines, whose direction angles are α_1, β_1, γ_1 and α_2, β_2, γ_2:

$$\cos\theta = \cos\alpha_1 \cos\alpha_2 + \cos\beta_1 \cos\beta_2 + \cos\gamma_1 \cos\gamma_2$$

Equation of a plane is of the first degree in x, y, and z:

$$Ax + By + Cz + D = 0$$

where A, B, C are proportional to the direction cosines of a normal or perpendicular to the plane.

Angle between two planes is the angle between their normals. Equations of a straight line are two equations of the first degree.

$$A_1x + B_1y + C_1z + D_1 = 0 \qquad A_2x + B_2y + C_2z + D_2 = 0$$

Equations of a straight line through the point $P_1(x_1, y_1, z_1)$ with direction cosines proportional to a, b, and c:

$$\frac{x - x_1}{a} = \frac{y - y_1}{b} = \frac{z - z_1}{c}$$

DIFFERENTIAL CALCULUS

DERIVATIVE RELATIONS

$$\text{If } x = f(y)\text{, then } \frac{dy}{dx} = 1 \div \frac{dx}{dy}$$

$$\text{If } x = f(t) \text{ and } y = F(t)\text{, then } \frac{dy}{dx} = \frac{dy}{dt} \div \frac{dx}{dt}$$

$$\text{If } y = f(u) \text{ and } u = F(x)\text{, then } \frac{dy}{dx} = \frac{dy}{du} \cdot \frac{du}{dx}$$

DIFFERENTIAL CALCULUS, CONTINUED

DERIVATIVES

Functions of x represented by u and v, constants by a, n, and e:

$$\frac{d}{dx}(x)=1 \quad \frac{d}{dx}(a)=0$$

$$\frac{d}{dx}(u \pm v \pm \cdots)=\frac{du}{dx} \pm \frac{dv}{dx} \pm \cdots$$

$$\frac{d}{dx}(au)=a\frac{du}{dx} \quad \frac{d}{dx}a^u=a^u \ln a\frac{du}{dx}$$

$$\frac{d}{dx}(uv)=u\frac{dv}{dx}+v\frac{du}{dx} \quad \frac{d}{dx}e^u=e^u\frac{du}{dx}$$

$$\frac{d}{dx}\left(\frac{u}{v}\right)=\left(v\frac{du}{dx}-u\frac{dv}{dx}\right)\bigg/v^2 \quad \frac{d}{dx}u^v=vu^{v-1}\frac{du}{dx}+u^v \ln u\frac{dv}{dx}$$

$$\frac{d}{dx}(u^n)=nu^{n-1}\frac{du}{dx} \quad \frac{d}{dx}\sin u=\cos u\frac{du}{dx}$$

$$\frac{d}{dx}\log_a u=\frac{\log_a e}{u}\frac{du}{dx}$$

$$\frac{d}{dx}\ln u=\frac{1}{u}\frac{du}{dx}$$

$$\frac{d}{dx}\cos u=-\sin u\frac{du}{dx} \quad \frac{d}{dx}\tan u=\sec^2 u\frac{du}{dx}$$

$$\frac{d}{dx}\cot u=-\csc^2 u\frac{du}{dx} \quad \frac{d}{dx}\sin^{-1} u=\frac{1}{\sqrt{1-u^2}}\frac{du}{dx}$$

(where $\sin^{-1} u$ lies between $-\pi/2$ and $+\pi/2$)

$$\frac{d}{dx}\cos^{-1} u=-\frac{1}{\sqrt{1-u^2}}\frac{du}{dx}$$

(where $\cos^{-1} u$ lies between 0 and π)

$$\frac{d}{dx}\tan^{-1} u=\frac{1}{1+u^2}\frac{du}{dx} \quad \frac{d}{dx}\cot^{-1} u=-\frac{1}{1+u^2}\frac{du}{dx}$$

$$\frac{d}{dx}\sec^{-1} u=\frac{1}{u\sqrt{u^2-1}}\frac{du}{dx}$$

(where $\sec^{-1} u$ lies between 0 and π)

$$\frac{d}{dx}\csc^{-1} u=-\frac{1}{u\sqrt{u^2-1}}\frac{du}{dx}$$

(where $\csc^{-1} u$ lies between $-\pi/2$ and $+\pi/2$)

DIFFERENTIAL CALCULUS, CONTINUED

nTH DERIVATIVE OF CERTAIN FUNCTIONS

$$\frac{d^n}{dx^n} e^{ax} = a^n e^{ax}$$

$$\frac{d^n}{dx^n} a^x = (\ln a)^n a^x$$

$$\frac{d^n}{dx^n} \ln x = \frac{(-1)^{n-1}(n-1)!}{x^n}, \text{ where } (n-1)! = 1 \cdot 2 \cdot 3 \cdots (n-1)$$

$$\frac{d^n}{dx^n} \sin ax = a^n \sin\left(ax + \frac{n\pi}{2} \right)$$

$$\frac{d^n}{dx^n} \cos ax = a^n \cos\left(ax + \frac{n\pi}{2} \right)$$

INTEGRAL CALCULUS

THEOREMS ON INTEGRALS AND SOME EXAMPLES

$$\int df(x) = f(x) + C$$

$$d\int f(x)dx = f(x)dx$$

$$\int [f_1(x) \pm f_2(x) \pm \cdots]\, dx = \int f_1(x)dx \pm \int f_2(x)dx \pm \cdots$$

$$\int af(x)dx = a\int f(x)dx \quad a \text{ any constant}$$

$$\int u^n du = \frac{u^{n+1}}{n+1} + C \quad (n \neq -1) \quad u \text{ any function of } x$$

$$\int \frac{du}{u} = \ln u + C \quad u \text{ any function of } x$$

$$\int u\, dv = uv - \int v\, du \quad u \text{ and } v \text{ any functions of } x$$

INTEGRAL CALCULUS, CONTINUED

SELECTED INTEGRALS

$$\int \mathrm{d}u = u + C$$

$$\int a\,\mathrm{d}u = a\int \mathrm{d}u = au + C$$

$$\int [f_1(u) + f_2(u) + \cdots + f_n(u)]\,\mathrm{d}u = \int f_1(u)\,\mathrm{d}u + \int f_2(u)\mathrm{d}u + \int f_n(u)\,\mathrm{d}u$$

$$\int u^n \mathrm{d}u = \frac{u^{n+1}}{n+1} + C \quad (n \neq -1)$$

$$\int \frac{\mathrm{d}u}{u} = \ell n\,|\,u\,| + C$$

$$\int a^u\,\mathrm{d}u = \frac{a^u}{\ell n\, a} + C \quad (a > 0, a \neq 1)$$

$$\int e^u \mathrm{d}u = e^u + C$$

$$\int \sin u\,\mathrm{d}u = -\cos u + C$$

$$\int \cos u\,\mathrm{d}u = \sin u + C$$

$$\int \sec^2 u\,\mathrm{d}u = \tan u + C$$

$$\int \csc^2 u\,\mathrm{d}u = -\cot u + C$$

$$\int \sec u \tan u\,\mathrm{d}u = \sec u + C$$

$$\int \csc u\,\cot u\,\mathrm{d}u = -\csc u + C$$

$$\int \tan u\,\mathrm{d}u = -\ell n\left|\cos u\right| + C = \ell n\left|\sec u\right| + C$$

INTEGRAL CALCULUS, CONTINUED

SELECTED INTEGRALS, CONTINUED

$$\int \cot u\, du = \ln|\sin u| + C$$

$$\int \sec u\, du = \ln|\sec u + \tan u| + C$$

$$\int \csc u\, du = \ln|\csc u - \cot u| + C$$

$$\int \frac{du}{\sqrt{a^2 - u^2}} = \arcsin\frac{u}{a} + C \quad (a > 0)$$

$$\int \frac{du}{\sqrt{a^2 + u^2}} = \frac{1}{a}\arctan\frac{u}{a} + C$$

$$\int u\, dv = uv - \int v\, du$$

$$\int \frac{du}{\sqrt{u^2 \pm a^2}} = \ln|u + \sqrt{u^2 \pm a^2}| + C$$

$$\int \frac{du}{u\sqrt{u^2 - a^2}} = \frac{1}{a}\operatorname{arcsec}\frac{u}{a} + C \quad (a > 0)$$

$$\int \sqrt{a^2 - u^2}\, du = \frac{u}{2}\sqrt{a^2 - u^2} + \frac{a^2}{2}\arcsin\frac{u}{a} + C \quad (a > 0)$$

$$\int \sqrt{u^2 \pm a^2}\, du = \frac{u}{2}\sqrt{u^2 \pm a^2} \pm \frac{a^2}{2}\ln|u + \sqrt{u^2 \pm a^2}| + C$$

SOME DEFINITE INTEGRALS

$$\int_0^a \sqrt{a^2 - x^2}\, dx = \frac{\pi a^2}{4} \qquad \int_0^a \sqrt{2ax - x^2}\, dx = \frac{\pi a^2}{4}$$

$$\int_0^\infty \frac{dx}{ax^2 + b} = \frac{\pi}{2\sqrt{ab}} \quad a \text{ and } b \text{ positive}$$

$$\int_0^\pi \sin^2 ax\, dx = \int_0^\pi \cos^2 ax\, dx = \frac{\pi}{2}$$

$$\int_0^\infty e^{-ax^2}\, dx = \frac{1}{2}\sqrt{\frac{\pi}{a}}$$

$$\int_0^\infty x^n e^{-ax}\, dx = \frac{n!}{a^{n-1}} \quad n \text{ a positive integer}$$

DIFFERENTIAL EQUATIONS

EQUATIONS OF FIRST ORDER AND FIRST DEGREE: Mdx + Ndy = 0

VARIABLES SEPARABLE

$$X_1Y_1\,\mathrm{d}x + X_2Y_2\,\mathrm{d}y = 0$$

$$\int \frac{X_1}{X_2}\,\mathrm{d}x + \int \frac{Y_2}{Y_1}\,\mathrm{d}y = C$$

HOMOGENEOUS EQUATION

$$\mathrm{d}y - f\left(\frac{y}{x}\right)\mathrm{d}x = 0$$

$$\frac{1}{x} = \int \frac{\mathrm{d}v}{f(v) - v} \quad \text{and} \quad v = \frac{y}{x}$$

$M \div N$ can be written so that x and y occur only as $y \div x$, if every term in M and N have the same degree in x and y.

LINEAR EQUATION

$$\mathrm{d}y + (X_1y - X_2)\,\mathrm{d}x = 0$$

$$y = e^{-\int X_1\,\mathrm{d}x}\left(\int X_2 e^{\int X_1 dx}\,\mathrm{d}x + C\right)$$

A similar solution exists for $\mathrm{d}x + (Y_1x - Y_2)\mathrm{d}y = 0$.

EXACT EQUATION

$$M\mathrm{d}x + N\mathrm{d}y = 0 \quad \text{where } \partial M/\partial y = \partial N/\partial x$$

$$\int M\mathrm{d}x + \int \left| N - \frac{\partial}{\partial y}\int M\mathrm{d}x \right| dy = C$$

for y constant when integrating with respect to x.

NONEXACT EQUATION

$$M\mathrm{d}x + N\mathrm{d}y = 0 \quad \text{where } \partial M/\partial y \neq \partial N/\partial x$$

Make equation exact by multiplying by an integrating factor $\mu(x, y)$—a form readily recognized in many cases.

COMPLEX QUANTITIES

PROPERTIES OF COMPLEX QUANTITIES

If z, z_1, z_2 represent complex quantities, then

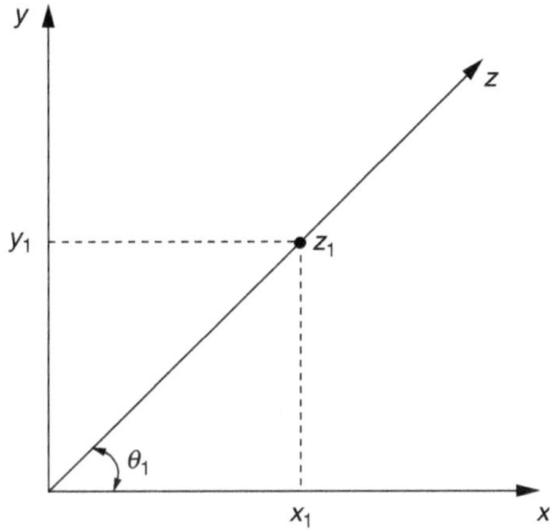

Sum or difference: $z_1 \pm z_2 = (x_1 \pm x_2) + j(y_1 \pm y_2)$

Product: $z_1 \cdot z_2 = r_1 r_2[\cos(\theta_1 + \theta_2) + j\sin(\theta_1 + \theta_2)]$

$$= r_1 r_2 e^{j(\theta_1+\theta_2)} = (x_1 x_2 - y_1 y_2) + j(x_1 y_2 + x_2 y_1)$$

Quotient: $z_1 / z_2 = r_1 / r_2[\cos(\theta_1 - \theta_2) + j\sin(\theta_1 - \theta_2)]$

$$= \frac{r_1}{r_2} e^{j(\theta_1+\theta_2)} = \frac{x_1 x_2 + y_1 y_2}{x_2^2 + y_2^2} + j\frac{x_2 y_1 - x_1 y_2}{x_2^2 + y_2^2}$$

Power: $z^n = r^n[\cos n\theta + j\sin n\theta] = r^n e^{jn\theta}$

Root: $\sqrt[n]{z} = \sqrt[n]{r}\left|\cos\frac{\theta + 2k\pi}{n} + j\sin\frac{\theta + 2k\pi}{n}\right| = \sqrt[n]{r} e^{j(\theta+2k\pi/n)}$

where k takes in succession the values 0, 1, 2, 3, …, $n - 1$.

If $z_1 = z_2$, then $x_1 = x_2$ and $y_1 = y_2$.

Periodicity: $z = r(\cos\theta + j\sin\theta)$

$$= r[\cos(\theta + 2k\pi) + j\sin(\theta + 2k\pi)]$$

or

$$z = re^{j\theta} = re^{j(\theta + 2k\pi)} \quad \text{and} \quad e^{i2k\pi} = 1$$

where k is any integer.

Exponential-trigonometric relations:

$$e^{jz} = \cos z + j\sin z \quad e^{-jz} = \cos z - j\sin z$$

$$\cos z = \frac{1}{2}(e^{jz} + e^{-jz}) \quad \sin z = \frac{1}{2j}(e^{jz} - e^{-jz})$$

SOME STANDARD SERIES

BINOMIAL SERIES

$$(a+x)^n = a^n + na^{n-1}x + \frac{n(n-1)}{2!}a^{n-2}x^2 + \frac{n(n-1)(n-2)}{3!}a^{n-3}x^3 + \cdots \quad (x^2 < a^2)$$

The number of terms becomes infinite when n is negative or fractional.

$$(a-bx)^{-1} = \frac{1}{a}\left(1 + \frac{bx}{a} + \frac{b^2x^2}{a^2} + \frac{b^3x^3}{a^3} + \cdots\right) \quad (b^2x^2 < a^2)$$

EXPONENTIAL SERIES

$$a^x = 1 + x\ln a + \frac{(x\ln a)^2}{2!} + \frac{(x\ln a)^3}{3!} + \cdots$$

$$e^x = 1 + x + \frac{x^2}{2!} + \frac{x^3}{3!} + \cdots$$

LOGARITHMIC SERIES

$$\ln x = (x-1) - \frac{1}{2}(x-1)^2 + \frac{1}{3}(x-1)^3 - \cdots \quad (0 < x < 2)$$

$$\ln x = \frac{x-1}{x} + \frac{1}{2}\left(\frac{x-1}{x}\right)^2 + \frac{1}{3}\left(\frac{x-1}{x}\right)^3 + \cdots \quad \left(x > \frac{1}{2}\right)$$

$$\ln x = 2\left[\frac{x-1}{x+1} \cdot \frac{1}{3}\left(\frac{x-1}{x+1}\right)^3 + \frac{1}{5}\left(\frac{x-1}{x+1}\right)^5 + \cdots\right] \quad (x \text{ positive})$$

$$\ln(1+x) = x - \frac{x^2}{2} + \frac{x^3}{3} - \frac{x^4}{4} + \cdots$$

TRIGONOMETRIC SERIES

$$\sin x = x - \frac{x^3}{3!} + \frac{x^5}{5!} - \frac{x^7}{7!} + \cdots$$

$$\cos x = 1 - \frac{x^2}{2!} + \frac{x^4}{4!} - \frac{x^6}{6!} + \cdots$$

SOME STANDARD SERIES, CONTINUED

TRIGONOMETRIC SERIES, CONTINUED

$$\tan x = x + \frac{x^3}{3} + \frac{2x^5}{15} + \frac{17x^7}{315} + \frac{62x^9}{2835} + \cdots \quad \left(x^2 < \frac{\pi^2}{4}\right)$$

$$\sin^{-1} x = x + \frac{1}{2}\frac{x^3}{3} + \frac{1\cdot 3}{2\cdot 4}\frac{x^5}{5} + \frac{1\cdot 3\cdot 5}{2\cdot 4\cdot 6}\frac{x^7}{7} + \cdots \quad \left(x^2 < 1\right)$$

$$\tan^{-1} x = x - \frac{1}{3}x^3 + \frac{1}{5}x^5 - \frac{1}{7}x^7 + \cdots \quad (x^2 \leqq 1)$$

OTHER SERIES

$$1 + 2 + 3 + 4 + \cdots + (n - 1) + n = n\,(n + 1)/2$$

$$p + (p + 1) + (p + 2) + \cdots + (q - 1) + q = (q + p)(q - p + 1)/2$$

$$2 + 4 + 6 + 8 + \cdots + (2n - 2) + 2n = n(n + 1)$$

$$1 + 3 + 5 + 7 + \cdots + (2n - 3) + (2n - 1) = n^2$$

$$1^2 + 2^2 + 3^3 + 4^2 + \cdots + (n - 1)^2 + n^2 = n(n + 1)(2n + 1)/6$$

$$1^3 + 2^3 + 3^3 + 4^3 + \cdots + (n - 1)^3 + n^3 = n^2(n + 1)^2/4$$

$$\left.\begin{aligned} \frac{1+2+3+4+5+\cdots+n}{n^2} &\rightarrow \frac{1}{2} \\ \frac{1+2^2+3^2+4^2+\cdots+n^2}{n^3} &\rightarrow \frac{1}{3} \\ \frac{1+2^3+3^3+4^3+\cdots+n^3}{n^4} &\rightarrow \frac{1}{4} \end{aligned}\right\} \text{as } n \rightarrow \infty$$

MATRIX OPERATIONS

Two matrices, **A** and **B**, can be added if the number of rows in **A** equals the number of rows in **B**.

$$\mathbf{A} \pm \mathbf{B} = \mathbf{C}$$

$$a_{ij} \pm b_{ij} = c_{ij} \quad i = 1, 2, \ldots, m \quad j = 1, 2, \ldots, n$$

Matrix Operations, Continued

Multiplying a matrix or vector by a scalar implies multiplication of each element by the scalar. If $\mathbf{B} = \gamma\mathbf{A}$, then $b_{ij} = \gamma a_{ij}$ for all elements.

Two matrices, **A** and **B**, can be multiplied if the number of columns in **A** equals the number of rows in **B**. For **A** of order $m \times n$ (m rows and n columns) and **B** of order $n \times p$, the product of two matrices, $\mathbf{C} = \mathbf{AB}$, will be a matrix of order $m \times p$ elements:

$$c_{ij} = \sum_{k=1}^{n} a_{ik} b_{kj}$$

Thus, c_{ij} is the scalar product of the i'th row vector of **A** and the j'th column vector of **B**.

In general, matrix multiplication is not commutative: $\mathbf{AB} \neq \mathbf{BA}$. Matrix multiplication is associative: $\mathbf{A(BC)} = \mathbf{(AB)C}$. The distributive law for multiplication and addition holds as in the case of scalars.

$$(\mathbf{A} + \mathbf{B})\mathbf{C} = \mathbf{AC} + \mathbf{BC}$$
$$\mathbf{C}(\mathbf{A} + \mathbf{B}) = \mathbf{CA} + \mathbf{CB}$$

For some applications, the term-by-term product of two matrices, **A** and **B**, of identical order is defined as $\mathbf{C} = \mathbf{A} \cdot \mathbf{B}$, where $c_{ij} = a_{ij} b_{ij}$.

$$(\mathbf{ABC})' = \mathbf{C}'\,\mathbf{B}'\,\mathbf{A}'$$
$$(\mathbf{ABC})^{\mathrm{H}} = \mathbf{C}^{\mathrm{H}}\mathbf{B}^{\mathrm{H}}\mathbf{A}^{\mathrm{H}}$$

If both **A** and **B** are symmetric, then $(\mathbf{AB})' = \mathbf{BA}$. The product of two symmetric matrices will usually not be symmetric.

Determinants

A determinant $|\mathbf{A}|$ or $\det(\mathbf{A})$ is a scalar function of a square matrix.

$$|\mathbf{A}||\mathbf{B}| = |\mathbf{AB}|$$

$$|\mathbf{A}| = \begin{vmatrix} a_{11} & a_{12} \\ a_{21} & a_{22} \end{vmatrix} = a_{11}a_{22} - a_{12}a_{21}$$

$$|\mathbf{A}| = |\mathbf{A}'|$$

$$|\mathbf{A}| = \begin{vmatrix} a_{11} & a_{12} & a_{13} \\ a_{21} & a_{22} & a_{23} \\ a_{31} & a_{32} & a_{33} \end{vmatrix} = \begin{matrix} a_{11}a_{22}a_{33} + a_{12}a_{23}a_{31} + a_{13}a_{21}a_{32} \\ -a_{13}a_{22}a_{31} - a_{11}a_{23}a_{32} - a_{12}a_{21}a_{33} \end{matrix}$$

DETERMINANTS, CONTINUED

$$|\mathbf{A}| = \begin{vmatrix} a_{11} & a_{12} & \cdots & a_{1n} \\ a_{21} & a_{22} & \cdots & a_{2n} \\ \cdot & \cdot & \cdots & \cdot \\ a_{n1} & a_{n2} & \cdots & a_{nm} \end{vmatrix} = \Sigma(-1)^{\delta} a_{1i}, a_{2i}, \ldots, a_{ni_n}$$

where the sum is over all permutations: $i_1 \neq i_2 \neq \cdots \neq i_n$, and δ denotes the number of exchanges necessary to bring the sequence (i_1, i_2, … , i_n) back into the natural order (1, 2, … , n).

CURVE FITTING

POLYNOMIAL FUNCTION

$$y = b_0 + b_1 x + b_2 x^2 + \cdots + b_m x^m$$

For a polynomial function fit by the method of least squares, obtain the values of b_0, b_1,…, b_m by solving the system of m + 1 normal equations.

$$nb_0 + b_1 \Sigma x_i + b_2 \Sigma x_i^2 + \cdots + b_m \Sigma x_i^m = \Sigma y_i$$

$$b \Sigma x_i + b_1 \Sigma x_i^2 + b_2 \Sigma x_i^3 + \cdots + b_m \Sigma x_i^{m+1} = \Sigma x_i y_i$$

$$b_0 \Sigma x_i^m + b_1 \Sigma x_i^{m+1} + b_2 \Sigma x_i^{m+2} + \cdots + b_m \Sigma x_i^{2m} = \Sigma x_i^m y_i$$

where y_i are the function values at points x_i.

STRAIGHT LINE

$$y = b_0 + b_1 x$$

For a straight line fit by the method of least squares, obtain the values b_0 and b_1 by solving the normal equations.

$$nb_0 + b_1 \Sigma x_i = \Sigma y_i$$

$$b_0 \Sigma x_i + b_1 \Sigma x_i^2 = \Sigma x_i y_i$$

Solutions for these normal equations:

$$b_1 = \frac{n \Sigma x_i y_i - (\Sigma x_i)(\Sigma y_i)}{n \Sigma x_i^2 - (\Sigma x_i)^2} \qquad b_0 = \frac{\Sigma y_i}{n} - b_1 \frac{\Sigma x_i}{n} = \bar{y} - b_1 \bar{x}$$

CURVE FITTING, CONTINUED

EXPONENTIAL CURVE

$$y = ab^x \quad \text{or} \quad \log y = \log a + (\log b)x$$

For an exponential curve fit by the method of least squares, obtain the values log a and log b by fitting a straight line to the set of ordered pairs $\{(x_i, \log y_i)\}$.

POWER FUNCTION

$$y = ax^b \quad \text{or} \quad \log y = \log a + b \log x$$

For a power function fit by the method of least squares, obtain the values log a and b by fitting a straight line to the set of ordered pairs $\{(\log x_i, \log y_i)\}$.

SMALL-TERM APPROXIMATIONS

This section lists some first approximations derived by neglecting all powers but the first of the small positive or negative quantity, $x = s$. The expression in brackets gives the next term beyond that used, and, by means of it, the accuracy of the approximation can be estimated.

$$\frac{1}{1+s} = 1 - s \quad [+s^2]$$

$$(1+s)^n = 1 + ns \quad \left[+\frac{n(n-1)}{2}s^2\right]$$

$$e^s = 1 + s \quad \left[+\frac{s^2}{2}\right]$$

$$\ln(1+s) = s \quad \left[-\frac{s^2}{2}\right]$$

$$\sin s = s \quad \left[-\frac{s^3}{6}\right]$$

$$\cos s = 1 \quad \left[-\frac{s^2}{2}\right]$$

$$(1+s_1)(1+s_2) = (1+s_1+s_2) \quad [+s_1 s_2]$$

SMALL-TERM APPROXIMATIONS, CONTINUED

The following expressions may be approximated by $1 + s$, where s is a small ($|s| < 1$) positive or negative quantity and n any positive number:

$$e^s \quad 2-e^{-s} \quad \cos\sqrt{-2s}$$

$$\left(1+\frac{s}{n}\right)^n \quad 1+\ln\sqrt{\frac{1+s}{1-s}}$$

$$\sqrt[n]{1+ns} \quad 1+n\sin\frac{s}{n}$$

$$\sqrt[n]{\frac{1+ns/2}{1-ns/2}} \quad 1+n\ln\left(1+\frac{s}{n}\right)$$

VECTOR EQUATIONS

A vector is a line segment that has both magnitude (length) and direction. Examples are velocity, acceleration, and force. A unit vector has length one. A scalar quantity has only a magnitude, e.g., mass, temperature, and energy.

PROPERTIES OF VECTORS

A vector may be expressed in terms of a set of unit vectors along specified directions and the magnitudes of the vector's components in those directions. For example, using a set of mutually perpendicular unit vectors ($\boldsymbol{i}, \boldsymbol{j}, \boldsymbol{k}$). $\boldsymbol{V} = a\boldsymbol{i} + b\boldsymbol{j}$ in two dimensions, or $\boldsymbol{V} = a\boldsymbol{i} + b\boldsymbol{j} + c\boldsymbol{k}$ in three dimensions.

The vector sum $\boldsymbol{V}$ of any number of vectors $\boldsymbol{V}_1$, $\boldsymbol{V}_2$, $\boldsymbol{V}_3$, where $\boldsymbol{V}_1 = a_1\boldsymbol{i} + b_1\boldsymbol{j} + c_1\boldsymbol{k}$, etc., is given by

$$\boldsymbol{V} = \boldsymbol{V}_1 + \boldsymbol{V}_2 + \boldsymbol{V}_3 + \cdots = (a_1 + a_2 + a_3 + \cdots)\boldsymbol{i} + (b_1 + b_2 + b_3 + \cdots)\boldsymbol{j} + (c_1 + c_2 + c_3 + \cdots)\boldsymbol{k}$$

*PRODUCT OF A VECTOR **V** AND A SCALAR QUANTITY s*

$$s\boldsymbol{V} = (sa)\boldsymbol{i} + (sb)\boldsymbol{j} + (sc)\boldsymbol{k}$$

$$(s_1 + s_2)\boldsymbol{V} = s_1\boldsymbol{V} + s_2\boldsymbol{V} \quad (\boldsymbol{V}_1 + \boldsymbol{V}_2)s = \boldsymbol{V}_1 s + \boldsymbol{V}_2 s$$

VECTOR EQUATIONS, CONTINUED

PRODUCT OF A VECTOR V AND A SCALAR QUANTITY S, CONTINUED

where $s\boldsymbol{V}$ has the same direction as $\boldsymbol{V}$, and its magnitude is s times the magnitude of $\boldsymbol{V}$.

SCALAR PRODUCT (DOT PRODUCT) OF TWO VECTORS, $\boldsymbol{V}_1 \cdot \boldsymbol{V}_2$

$$V_1 \cdot V_2 = |V_1||V_2| \cos\phi$$

where ϕ is the angle between $\boldsymbol{V}_1$ and $\boldsymbol{V}_2$.

$$\boldsymbol{V}_1 \cdot \boldsymbol{V}_1 = |\boldsymbol{V}_1|^2 \quad \text{and} \quad \boldsymbol{i} \cdot \boldsymbol{i} = \boldsymbol{j} \cdot \boldsymbol{j} = \boldsymbol{k} \cdot \boldsymbol{k} = 1$$

because $\cos\phi = \cos(0) = 1$. On the other hand,

$$\boldsymbol{i} \cdot \boldsymbol{j} = \boldsymbol{j} \cdot \boldsymbol{k} = \boldsymbol{k} \cdot \boldsymbol{i} = 0$$

because $\cos\phi = \cos(90°) = 0$.

Scalar products behave much like products in normal (scalar) arithmetic.

$$\boldsymbol{V}_1 \cdot \boldsymbol{V}_2 = \boldsymbol{V}_2 \cdot \boldsymbol{V}_1 \quad (\boldsymbol{V}_1 + \boldsymbol{V}_2) \cdot \boldsymbol{V}_3 = \boldsymbol{V}_1 \cdot \boldsymbol{V}_3 + \boldsymbol{V}_2 \cdot \boldsymbol{V}_3$$

In a plane,

$$\boldsymbol{V}_1 \cdot \boldsymbol{V}_2 = a_1 a_2 + b_1 b_2$$

In space,

$$\boldsymbol{V}_1 \cdot \boldsymbol{V}_2 = a_1 a_2 + b_1 b_2 + c_1 c_2$$

An example of the scalar product of two vectors $\boldsymbol{V}_1 \cdot \boldsymbol{V}_2$ is the work done by a constant force of magnitude $|\boldsymbol{V}_1|$ acting through a distance $|\boldsymbol{V}_2|$, where ϕ is the angle between the line of force and the direction of motion.

VECTOR PRODUCT (CROSS PRODUCT) OF TWO VECTORS, $\boldsymbol{V}_1 \times \boldsymbol{V}_2$

$$|V_1 \times V_2| = |V_1||V_2| \sin\phi$$

where ϕ is the angle between $\boldsymbol{V}_1$ and $\boldsymbol{V}_2$.

Among unit vectors, the significance of the order of vector multiplication can be seen.

$$\boldsymbol{i} \times \boldsymbol{j} = \boldsymbol{k} \quad \boldsymbol{j} \times \boldsymbol{k} = \boldsymbol{i} \quad \boldsymbol{k} \times \boldsymbol{i} = \boldsymbol{j}$$

but reversing the order changes the sign, so that

$$\boldsymbol{j} \times \boldsymbol{i} = -\boldsymbol{k}, \text{ etc.}$$

and

VECTOR EQUATIONS, CONTINUED

VECTOR PRODUCT (CROSS PRODUCT) OF TWO VECTORS, $\boldsymbol{V}_1 \times \boldsymbol{V}_2$, CONTINUED

$$\boldsymbol{V}_1 \times \boldsymbol{V}_2 = -\boldsymbol{V}_2 \times \boldsymbol{V}_1$$

Also

$$\boldsymbol{i} \times \boldsymbol{i} = \boldsymbol{j} \times \boldsymbol{j} = \boldsymbol{k} \times \boldsymbol{k} = 0$$

and always

$$\boldsymbol{V}_1 \times \boldsymbol{V}_1 = 0$$

because $\sin\phi = \sin(0) = 0$.

An example of a vector product is the moment $\boldsymbol{M}_0$ about a point 0 of a force $\boldsymbol{F}_P$ applied at point P, where $\boldsymbol{r}_{P/0}$ is the position vector of the point of force application with respect to 0.

$$\boldsymbol{M}_0 = \boldsymbol{r}_{P/0} \times \boldsymbol{F}_P$$

SECTION 2

CONVERSION FACTORS

LENGTH (*l*)

Item	mm	m	km	in.	ft	statute mile
1 millimeter (mm)	1	10^{-3}	10^{-6}	3.937×10^{-2}	3.281×10^{-3}	6.214×10^{-7}
1 meter[a] (m)	1000	1	10^{-3}	39.37	3.281	6.214×10^{-4}
1 kilometer (km)	10^{6}	1000	1	3.937×10^{4}	3281	0.6214
1 inch (in.)	25.40	2.540×10^{-2}	2.540×10^{-5}	1	8.333×10^{-2}	1.578×10^{-5}
1 foot (ft)	304.8	0.3048	3.048×10^{-4}	12	1	1.894×10^{-4}
1 statute mile	1.609×10^{6}	1609	1.609	6.336×10^{4}	5280	1

[a]Denotes SI units.

1 foot = 1200/3937 m
1 meter = 3937/1200 ft
1 angstrom (Å) = 10^{-10} m
1 X-unit = 10^{-13} m
1 micron (μ) = 10^{-6} m
1 nanometer = 10^{-9} m

1 light year = 9.460×10^{12} km
1 parsec = 3.086×10^{13} km
1 fathom = 6 ft
1 yard = 3 ft
1 rod = 16.5 ft
1 mil = 10^{-3} in.

1 nautical mile (n mile) = 1852 m = 1.1508 statute miles (st mile) = 6076.10 ft

AREA (A)

Item	m^2	mm^2	ft^2	in.^{2}	circ mil
1 square meter[a] (m^2)	1	10^6	10.76	1550	1.974×10^9
1 square mm	10^{-6}	1	1.076×10^{-5}	1.550×10^{-3}	1.974×10^3
1 square ft	9.290×10^{-2}	9.290×10^4	1	144	1.833×10^8
1 square in.	6.452×10^{-4}	645.2	6.944×10^{-3}	1	1.273×10^6
1 circular mil	5.067×10^{-10}	5.067×10^{-4}	5.454×10^{-9}	7.854×10^{-7}	1[b]

[a]Denotes SI units.
[b]I circular mil = $\pi (d)^2/4$, where d is measured in units of 0.001 in. = 1 mil.

1 square mile = 27,878,400 ft^2 = 640 acres; 1 acre = 43,560 ft^2; 1 barn = 10^{-28} m^2

VOLUME (V)

Item	m^3	mm^3	L	ft^3	in.^{3}
1 cubic m	1	10^9	1000	35.31	6.102×10^4
1 cubic mm	10^{-9}	1	1.000×10^{-6}	3.531×10^{-8}	6.102×10^{-5}
1 liter[a] L	1.000×10^{-3}	1.000×10^6	1	3.531×10^{-2}	61.02
1 cubic ft	2.832×10^{-2}	2.832×10^7	28.32	1	1728
1 cubic in.	1.639×10^{-5}	16.39×10^3	1.639×10^{-2}	5.787×10^{-4}	1

[a]Denotes SI units.

1 U.S. fluid gallon = 4 U.S. fluid quarts = 8 U.S. fluid pints = 128 U.S. fluid ounces = 231 in.^{3}; 1 British imperial gallon = 277.42 in.^{3} (volume of 10 1b H_2O at 62°F)

PLANE ANGLE (θ)

Item	deg	min	s	rad	rev
1 degree (deg)	1	60	3600	1.745×10^{-2}	2.778×10^{-3}
1 minute (min or ′)	1.667×10^{-2}	1	60	2.909×10^{-4}	4.630×10^{-5}
1 second[a] (s or ″)	2.778×10^{-4}	1.667×10^{-2}	1	4.848×10^{-6}	7.716×10^{-7}
1 radian[a] (rad)	57.30	3438	2.063×10^5	1	0.1592
1 revolution (rev)	360	2.16×10^4	1.296×10^5	6.283	1

[a]Denotes SI units.

$$1 \text{ rev} = 2\pi \text{ rad} = 360 \text{ deg}; 1 \text{ deg} = 60' = 3600''$$

SOLID ANGLE

$$1 \text{ sphere} = 4\pi \text{ steradians} = 12.57 \text{ steradians}$$

MASS (m)

Item	g	kg	slug	amu	oz	lb	ton
1 gram (g)	1	0.001	6.852×10^{-5}	6.024×10^{23}	3.527×10^{-2}	2.205×10^{-3}	1.102×10^{-6}
1 kilogram[a] (kg)	1000	1	6.852×10^{-2}	6.024×10^{26}	35.27	2.205	1.102×10^{-3}
1 slug	1.459×10^{4}	14.59	1	8.789×10^{27}	514.8	32.17	1.609×10^{-2}
1 atomic mass unit (amu)	1.661×10^{-24}	1.661×10^{-27}	1.138×10^{-28}	1	5.857×10^{-26}	3.661×10^{-27}	1.830×10^{-30}
1 ounce (avoirdupois)	28.35	2.835×10^{-2}	1.943×10^{-3}	1.707×10^{25}	1	6.250×10^{-2}	3.125×10^{-5}
1 pound (avoirdupois)	453.6	0.4536	3.108×10^{-2}	2.732×10^{26}	16	1	5.000×10^{-4}
1 ton	9.072×10^{5}	907.2	62.16	5.464×10^{29}	3.200×10^{4}	2000	1

[a]Denotes SI units.

Note: Portion of table enclosed in the box must be used with caution because those units are not properly mass units but weight equivalents, which depend on standard terrestrial acceleration due to gravity g, where $g = 9.80665\ m/s^2 = 32.174\ ft/s^2$ on Earth at sea level. The conversion between the equivalent weight (lb) and the true mass parameter (lbm = pound mass) is given by $1.0\ lb = 32.174\ lbm{\cdot}ft/s^2$.

1 metric ton = 1000 kg

DENSITY (ρ)

Item	slug/ft^3	kg/m^3	g/cm^3	lb/ft^3	lb/in.^{3}
1 slug per ft^3	1	515.4	0.5154	32.17	1.862×10^{-2}
1 kg per m^3	1.940×10^{-3}	1	0.001	6.243×10^{-2}	3.613×10^{-5}
1 g per cm^3	1.940	1000	1	62.43	3.613×10^{-2}
1 lb per ft^3	3.108×10^{-2}	16.02	1.602×10^{-2}	1	5.787×10^{-4}
1 lb per in.^{3}	53.71	2.768×10^4	27.68	1728	1

Note: Portion of table enclosed in the box must be used with caution because those units are not mass-density units but weight-density units, which depend on g.

TIME (t)

Item	yr	day	h	min	s
1 year (yr)	1	365.2[a]	8.766×10^3	5.260×10^5	3.156×10^7
1 solar day	2.738×10^{-3}	1	24	1440	8.640×10^4
1 hour (h)	1.4108×10^{-4}	4.167×10^{-2}	1	60	3600
1 minute (min)	1.901×10^{-6}	6.944×10^{-4}	1.667×10^{-2}	1	60
1 second[b] (s)	3.169×10^{-8}	1.157×10^{-5}	2.778×10^{-4}	1.667×10^{-2}	1

[a]1 year = 365.24219879 days.
[b]Denotes SI units.

VELOCITY (v)

Item	ft/s	km/h	m/s	mile/h	mm/s	kn
1 ft per s	1	1.097	0.3048	0.6818	304.8	0.5925
1 km per h	0.9113	1	0.2778	0.6214	277.8	0.5400
1 m per s[a]	3.281	3.600	1	2.237	1000	1.944
1 mile per h	1.467	1.609	0.4470	1	447.0	.8690
1 mm per s	3.281×10^{-3}	3.600×10^{-3}	10^{-3}	2.237×10^{-3}	1	1.944×10^{-3}
1 knot (kn)	1.688	1.852	0.5144	1.151	514.4	1

[a]Denotes SI units.

1 knot = 1 n mile/h; 1 mile/min = 88 ft/s = 60 mile/h

FORCE (*F*)

Item	dyne	N	lb	poundal	gf	kgf
1 dyne	1	10^{-5}	2.248×10^{-6}	7.233×10^{-5}	1.020×10^{-3}	1.020×10^{-6}
1 newton (N)	10^{5}	1	0.2248	7.233	102.0	0.1020
1 pound (lb)	4.448×10^{5}	4.448	1	32.17	453.6	0.4536
1 poundal	1.383×10^{4}	0.1383	3.108×10^{-2}	1	14.10	1.410×10^{-2}
1 gram-force (gf)	980.7	9.807×10^{-3}	2.205×10^{-3}	7.093×10^{-2}	1	10^{-3}
1 kilogram-force (kgf)	9.807×10^{5}	9.807	2.205	70.93	1000	1

Note: Portion of table enclosed in the box must be used with caution because those units are not force units but weight equivalents of mass, which depend on *g*.

$$1 \text{ kgf} = 9.80665 \text{ N};\ 1 \text{ lb} = 32.174 \text{ poundal}$$

PRESSURE (*p*)

Item	atm	dyne/cm^2	in. H_2O	cm Hg	N/m^2 (Pa)	lb/in.^{2}	lb/ft^2
1 atmosphere (atm)	1	1.013×10^{6}	406.8	76	1.013×10^{5}	14.70	2116
1 dyne per cm^2	9.869×10^{-7}	1	4.015×10^{-4}	7.501×10^{-5}	0.1000	1.450×10^{-5}	2.089×10^{-3}
1 in. of water at 4°C[a]	2.458×10^{-3}	2491	1	0.1868	249.1	3.613×10^{-2}	5.202
1 cm of mercury at 0°C[a]	1.316×10^{-2}	1.333×10^{4}	5.353	1	1333	0.1934	27.85
1 N per m^2 (pascal)	9.869×10^{-6}	10	4.015×10^{-3}	7.501×10^{-4}	1	1.450×10^{-4}	2.089×10^{-2}
1 lb per in.^{2}	6.805×10^{-2}	6.895×10^{4}	27.68	5.171	6.895×10^{3}	1	144
1 lb per ft^2	4.725×10^{-4}	478.8	0.1922	3.591×10^{-2}	47.88	6.944×10^{-3}	1

[a]Where the acceleration of gravity has the standard value 9.80665 m/s^2.

$$1 \text{ bar} = 10^6 \text{ dyne/cm}^2 = 100 \text{ kPa}$$

$$1 \text{ in. of Hg} = 0.4912 \text{ lb/in.}^2$$

ENERGY, WORK, HEAT (W)

Item	Btu	erg	ft-lb	hp-h	J
1 British thermal unit (Btu)	1	1.055×10^{10}	778.2	3.930×10^{-4}	1055
1 erg	9.478×10^{-11}	1	7.376×10^{-8}	3.725×10^{-14}	10^{-7}
1 foot-pound (ft-lb)	1.285×10^{-3}	1.356×10^{7}	1	5.051×10^{-7}	1.356
1 horsepower-hour (hp-h)	2544	2.685×10^{13}	1.980×10^{6}	1	2.685×10^{6}
1 joule[a] (J)	9.478×10^{-4}	10^{7}	0.7376	3.725×10^{-7}	1
1 calorie (cal)	3.968×10^{-3}	4.186×10^{7}	3.087	1.559×10^{-6}	4.187
1 kilowatt-hour (kW-h)	3412	3.600×10^{13}	2.655×10^{6}	1.341	3.600×10^{6}
1 electron volt (eV)	1.519×10^{-22}	1.602×10^{-12}	1.182×10^{-19}	5.967×10^{-26}	1.602×10^{-19}
1 million electron volts	1.519×10^{-16}	1.602×10^{-6}	1.182×10^{-13}	5.967×10^{-20}	1.602×10^{-13}
1 kg	8.521×10^{13}	8.987×10^{23}	6.629×10^{16}	3.348×10^{10}	8.987×10^{16}
1 amu	1.415×10^{-3}	1.492×10^{-3}	1.100×10^{-10}	5.558×10^{-17}	1.492×10^{-10}

(continued)

[a]Denotes SI units.

Notes: The electron volt is the kinetic energy an electron gains from being accelerated through the potential difference of one volt in an electric field. The units enclosed in the box are not properly energy units; they arise from the relativistic mass-energy equivalent formula, $E = mc^2$.

1 m-kgf = 9.807 J; 1 W-s = 1 J = 1 N-m, 1 dyne-cm = 1 erg

ENERGY, WORK, HEAT (W), CONTINUED

Item	cal	kW-h	eV	MeV	kg	amu
1 British thermal unit (Btu)	252.0	2.930×10^{-4}	6.585×10^{21}	6.585×10^{15}	1.174×10^{-14}	7.074×10^{12}
1 erg	2.389×10^{-8}	2.778×10^{-14}	6.242×10^{11}	6.242×10^{5}	1.113×10^{-24}	670.5
1 foot-pound (ft-lb)	0.3239	3.766×10^{-7}	8.463×10^{18}	8.463×10^{12}	1.509×10^{-17}	9.092×10^{9}
1 horsepower-hour (hp-h)	6.414×10^{5}	0.7457	1.676×10^{25}	1.676×10^{19}	2.988×10^{-11}	1.800×10^{16}
1 joule[a] (J)	0.2389	2.778×10^{-7}	6.242×10^{18}	6.242×10^{12}	1.113×10^{-17}	6.705×10^{7}
1 calorie (cal)	1	1.163×10^{-6}	2.613×10^{19}	2.613×10^{13}	4.659×10^{-17}	2.807×10^{10}
1 kilowatt-hour (kW-h)	8.601×10^{5}	1	2.247×10^{25}	2.247×10^{19}	4.007×10^{-11}	2.414×10^{16}
1 electron volt (eV)	3.827×10^{-20}	4.450×10^{-26}	1	10^{-6}	1.783×10^{-36}	1.074×10^{-9}
1 million electron volts	3.827×10^{-14}	4.450×10^{-20}	10^{6}	1	1.783×10^{-30}	1.074×10^{-3}
1 kg	2.147×10^{16}	2.497×10^{10}	5.610×10^{35}	5.610×10^{29}	1	6.025×10^{26}
1 amu	3.564×10^{-11}	4.145×10^{-17}	9.310×10^{8}	931.0	1.660×10^{-27}	1

[a]Denotes SI units. Notes: The electron volt is the kinetic energy an electron gains from being accelerated through the potential difference of one volt in an electric field. The units enclosed in the box are not properly energy units; they arise from the relativistic mass-energy equivalent formula, $E = mc^2$.

POWER (*P*)

Item	Btu/h	Btu/s	ft-lb/min	ft-lb/s	hp	cal/s	kW	W
1 Btu/h	1	2.778×10^{-4}	12.97	0.2162	3.930×10^{-4}	7.000×10^{-2}	2.931×10^{-4}	0.2931
1 Btu/s	3600	1	4.669×10^{4}	778.2	1.415	252.0	1.055	1.055×10^{-3}
1 ft-lb/min	7.710×10^{-2}	2.142×10^{-5}	1	1.667×10^{-2}	3.030×10^{-5}	5.397×10^{-3}	2.260×10^{-5}	2.260×10^{-2}
1 ft-lb/s	4.626	1.286×10^{-3}	60.00	1	1.818×10^{-3}	0.3238	1.356×10^{-3}	1.356
1 hp	2544	0.7068	3.3×10^{4}	550.0	1	178.1	0.7457	745.7
1 cal/s	14.29	3.968×10^{-3}	185.3	3.088	5.615×10^{-3}	1	4.187×10^{-3}	4.187
1 kW	3412	0.9478	4.425×10^{4}	737.6	1.341	238.8	1	1000
1 W	3.412	9.478×10^{-4}	44.25	0.7376	1.341×10^{-3}	0.2388	0.0010	1

1 W = 1 J/s

TEMPERATURE (*T*)

Item	Conversion
Kelvin, K	5/9 R
Rankine, °R	9/5 K
Celsius, °C	5/9(F-32), K-273.15
Fahrenheit, °F	9/5C + 32, R-459.67

THERMAL CONDUCTIVITY (k)

Item	cal/s-cm °C	W/m K	W/in. °C	Btu/h-ft °F	Btu/s-in. °F	hp/ft °F
1 cal per s per cm per °C	1	418.7	10.63	241.9	5.603×10^{-3}	9.507×10^{-2}
1 W per m per K[a]	2.389×10^{-3}	1	2.540×10^{-2}	.5778	1.338×10^{-5}	2.271×10^{-4}
1 W per in. per °C	9.410×10^{-2}	39.37	1	22.75	5.266×10^{-4}	8.940×10^{-3}
1 Btu per h per ft per °F	4.134×10^{-3}	1.730	4.396×10^{-2}	1	2.315×10^{-5}	3.930×10^{-4}
1 Btu per s per in. per °F	1.785×10^{2}	7.472×10^{4}	1.898×10^{3}	4.320×10^{4}	1	16.98

[a]Denotes SI units.

ABSOLUTE OR DYNAMIC VISCOSITY (μ)

Item	Pa-s	poise	kgf-s/m^2	lb-s/ft^2	kg/m-s	lbm/ft-s
1 Pa-s[a]	1	10	1.020×10^{-1}	2.089	10^{-3}	0.6720
1 poise	10^{-1}	1	1.020×10^{-2}	2.089×10^{-3}	10^{-1}	6.720×10^{-2}
1 kg (force)-s per m^2	9.807	98.07	1	0.2048	9.807	6.590
1 lb (force)-s per ft^2	47.88	4.788×10^2	4.882	1	47.88	32.17
1 kg per m-s	1	10	0.1020	2.089×10^{-2}	1	0.6720
1 lb (mass) per ft-s	1.488	14.88	0.1518	3.108×10^{-2}	1.488	1

[a]Denotes SI units.

Notes: The absolute viscosity μ is properly expressed in force units according to its definition. In heat transfer and fluid mechanics it is usually expressed in mass-equivalent units to avoid the use of a conversion factor in Reynolds number. Mass equivalent units have been used in the portion of the table enclosed in the box. The poise, the cgs absolute viscosity unit, is defined as 1 dyne-s/cm^2.

KINEMATIC VISCOSITY ($\nu = \mu/\rho$)

Item	centistoke	stoke	m^2/s	ft^2/s
1 centistoke	1	10^{-2}	10^{-6}	1.076×10^{-5}
1 stoke	100	1	10^{-4}	1.076×10^{-3}
1 m^2/s[a]	10^6	10^4	1	10.76
1ft^2/s	9.290×10^4	929.0	9.290×10^{-2}	1

[a]Denotes SI units.

$$1 \text{ stoke} = 1 \text{ cm}^2/\text{s}$$

INDUCTANCE (*L*)

Item	abhenry	henry	μH	stathenry
1 abhenry[a] (1 emu)	1	10^{-9}	0.001	1.113×10^{-21}
1 henry[b] (H)	10^{9}	1	10^{6}	1.113×10^{-12}
1 microhenry (μH)	10^{3}	10^{-6}	1	1.113×10^{-18}
1 stathenry[a] (1 esu)	8.988×10^{20}	8.988×10^{11}	8.988×10^{17}	1

[a]These units are listed for historical completeness only. They are no longer used.
[b]Denotes SI units.

CAPACITANCE (*C*)

Item	abF	F	*μ*F	statF
1 abfarad[a] (1 emu)	1	10^{9}	10^{15}	8.988×10^{20}
1 farad[b] (F)	10^{-9}	1	10^{6}	8.988×10^{11}
1 microfarad (μF)	10^{-15}	10^{-6}	1	8.988×10^{5}
1 statfarad[a] (1 esu)	1.113×10^{-21}	1.113×10^{-12}	1.113×10^{-6}	1

[a]These units are listed for historical completeness only. They are no longer used.
[b]Denotes SI units.

ELECTRICAL RESISTANCE (*R*)

Item	abohm	ohm	statohm
1 abohm[a] (1 emu)	1	10^{-9}	1.113×10^{-21}
1 ohm[b]	10^{9}	1	1.113×10^{-12}
1 statohm[a] (1 esu)	8.988×10^{20}	8.988×10^{11}	1

[a]These units are listed for historical completeness only. They are no longer used.
[b]Denotes SI units.

ELECTRICAL RESISTIVITY, RECIPROCAL CONDUCTIVITY (ρ)

Item	abohm-cm	ohm-cm	ohm-m	statohm-cm	ohm-circ mil/ft[a]
1 abohm-cm (1 emu)	1	10^{-9}	10^{-11}	1.113×10^{-21}	6.015×10^{-3}
1 ohm-cm	10^{9}	1	0.0100	1.113×10^{-12}	6.015×10^{6}
1 ohm-m[b]	10^{11}	100	1	1.113×10^{-10}	6.015×10^{8}
1 statohm-cm (1 esu)	8.988×10^{20}	8.988×10^{11}	8.988×10^{9}	1	5.406×10^{18}
1 ohm-circular mil per ft	166.2	1.662×10^{-7}	1.662×10^{-9}	1.850×10^{-19}	1

[a]1 circular mil = $\pi (d)^2/4$, where d is measured in units of 0.001 in. = 1 mil.
[b]Denotes SI units.

MAGNETIC FIELD INTENSITY (H)

Item	abamp-turn/cm	amp-turn/cm	amp-turn/m	amp-turn/in.	oersted
1 abamp-turn per cm[a]	1	10	1000	25.40	12.57
1 amp-turn per cm	0.100	1	100	2.54	1.257
1 amp-turn per m[b]	10^{-3}	10^{-2}	1	2.540×10^{-2}	1.257×10^{-2}
1 amp-turn per in.	3.937×10^{-2}	0.3937	39.37	1	0.4947
1 oersted	7.958×10^{-2}	0.7958	79.58	2.021	1

[a]These units are listed for historical completeness only. They are no longer used.
[b]Denotes SI units.

1 oersted = 1 gilbert/cm; 1 esu = 2.655×10^{-9} amp-turn/m; 1 praoersted = 4π amp-turn/m

MAGNETOMOTIVE FORCE

Item	abamp-turn	amp-turn	gilbert
1 abamp-turn[a]	1	10	12.57
1 amp-turn[b]	0.100	1	1.257
1 gilbert	7.958×10^{-2}	0.7958	1

[a]These units are listed for historical completeness only. They are no longer used.
[a]Denotes SI units.

1 pragilbert = 4π amp-turn
1 esu = 2.655×10^{-11} amp-turn

MAGNETIC FLUX (ϕ)

Item	maxwell	kiloline	Wb
1 maxwell (1 line or 1 emu)	1	0.001	10^{-8}
1 kiloline	1000	1	10^{-5}
1 weber[a] (Wb)	10^{8}	10^{5}	1

[a]Denotes SI units.

1 esu = 2.998 Wb

MAGNETIC FLUX DENSITY (B)

Item	gauss	kiloline/in.^{2}	Wb/m^2	milligauss	gamma
1 gauss (line per cm^2)	1	6.452×10^{-3}	10^{-4}	1000	10^{5}
1 kiloline per in.^{2}	155.0	1	1.550×10^{-2}	1.550×10^{5}	1.550×10^{7}
1 Wb per m^2[a] (T)	10^{4}	64.52	1[b]	10^{7}	10^{9}
1 milligauss	10^{-3}	6.452×10^{-6}	10^{-7}	1	100
1 gamma	10^{-5}	6.452×10^{-8}	10^{-9}	10^{-2}	1

[a] Denotes SI units.
[b]1 Wb/m^2 = 1 tesla (T).

1 esu = 2.998×10^{6} Wb/m^2

FUNDAMENTAL CONSTANTS

Quantity	Symbol	Value
Absolute zero		–459.67°F –273.15°C
Acceleration of gravity	g	9.80665 m/s^2 32.17405 ft/s^2
Avogadro constant	N_A, L	$6.02214129 \times 10^{23}$ mol^{-1}
Bohr magneton, $eh/2m_e$	μ_B	$9.27400968 \times 10^{-24}$ J/T
Bohr radius, $\alpha/4\pi^2R_\infty$	a_o	$0.5291771092 \times 10^{-10}$ m
Boltzmann constant, R/N_A	k	$1.3806488 \times 10^{-23}$ J/K
Classical electron radius, $\alpha^2\alpha_0$	r_e	$2.8179403267 \times 10^{-15}$ m
Compton wavelength, h/m_ec	λ_c	$2.4263102389 \times 10^{-12}$ m
Conductance quantum, $2e^2/h$	G_o	$7.7480917346 \times 10^{-5}$ S
Deuteron-electron mass ratio	m_d/m_e	3670.4829652
Deuteron mass	m_d	$3.34358348 \times 10^{-27}$ kg
Deuteron-proton magnetic moment ratio	μ_d/μ_p	0.3070122070
Deuteron-proton mass ratio	m_d/m_p	1.99900750097
Electric constant $1/\mu_oc^2$	ε_o	$8.854187817 \times 10^{-12}$ F/m
Electron magnetic moment	μ_e	$-928.476430 \times 10^{-26}$ J/T
in Bohr magnetons	μ_e/μ_B	–1.00115965218076
Electron mass	m_e	$9.10938291 \times 10^{-31}$ kg
Electron mass in atomic mass units		$5.4857990946 \times 10^{-4}$ u
Elementary charge	e	$1.602176565 \times 10^{-19}$ C
Faraday constant, N_Ae	F	96,485.3365 C/mol
Fine structure constant	α	$7.2973525968 \times 10^{-3}$
First radiation constant, $2\pi hc^2$	c_1	$3.74177153 \times 10^{-16}$ Wm^2
Josephson constant, 2 e/h	K_J	$4.83597870 \times 10^{14}$ Hz/V
Magnetic constant ($4\pi \times 10^{-7}$)	μ_0	$12.566370614 \times 10^{-7}$ NA^{-2}
Magnetic flux quantum, $h/2e$	θ_o	$2.067833758 \times 10^{-15}$ Wb
Molar gas constant	R	8.3144621 J/mol/K
Molar volume (ideal gas), RT/p $T = 273.15$ K, $p = 101.325$ kPa	V_m	0.022413968 m^3/mol
Muon-electron mass ratio	m_μ/m_e	206.7682843

(continued)

FUNDAMENTAL CONSTANTS, CONTINUED

Quantity	Symbol	Value
Muon magnetic moment	μ_μ	$-4.49044807 \times 10^{-26}$ J/T
Muon magnetic moment anomaly, $[\mu_\mu/(eh/2m_\mu)] - 1$	a_μ	0.00116592091
Muon mass	m_μ	$1.883531475 \times 10^{-28}$ kg
Muon-proton magnetic moment ratio	μ_μ/μ_p	−3.183345107
Neutron-electron mass ratio	m_n/m_e	1838.6836605
Neutron magnetic moment	μ_n	$-0.96623647 \times 10^{-26}$ J/T
in nuclear magnetons	μ_n/μ_N	−1.91304272
Neutron mass	m_n	$1.674927351 \times 10^{-27}$ kg
Neutron-proton magnetic moment ratio	μ_n/μ_p	−0.68497934
Neutron-proton mass ratio	m_n/m_p	1.00137841917
Newtonian constant of gravitation	G	6.67384×10^{-11} m^3 kg^{-1} s^{-2}
Nuclear magneton, $eh/2m_p$	μ_N	$5.05078353 \times 10^{-27}$ J/T
Planck constant	h	$6.62606957 \times 10^{-34}$ J-s
Planck constant, molar	$N_A h$	$3.9903127176 \times 10^{-10}$ J-s/mol
	$N_A hc$	0.11962656472 Jm/mol
Proton-electron mass ratio	m_p/m_e	1836.15267245
Proton magnetic moment	μ_p	$1.410606743 \times 10^{-26}$ J/T
in Bohr magnetons	μ_p/μ_B	$1.521032210 \times 10^{-3}$
in nuclear magnetons	μ_p/μ_N	2.792847356
Proton mass	m_p	$1.672621777 \times 10^{-27}$ kg
Rydberg constant, $m_e c\alpha^2/2h$	R_∞	10,973,731.568527 m^{-1}
Second radiation constant, hc/k	c_2	0.014387752 mK
Speed of light in vacuum	c	299,792,458 m/s
Stefan–Boltzmann constant	σ	5.670373×10^{-8} W/m^{2}K^4
Thomson cross section, $(8\pi / 3)r_o^2$	σ_e	$0.6652458734 \times 10^{-28}$ m^2
Von Klitzing constant, h/e^2	R_k	25,812.8074434 Ω

COMMON DERIVED UNITS

Quantity	Unit	Symbol
Acceleration	meter per second squared, m s^{-2}	a
	feet per second squared, ft s^{-2}	a
Acceleration, angular	radian per second squared, rad s^{-2}	α
Activity of a radionuclide	becquerel, (one per second) s^{-1}	Bq
Angular velocity	radian per second, rad s^{-1}	ω
Area	square meter, m^{2}	A
	square feet, ft^{2}	A
Capacitance	farad, (coulomb per volt) C V^{-1}	F
Concentration	mole per cubic meter, mol m^{-3}	
	mole per cubic foot, mol ft^{-3}	
Density	kilogram per cubic meter, kg m^{-3}	ρ
	pound per cubic inch, lb in.$^{-3}$	
Dose, absorbed	gray, (joule per kilogram) J kg^{-1}	ρ
Dose, equivalent	sievert, (joule per kilogram) J kg^{-1}	Gy
Electric conductance	siemens, (one per ohm) Ω^{-1}	Sv
Electric field strength	volt per meter, V m^{-1}	S
	volt per foot, V ft^{-1}	
Electric resistance	ohm, (volt per ampere) V amp^{-1}	Ω
Energy, work, quantity	joule, (newton meter) N-m	J
of heat	British thermal unit	Btu
	foot pound, ft-lb	
Force	newton, (meter kilogram per	N
	second squared) m kg s^{-2}	
	pound, lb	
Frequency	hertz, (one per second) s^{-1}	Hz
Heat capacity, entropy	joule per degree kelvin, J K^{-1}	
	Btu per degree Fahrenheit, Btu °F^{-1}	
Heat flux density,	watt per square meter, W m^{-2}	
irradiance	watt per square foot, W ft^{-2}	
Illuminance	lux, (lumen per square meter) lm m^{-2}	lx
Inductance	henry, (webers per ampere) Wb amp^{-1}	H

(continued)

COMMON DERIVED UNITS, CONTINUED

Quantity	Unit	Symbol
Luminance	candela per square meter, cd m^{-2}	
	candela per square foot, cd ft^{-2}	
Luminance flux	lumen, cd sr	lm
Magnetic field strength	ampere per meter, amp m^{-1}	H
Magnetic flux	weber, (volt second) V s	Wb
Magnetic flux density	tesla, (weber per square meter) Wb m^{-2}	T
Magnetomotive force	ampere	amp
Molar energy	joule per mole, J mol^{-1}	
	Btu per mole, Btu mol^{-1}	
Molar entropy, molar heat capacity	joule per mole per kelvin, J mol^{-1} K^{-1}	
	Btu per mole per degree Fahrenheit, Btu mol^{-1} $°F^{-1}$	
Potential difference, electromotive force	volt, (watt per ampere) W amp^{-1}	V
Power, radiant flux	watt, (joule per second) J s^{-1}	W
	watt, (Btu per second) Btu s^{-1}	W
Pressure, stress	pascal, (newton per square meter) N m^{-2}	Pa
	atmosphere, (pounds per square inch) lb $in.^{-2}$	P
Quantity of electricity	coulomb, (ampere second) amp s	C
Specific heat capacity, specific entropy	joule per kilogram per kelvin, J kg^{-1} K^{-1}	
	Btu per pound per degree Fahrenheit, Btu lb^{-1} $°F^{-1}$	
Thermal conductivity	watt per meter per kelvin, W m^{-1} K^{-1}	
	watt per foot per degree Fahrenheit, W ft^{-1} $°F^{-1}$	
Velocity	meter per second, m s^{-1}	v
	feet per second, ft s^{-1}	v
	miles per hour, mph, mi hr^{-1}	v
Viscosity, dynamic	pascal second, Pa s	μ
	centipoise	μ
Viscosity, kinematic	meter squared per second, m^2 s^{-1}	υ
	foot squared per second, ft^2 s^{-1}	υ
Volume	cubic meter, m^3	V
	cubic inch, $in.^3$	V
Wave number	one per meter, m^{-1}	
	one per inch, $in.^{-1}$	

SECTION 3

MATERIALS AND SPECIFICATIONS

METALLIC MATERIALS

Material	F_{tu} [a] MPa	F_{tu} (ksi)	F_{ty} [b] MPa	F_{ty} (ksi)	F_{cy} [c] (MPa)	F_{cy} (ksi)	E_t [d] GPa	E_t (psi/10^6)	w [e] kg m^{-3}	w (lb/in.^{3})	Characteristics
Alloy steels											
4130 normalized sheet, strip, plate, and tubing	655	95	517	75	517	75	200	29	7833	0.283	≤ 0.187 thick } weldable
	621	90	483	70	483	70	200	29	7833	0.283	> 0.187 thick } weldable
4130 wrought forms (180 H.T.)	1241	180	1124	163	1193	173	200	29	7833	0.283	<0.5 equiv. diam, weldable
4330 wrought forms (220 H.T.)	1517	220	1282	186	1338	194	200	29	7833	0.283	<2.5 equiv. diam
DGAC wrought forms (220 H.T.)	1517	220	1310	190	1365	198	200	29	7833	0.283	<5.0 equiv. diam
300 M bars, forgings, and tubing (280 H.T.)	1931	280	1586	230	1703	247	200	29	7833	0.283	<5.0 equiv. diam
4340 bars, forgings, and tubing (260 H.T.)	1793	260	1482	215	1655	240	200	29	7833	0.283	<3.5 equiv. diam
Stainless steels											
301 (full hard) sheet and strip	1276	185	965	140	676*	98*	179	26	7916	0.286	Weldable *Longitudinal grain direction
15-5 PH bars and forgings	793	115	517	75	683	99					Readily forged and welded
	1310	190	1172	170	986	143	197	28.5	7833	0.283	Available in range of H.T.
PH15-7 Mo sheet, strip, and plate	1310	190	1172	170	1234	179	200	29	7667	0.277	Readily cold formed and cold drawn
bars and forgings	1241	180	1103	160	1158	168	200	29	7667	0.277	Readily cold formed and cold drawn

(continued)

METALLIC MATERIALS, CONTINUED

	F_{tu} [a]		F_{ty} [b]		F_{cy} [c]		E_t [d]		w [e]		
Material	MPa	(ksi)	MPa	(ksi)	(MPa)	(ksi)	GPa	(psi/10^6)	kg m^{-3}	(lb/in.^{3})	Characteristics
Stainless steels, continued											
17-4 PH sheet, strip,	931	135	724	105					7861	0.284	
and plate	1310	190	1172	170			197	28.5	7806	0.282	Readily forged, welded, and brazed
bars	1345	195	517	75					7861	0.284	Can be sand or investment molded or centrifugally cast
	1310	190	1172	170			197	28.5	7806	0.282	
castings	896	130	827	120							
	1241	180	1103	160			197	28.5	7806	0.282	
Heat resistant steels											
A286 sheet, strip, and plate	965	140	655	95	655	95	201	29.1	7944	0.287	High strength up to 1300 °F Weldable
bars, forgings, tubing	896	130	586	85	586	85	201	29.1	7944	0.287	
Inconel 600											
sheet, strip, plate, tubing, and forgings	552	80	207	30	207	30	207	30	8415	0.304	Annealed — For low stressed parts up to 2000°F—weldable
	655	95	483	70							
bars and rods	827	120	621	90			207	30	8415	0.304	Cold drawn — For low stressed parts up to 2000°F—weldable
Inconel 718											
sheet, plate, and tubing	1172	170	1000	145			204	29.6	8221	0.297	High strength and creep resistant to 1300°F—can be cast (lower values)
bars and forgings	1241	180	1034	150			204	29.6	8221	0.297	

(continued)

[a]F_{tu} = strength, tensile ultimate. [b]F_{ty} = strength, tensile yield. [c]F_{cy} = strength, compressive yield. [d]E_t = tangent modulus. [e]w = density. $E = 2G(1+\mu)$ where Poisson's ratio $\mu = 0.31–0.33$ for most metals.

Source: MMPDS-05, "Metallic Materials Properties Development and Standards Handbook."

METALLIC MATERIALS, CONTINUED

Material	F_{tu} [a] MPa	F_{tu} [a] (ksi)	F_{ty} [b] MPa	F_{ty} [b] (ksi)	F_{cy} [c] MPa	F_{cy} [c] (ksi)	E_t [d] GPa	E_t [d] (psi/10^6)	w [e] kg m^{-3}	w [e] (lb/in.$^{-3}$)	Characteristics
Aluminum alloy sheet											
2024-T3 (bare)	434	63	290	42	310	45	74	10.7	2768	0.100	Common use—low cost
2024-T3 (clad)	400	58	269	39	290	42	72	10.5	2768	0.100	Good strength/weight
2219-T87	427	62	345	50	345	50	72	10.5	2823	0.102	High strength—creep resistant
5456-H343	365	53	283	41	269	39	70	10.2	2657	0.096	Corrosion resistant—good weldability
6061-T6	290	42	248	36	241	35	68	9.9	2713	0.098	Low cost—formable—weldable
7075-T6	524	76	455	66	462	67	71	10.3	2796	0.101	High strength/weight
7075-T73	462	67	386	56	379	55	71	10.3	2796	0.101	Stress, corrosion resistant
7178-T6	572	83	503	73	503	73	71	10.3	2823	0.102	High strength/weight
Plate											
2024-T351	393	57	283	41	248	36	74	10.7	2768	0.100	Common use—low cost
2219-T87	427	62	345	50	345	50	72	10.5	2823	0.102	High strength—creep resistant
5456-H343	365	53	283	41	269	39	70	10.2	2657	0.096	Corrosion resistant—good weldablility
6061-T651	290	42	248	36	241	35	68	9.9	2713	0.098	Low cost—formable—weldable
7050-T73651	490	71	427	62	414	60	71	10.3	2823	0.102	Good fracture toughness
Extrusions											
2024-T4	414	60	303	44	269	39	74	10.7	2768	0.100	Common use—low cost
6061-T6	262	38	241	35	234	34	70	10.1	2713	0.098	Low cost—corrosion resistant—weldable
7050-T6510/1	469	68	407	59	441	64	71	10.3	2823	0.102	High stress/corrosion resistance
7075-T6	558	81	503	73	510	74	72	10.5	2796	0.101	High strength/weight

(continued)

[a]F_{tu} = strength, tensile ultimate. [b]F_{ty} = strength, tensile yield. [c]F_{cy} = strength, compressive yield. [d]E_t = tangent modulus. [e]w = density. $E = 2G(1 + \mu)$ where Poisson's ratio $\mu = 0.31$–0.33 for most metals.

Source: MMPDS-05, "Metallic Materials Properties Development and Standards Handbook."

METALLIC MATERIALS, CONTINUED

Material	F_{tu}[a] MPa	F_{tu}[a] (ksi)	F_{ty}[b] MPa	F_{ty}[b] (ksi)	F_{cy}[c] MPa	F_{cy}[c] (ksi)	E_t[d] GPa	E_t[d] (psi/10^6)	w[e] kg m^{-3}	w[e] (lb/$in.^{-3}$)	Characteristics
Aluminum alloy sheet, continued											
7075-T73	455	66	400	58	400	58	72	10.5	2796	0.101	Good stress/corrosion resistance
7178-T6	607	88	545	79	545	79	72	10.5	2823	0.102	High strength/weight
Tubing											
2024-T3	441	64	290	42	290	42	72	10.5	2768	0.100	Low cost—common use
6061-T6	290	42	241	35	234	34	70	10.1	2713	0.098	Weldable—corrosion resistant
Forgings											
2014-T6	448	65	379	55	379	55	74	10.7	2796	0.101	Common use
7050-T736	483	70	372	54	393	57	70	10.2	2823	0.102	Good fracture toughness
7075-T73	421	61	359	52	372	54	69	10.0	2796	0.101	Good stress/corrosion resistance
Castings											
356-T6	207	30	138	20	138	20	72	10.4	2685	0.097	Easy sand and investment
A356-T61	262	38	193	28	193	28	72	10.4	2685	0.097	Good corrosion resistance
A357-T61	345	50	276	40	276	40	72	10.4	2685	0.097	Premium castings
Titanium alloys											
6AL-4V (S.T.A.)											
sheet, strip, and plate	1103	160	1000	145	1034	150	110	16.0	4429	0.160	Can be spot and fusion welded
forgings (aww)	896	130	827	120			110	16.0	4429	0.160	Corrosion resistant
bars	1000	145	931	135			110	16.0	4429	0.160	High strength
6AL-6V-2SN											
sheet, strip, and plate	1172	170	1103	160	1172	170	117	17.0	4540	0.164	High strength
forgings	1034	150	965	140			117	17.0	4540	0.164	Good formability
bars	1172	170	1069	155			117	17.0	4540	0.164	Corrosion resistant

[a]F_{tu} = strength, tensile ultimate. [b]F_{ty} = strength, tensile yield. [c]F_{cy} = strength, compressive yield. [d]E_t = tangent modulus. [e]w = density. $E = 2G(1+\mu)$ where Poisson's ratio $\mu = 0.31$–0.33 for most metals.

Source: MMPDS-05, "Metallic Materials Properties Development and Standards Handbook."

WEIGHTS OF LIQUIDS

Liquid	Specific Gravity	at °C	Specific Weight kg L^{-1}	(lb/U.S. gal)	kg m^{-3}	(lb/ft^3)
Alcohol (methyl)	0.790	20	0.790	(6.59)	790	(49.3)
Benzine	0.899	0	0.899	(7.50)	899	(56.1)
Carbon tetrachloride	1.595	20	1.595	(13.31)	1595	(99.6)
Ethylene glycol	1.12		1.120	(9.35)	1120	(69.9)
Gasoline	0.72		0.720	(6.01)	720	(44.9)
Glycerine	1.261	20	1.261	(10.52)	1261	(78.7)
JP1	0.80		0.800	(6.68)	800	(49.9)
JP3	0.775		0.775	(6.47)	775	(48.4)
JP4	0.785		0.785	(6.55)	785	(49.0)
JP5	0.817	15	0.817	(6.82)	817	(51.0)
JP6	0.810		0.810	(6.76)	810	(50.6)
Kerosene	0.82		0.820	(6.84)	820	(51.2)
Mercury	13.546	20	13.546	(113.05)	13546	(845.6)
Oil	0.89	15	0.890	(7.43)	890	(55.6)
Sea water	1.025	15	1.025	(8.55)	1025	(64.0)
Synthetic oil	0.928	15	0.928	(7.74)	928	(57.9)
Water	1.000	4	1.000	(8.35)	1000	(62.4)

WEIGHTS OF GASES

Gas	Specific wt.,[a] kg m^{-3}	(lb/ft^3)	
Air	1.2256	(0.07651)	(at 59.0°F)
Air	1.2928	(0.08071)	
Carbon dioxide	1.9768	(0.12341)	
Carbon monoxide	1.2504	(0.07806)	
Helium	0.1784	(0.01114)	
Hydrogen	0.0899	(0.00561)	
Nitrogen	1.2506	(0.07807)	
Oxygen	1.4290	(0.08921)	

[a]At atmospheric pressure and 0 °C.

COMPOSITE VS METAL ALLOY

Style of Material	Maximum Service Temp.		Density		Ultimate Tensile Strength		Tensile Modulus		Ultimate Flexural Strength		Flexural Modulus of Elasticity		Specific Tensile Modulus		Ultimate Compression Strength		Cured Ply Thickness		Coefficient of Thermal Expansion, $\times10^6$		Thermal Conductivity	
Cure Class	°C	(°F)	kg m^{-3}	(lb/in.^{3})	MPa	(ksi)	GPa	(psi $\times10^6$)	MPa	(ksi)	GPa	(psi $\times10^6$)	GPa	(psi $\times10^6$)	MPa	(ksi)	mm	mils	m m^{-1} K^{-1}	(in./in./°F)	W m^{-1} K^{-1}	(Btu/°F/h/ft)
Aluminum 7075-T6	149+	(300+)	2768	(0.100)	572	(83)	69	(10)	572	(83)	69	(10)	689	(100)	572	(83)	–	–	24.30	(13.5)	121.1	(70)
Carbon steel 4310	316	(600)	7999	(0.289)	1310	(190)	207	(30)	1310	(190)	207	(30)	717	(104)	1310	(190)	–	–	11.34	(6.3)	38.1	(22)
Stainless steel 316	538+	(1000+)	8027	(0.290)	579	(84)	193	(28)	579	(84)	193	(28)	665	(96.5)	579	(84)	–	–	16.02	(8.9)	16.3	(9.4)
Titanium 6AL 4VA	538	(1000)	4429	(0.160)	1103	(160)	114	(16.5)	1103	(160)	114	(16.5)	710	(103)	1103	(160)	–	–	9.54	(5.3)	7.3	(4.2)
Magnesium HK 31 A-H2A	93	(200)	1799	(0.065)	200	(29)	44	(6.4)	200	(29)	44	(6.4)	679	(98.5)	200	(29)	–	–	25.20	(14)	114.2	(66)
Low temp epoxy fiberglass 250°F cure 7781 cloth	93	(200)	1827	(0.066)	434	(63)	23	(3.4)	538	(78)	22	(3.2)	355	(51.5)	421	(61)	0.23–0.25	9–10	4.86	(2.7)	0.4	(0.26)
High temp epoxy fiberglass 350°F cure 7781 cloth	177	(350)	1827	(0.066)	483	(70)	31	(4.5)	607	(88)	24	(3.5)	469	(68)	490	(71)	0.23–0.25	9–10	5.04	(2.8)	0.4	(0.25)
Epoxy/fiberglass UD tape 350°F cure	177	(350)	1827	(0.066)	1103	(160)	48	(7.0)	1269	(184)	48	(6.9)	731	(106.0)	607	(88)	0.13	5	14.94	(8.3)	0.3	(0.2)
Phenolic/fiberglass 7781 cloth	177	(350)	1827	(0.066)	303	(44)	21	(3.1)	455	(66)	24	(3.5)	328	(47.5)	310	(45)	0.23–0.25	9–10	10.80	(6.0)	0.3	(0.18)

(continued)

Note: The reported property values are nominal laminate averages and can change due to a number of variables. The data are offered for general information only, subject to your verification.

COMPOSITE VS METAL ALLOY, CONTINUED

Style of Material	Maximum Service Temp.		Density		Ultimate Tensile Strength		Tensile Modulus		Ultimate Flexural Strength		Flexural Modulus of Elasticity		Specific Tensile Modulus		Ultimate Compression Strength		Cured Ply Thickness		Coefficient of Thermal Expansion, $\times 10^6$		Thermal Conductivity	
Cure Class	°C	(°F)	kg m^{-3}	(lb/in.^{3})	MPa	(ksi)	GPa	(psi $\times 10^6$)	MPa	(ksi)	GPa	(psi $\times 10^6$)	GPa	(psi $\times 10^6$)	MPa	(ksi)	mm	mils	m m^{-1} K^{-1}	(in./in./°F)	W m^{-1} K^{-1}	(Btu/°F/h/ft)
Polyimide/fiberglass 7781 cloth condensation type	316	(600)	1855	(0.067)	400	(58)	28	(4.0)	538	(78)	24	(3.5)	414	(60.0)	400	(58)	0.23–0.25	9–10	10.80	(6.0)	0.31	(0.18)
Polyimide/fiberglass 7781 cloth PMR-15 600°F cure	316	(600)	1993	(0.072)	434	(63)	26	(3.8)	579	(84)	28	(4.0)	365	(53.0)	483	(70)	0.23–0.25	9–10	9.00	(5.0)	0.22	(0.125)
Epoxy/Kevlar cloth 350°F cure 285 style Kevlar 49	177	(350)	1439	(0.052)	496	(72)	24	(3.5)	400	(58)	23	(3.3)	462	(67.0)	193	(28)	0.23–0.25	9–10	0.00	(0.001)	0.90	(0.52)
Epoxy/graphite cloth 350°F cure HM woven graphite	177	(350)	1605	(0.058)	552	(80)	69	(10.0)	745	(108)	62	(9.0)	1186	(172)	586	(85)	0.20–0.23	8–9	0.01	(0.008)	48.43	(28)
Epoxy/graphite HM UD tape 350°F cure	177	(350)	1578	(0.057)	1586	(230)	131	(19.0)	1751	(254)	118	(17.1)	2296	(333)	1310	(190)	0.13	5	0.00	(±0)	69.18	(40)
BMI/graphite HM cloth	232	(450)	1522	(0.055)	648	(94)	76	(11.0)	855	(124)	59	(8.5)	1379	(200)	731	(106)	0.20–0.23	8–9	0.02	(0.01)	51.89	(30)
BMI/graphite HM UD tape	232	(450)	1522	(0.055)	1731	(251)	148	(21.5)	1986	(288)	127	(18.4)	2696	(391)	1641	(238)	0.13	5	0.00	(±0)	72.64	(42)
Polyimide graphite HM cloth PMR-15	316	(600)	1605	(0.058)	731	(106)	66	(9.5)	841	(122)	83	(12.0)	1131	(164)	717	(104)	0.20	8	0.02	(0.01)	51.89	(30)

Note: The reported property values are nominal laminate averages and can change due to a number of variables. The data are offered for general information only, subject to your verification.

COMPOSITE MATERIAL BEHAVIOR

Composite materials have been used in the design of structural elements for years because of the ability to tailor the laminate and sandwich construction to suit the loading requirements and the environment in which the materials operate.

Composites are not isotropic materials. The stiffness or compliance of an isotropic material is defined by three materials properties or engineering constants E, G, and ν, of which only two are independent since $G = E/2(1+\nu)$.

The stiffness of a single ply of unidirectional tape is orthotropic and is defined by the material properties E_x, E_y, ν_x, ν_y, and E_s, of which four are independent since $\nu_x/\nu_y = E_x/E_y$. x is defined as the axis aligned with the fiber direction and is called "longitudinal," where y is "transverse."

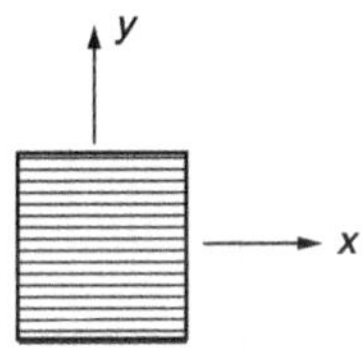

The on-axis stress–strain relationship for a unidirectional ply is defined by

	σ_x	σ_y	σ_s
ε_x	$\dfrac{1}{E_x}$	$-\dfrac{\nu_y}{E_y}$	
ε_y	$-\dfrac{\nu_y}{E_x}$	$\dfrac{1}{E_y}$	
ε_s			$\dfrac{1}{E_s}$

or

	σ_x	σ_y	σ_s
ε_x	$\dfrac{1}{E_x}$	$-\dfrac{\nu_y}{E_y}$	
ε_y	$-\dfrac{\nu_y}{E_x}$	$\dfrac{1}{E_y}$	
ε_s			$\dfrac{1}{E_s}$

COMPOSITE MATERIAL BEHAVIOR, CONTINUED

The 3 × 3 matrix, **S**, is the compliance matrix. The inverse below, **Q**, is the stiffness matrix, used to calculate stress from strain.

	σ_x	σ_y	σ_s
ε_x	S_{xx}	S_{xy}	
ε_y	S_{yx}	S_{yy}	
ε_s			S_{ss}

Evaluating a laminate of varying ply orientations makes it necessary to transform the *x–y* on-axis orientation to an off-axis orientation identified as the 1–2 axis, with the in-plane shear identified as the "6" direction.

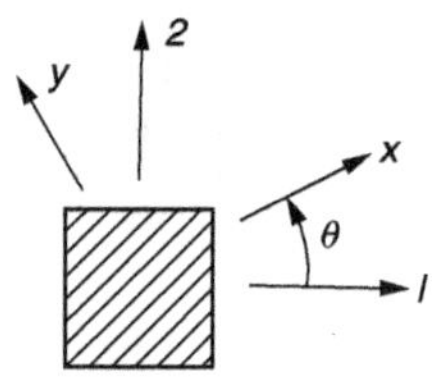

Noting the + or – rotational direction is important. Transforming from the 1, 2, 6 axis system to the *x*, *y*, *s* axis system is shown here for the three in-plane stress components:

	σ_1	σ_2	σ_6
σ_x	m^2	n^2	$2mn$
σ_y	n^2	m^2	$-2mn$
σ_s	$-mn$	mn	$m^2 - n^2$

where

$$m^2 = \cos^2\theta = \frac{1}{2} + \frac{1}{2}\cos 2\theta$$

$$n^2 = \sin^2\theta = \frac{1}{2} + \frac{1}{2}\cos 2\theta$$

$$2mn = \sin 2\theta$$

$$m^2 - n^2 = \cos 2\theta$$

COMPOSITE MATERIAL BEHAVIOR, CONTINUED

Similarly, for strain transformations,

	ε_1	ε_2	ε_6
ε_x	m^2	n^2	mn
ε_y	n^2	m^2	$-mn$
ε_s	$-2mn$	$2mn$	$m^2 - n^2$

The off-axis in-plane stress–strain relationship is defined by Q_{ij} $(i, j = 1, 2, 6)$.

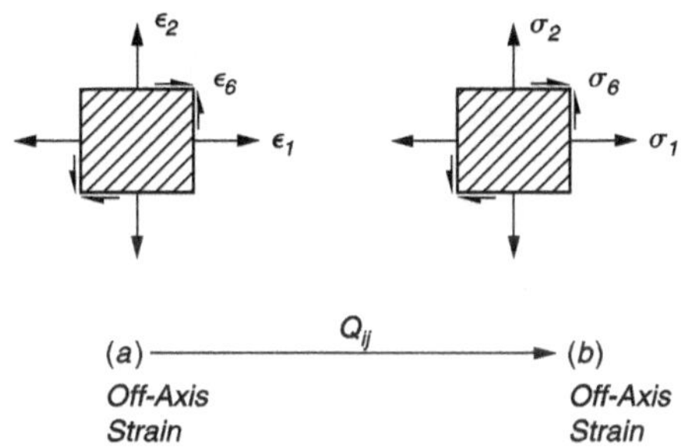

	ε_1	ε_2	ε_6
σ_1	Q_{11}	Q_{12}	Q_{16}
σ_2	Q_{21}	Q_{22}	Q_{26}
σ_6	Q_{61}	Q_{62}	Q_{66}

The transformation of on-axis unidirectional properties to off-axis is given by

	Q_{xx}	Q_{yy}	Q_{xy}	Q_{ss}
Q_{11}	m^4	n^4	$2m^2n^2$	$4m^2n^2$
Q_{22}	n^4	m^4	$2m^2n^2$	$4m^2n^2$
Q_{12}	m^2n^2	m^2n^2	$m^4 + n^4$	$-4m^2n^2$
Q_{66}	m^2n^2	m^2n^2	$-2m^2n^2$	$(m^2 - n^2)^2$
Q_{16}	m^3n	$-mn^3$	$mn^3 - m^3n$	$2(mn^3 - m^3n)$
Q_{26}	mn^3	$-m^3n$	$m^3n - mn^3$	$2(m^3n - mn^3)$

$$m = \cos\theta, \quad n = \sin\theta$$

COMPOSITE MATERIAL BEHAVIOR, CONTINUED

Shown here is a similarly transformed compliance matrix for on-axis unidirectional composites:

	S_{xx}	S_{yy}	S_{xy}	S_{ss}
S_{11}	m^4	n^4	$2m^2n^2$	m^2n^2
S_{22}	n^4	m^4	$2m^2n^2$	m^2n^2
S_{12}	m^2n^2	m^2n^2	$m^4 + n^4$	$-m^2n^2$
S_{66}	$4m^2n^2$	$4m^2n^2$	$-8m^2n^2$	$(m^2 - n^2)^2$
S_{16}	$2m^3n$	$-2mn^3$	$2(mn^3 - m^3n)$	$mn^3 - m^3n$
S_{26}	$2mn^3$	$-2m^3n$	$2(m^3n - mn^3)$	$m^3n - mn^3$

$$m = \cos\theta, \quad n = \sin\theta$$

IN-PLANE LAMINATE THEORY

Laminates add the third dimension of "z" into the equations, with $z = 0$ located at the mid plane of the laminate. With the laminate thickness of h, z varies from $-h/2$ to $+h/2$. Stiffness of a laminate is based on a contribution from each individual ply and the average or "equivalent modulus" is used. The following assumes that the laminate is symmetric about its mid plane; the general case is shown at the end of this section.

	ε_1^0	ε_2^0	ε_6^0
N_1	A_{11}	A_{12}	A_{16}
N_2	A_{21}	A_{22}	A_{26}
N_6	A_{61}	A_{62}	A_{66}

The superscript zero for the strain terms signifies the assumption that the strain remains constant across the laminate thickness. The A_{ij} terms have the units of lb/in. or N/m and include the thickness h of the laminate. N_1 also has the units of lb/in. or N/m and represents the force per unit width of the laminate with thickness h.

$$A_{11} = \int Q_{11}\,dz, \quad A_{22} = \int Q_{22}\,dz, \quad A_{12} = A_{21}$$
$$A_{12} = \int Q_{12}\,dz, \quad A_{66} = \int Q_{66}\,dz, \quad A_{16} = A_{61}$$
$$A_{16} = \int Q_{16}\,dz, \quad A_{26} = \int Q_{26}\,dz, \quad A_{26} = A_{62}$$

Q_{ij} has the units of lb/in.^{2} or N/m^2.

COMPOSITE MATERIAL BEHAVIOR, CONTINUED

FLEXURAL BEHAVIOR

For flexural behavior, the moment–curvature relationship is the counterpart of the stress-strain relationship for in-plane behavior. Simplifying for a symmetric laminate and assuming a linear strain distribution across the thickness of the laminate,

$$\varepsilon_1(z) = zk_1$$
$$\varepsilon_2(z) = zk_2$$
$$\varepsilon_6(z) = zk_6$$

where k is curvature with units of in.$^{-1}$ or m^{-1}.

	k_1	k_2	k_6
M_1	D_{11}	D_{12}	D_{16}
M_2	D_{21}	D_{22}	D_{26}
M_6	D_{61}	D_{62}	D_{66}

$$D_{11} = \int Q_{11} z^2 \, dz, \quad D_{22} = \int Q_{22} z^2 \, dz, \quad D_{12} = A_{21}$$
$$D_{12} = \int Q_{12} z^2 \, dz, \quad D_{66} = \int Q_{66} z^2 \, dz, \quad D_{16} = A_{61}$$
$$D_{16} = \int Q_{16} z^2 \, dz, \quad D_{26} = \int Q_{26} z^2 \, dz, \quad D_{26} = A_{62}$$

GENERAL LAMINATES

For general laminates, coupling exists between flexure and in-plane loading; this is seen by the addition of B_{ij} terms:

	ε_1^0	ε_2^0	ε_6^0	k_1	k_2	k_6
N_1	A_{11}	A_{12}	A_{16}	B_{11}	B_{12}	B_{16}
N_2	A_{21}	A_{22}	A_{26}	B_{21}	B_{22}	B_{26}
N_6	A_{61}	A_{62}	A_{66}	B_{61}	B_{62}	B_{66}
M_1	B_{11}	B_{12}	B_{16}	D_{11}	D_{12}	D_{16}
M_2	B_{21}	B_{22}	B_{26}	D_{21}	D_{22}	D_{26}
M_6	B_{61}	B_{62}	B_{66}	D_{61}	D_{62}	D_{66}

	N_1	N_2	N_6	M_1	M_2	M_6
ε_1^0	α_{11}	α_{12}	α_{16}	β_{11}	β_{12}	β_{16}
ε_2^0	α_{21}	α_{22}	α_{26}	β_{21}	β_{22}	β_{26}
ε_6^0	α_{61}	α_{62}	α_{66}	β_{61}	β_{62}	β_{66}
k_1	β_{11}	β_{21}	β_{61}	σ_{11}	σ_{12}	σ_{16}
k_2	β_{12}	β_{22}	β_{62}	σ_{21}	σ_{22}	σ_{26}
k_6	β_{16}	β_{26}	β_{66}	σ_{61}	σ_{62}	σ_{66}

COMPOSITE MATERIAL BEHAVIOR, CONTINUED

GENERAL LAMINATES, CONTINUED

For the symmetric anisotropic laminates, no coupling exists.

	ε_1^0	ε_2^0	ε_6^0	k_1	k_2	k_6
N_1	A_{11}	A_{12}	A_{16}			
N_2	A_{21}	A_{22}	A_{26}			
N_6	A_{61}	A_{62}	A_{66}			
M_1				D_{11}	D_{12}	D_{16}
M_2				D_{21}	D_{22}	D_{26}
M_6				D_{61}	D_{62}	D_{66}

	N_1	N_2	N_6	M_1	M_2	M_6
ε_1^0	σ_{11}	σ_{12}	σ_{16}			
ε_2^0	σ_{21}	σ_{22}	σ_{26}			
ε_6^0	σ_{61}	σ_{62}	σ_{66}			
k_1				d_{11}	d_{12}	d_{16}
k_2				d_{21}	d_{22}	d_{26}
k_6				d_{61}	d_{62}	d_{66}

For a homogeneous anisotropic laminate, the flexure components are directly related to the in-plane components.

	ε_1^0	ε_2^0	ε_6^0	k_1	k_2	k_6
N_1	A_{11}	A_{12}	A_{16}			
N_2	A_{21}	A_{22}	A_{26}			
N_6	A_{61}	A_{62}	A_{66}			
M_1						
M_2					$\frac{h^2}{12}A_{ij}$	
M_6						

	N_1	N_2	N_6	M_1	M_2	M_6
ε_1^0	σ_{11}	σ_{12}	σ_{16}			
ε_2^0	σ_{21}	σ_{22}	σ_{26}			
ε_6^0	σ_{61}	σ_{62}	σ_{66}			
k_1						
k_2					$\frac{12}{h^2}\sigma_{ij}$	
k_6						

Or, in matrix notation,

$$\boldsymbol{N} = \boldsymbol{A}\boldsymbol{\varepsilon}^0 + \boldsymbol{B}\boldsymbol{k}$$
$$\boldsymbol{M} = \boldsymbol{B}\boldsymbol{\varepsilon}^0 + \boldsymbol{D}\boldsymbol{k}$$

GALVANIC SERIES

GALVANIC SERIES OF SOME COMMERCIAL METALS AND ALLOYS IN SEAWATER

↑	Platinum
	Gold
Noble or cathodic	Graphite
	Titanium
	Silver
	Hastelloy C (62 Ni, 17 Cr, 15 Mo)
	18-8 stainless steel (passive)
	Chromium stainless steel 11–30% Cr (passive)
	Inconel (passive)
	Nickel (passive)
	Silver solder
	Monel
	Cupronickels (60–90 Cu, 40–10 Ni)
	Bronzes (Cu-Sn)
	Copper
	Brasses (Cu-Zn)
	Hastelloy B
	Inconel (active)
	Nickel (active)
	Tin
	Lead
	Lead–in solders
	18-8 stainless steel (active)
	Ni-Resist (high Ni cast iron)
	Chromium stainless steel, 13% Cr (active)
	Cast iron
	Steel or iron
	2024 aluminum
	Cadmium
Active or anodic	Commercially pure aluminum
	Zinc
↓	Magnesium and magnesium alloys

SECTION 4

SECTION PROPERTIES

PLANE AREAS

Figure	General Properties	Moment of Inertia: Area	Moment of Inertia: Weight	Radius of Gyration, ρ
Square 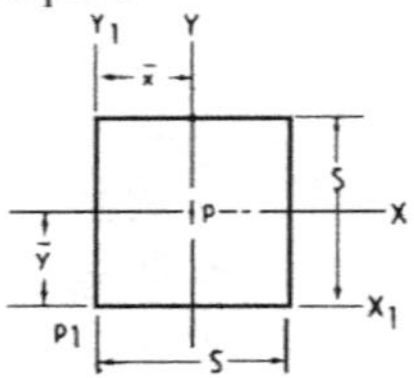	Area: $A = S^2$ Centroid: $\bar{x} = \bar{y} = \frac{S}{2}$	$I_x = I_y = \frac{S^4}{12}$ $I_{x_1} = I_{y_1} = \frac{S^4}{3}$ $I_p = I_x + I_y = \frac{S^4}{6}$ $I_{p_1} = I_{x_1} + I_{y_1} = \frac{2S^4}{3}$	$I_x = I_y = \frac{WS^2}{12}$ $I_{x_1} = I_{y_1} = \frac{WS^2}{3}$ $I_p = I_x + I_y = \frac{WS^2}{6}$ $I_{p_1} = I_{x_1} + I_{y_1} = \frac{2WS^2}{3}$	$\rho_x = \rho_y = 0.289S$ $\rho_{x_1} = \rho_{y_1} = 0.577S$ $\rho_p = 0.408S$ $\rho_{p_1} = 0.816S$
Rectangle 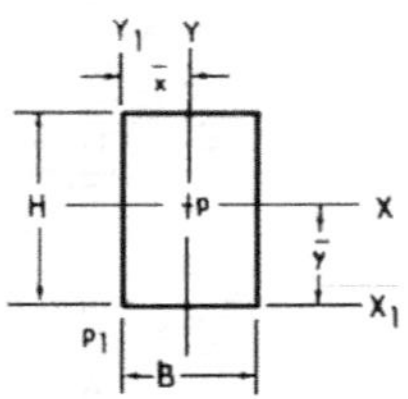	Area: $A = BH$ Centroid: $\bar{x} = \frac{B}{2}$ $\bar{y} = \frac{H}{2}$	$I_x = \frac{BH^3}{12}$ $I_y = \frac{HB^3}{12}$ $I_p = \frac{BH}{12}(H^2 + B^2)$ $I_{x_1} = \frac{BH^3}{3}$ $I_{y_1} = \frac{HB^3}{3}$ $I_{p_1} = \frac{BH}{3}(H^2 + B^2)$	$I_x = \frac{WH^2}{12}$ $I_y = \frac{WB^2}{12}$ $I_p = \frac{W}{12}(H^2 + B^2)$ $I_{x_1} = \frac{WH^2}{3}$ $I_{y_1} = \frac{WB^2}{3}$ $I_{p_1} = \frac{W}{3}(H^2 + B^2)$	$\rho_x = 0.289H$ $\rho_y = 0.289B$ $\rho_p = 0.289\sqrt{H^2 + B^2}$ $\rho_{x_1} = 0.577H$ $\rho_{y_1} = 0.577B$ $\rho_{p_1} = 0.577\sqrt{H^2 + B^2}$

(continued)

PLANE AREAS, CONTINUED

Figure	General Properties	Moment of Inertia: Area	Moment of Inertia: Weight	Radius of Gyration, ρ
Hollow square 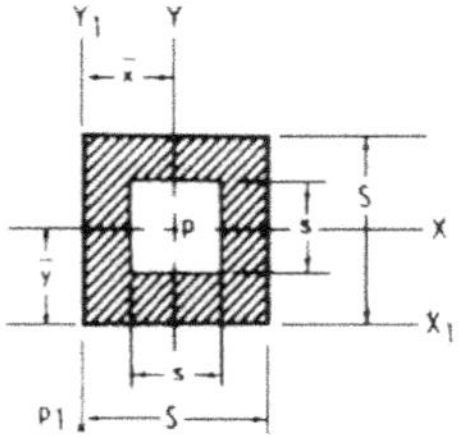	Area: $A = S^2 - s^2$ Centroid: $\bar{x} = \bar{y} = \frac{S}{2}$	$I_x = I_y = \frac{S^4 - s^4}{12}$ $I_p = \frac{S^4 - s^4}{6}$ $I_{x_1} = I_{y_1} = \frac{4S^4 - 3S^2s^2 - s^4}{12}$ $I_{p_1} = \frac{4S^4 - 3S^2s^2 - s^4}{6}$	$I_x = I_y = \frac{W(S^2 + s^2)}{12}$ $I_{x_1} = I_{y_1} = \frac{W(4S^2 + s^2)}{12}$ $I_p = \frac{W(S^2 + s^2)}{6}$ $I_{p_1} = \frac{W(4S^2 + s^2)}{6}$	$\rho_x = \rho_y = \sqrt{\frac{S^2 + s^2}{12}}$ $\rho_{x_1} = \rho_{y_1} = \sqrt{\frac{(4S^2 + s^2)}{12}}$ $\rho_p = 0.408\sqrt{S^2 + s^2}$ $\rho_{p_1} = \sqrt{\frac{(4S^2 + s^2)}{6}}$
Hollow rectangle 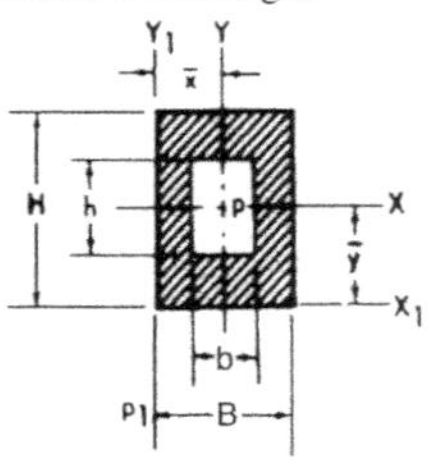	Area: $A = BH - bh$ Centroid: $\bar{x} = \frac{B}{2}$ $\bar{y} = \frac{H}{2}$	$I_x = \frac{BH^3 - bh^3}{12}$ $I_y = \frac{HB^3 - hb^3}{12}$ $I_{x_1} = \frac{BH^3}{3} - \frac{bh(3H^2 + h^2)}{12}$ $I_{y_1} = I_y + \frac{(BH - bh)B^2}{4}$ $I_p = I_x + I_y$ $I_{p_1} = I_{x_1} + I_{y_1}$	$I_x = \frac{W}{12}\left[\frac{BH^3 - bh^3}{BH - bh}\right]$ $I_y = \frac{W}{12}\left[\frac{HB^3 - hb^3}{BH - bh}\right]$ $I_{x_1} = I_x + \frac{WH^2}{4}$ $I_{y_1} = I_y + \frac{WB^2}{4}$ $I_p = I_x + I_y$ $I_{p_1} = I_{x_1} + I_{y_1}$	$\rho_x = \sqrt{\frac{BH^3 - bh^3}{12(BH - bh)}}$ $\rho_y = \sqrt{\frac{HB^3 - bh^3}{12(BH - bh)}}$ $\rho_{x_1} = \sqrt{\frac{I_{x_1}}{BH - bh}}$ $\rho_{y_1} = \sqrt{\frac{I_{y_1}}{BH - bh}}$ $\rho_{p_1} = \sqrt{\frac{I_{p_1}}{BH - bh}}$ $\rho_p = \sqrt{\frac{I_p}{BH - bh}}$

(continued)

PLANE AREAS, CONTINUED

Figure	General Properties	Moment of Inertia: Area	Moment of Inertia: Weight	Radius of Gyration, ρ
Oblique triangle 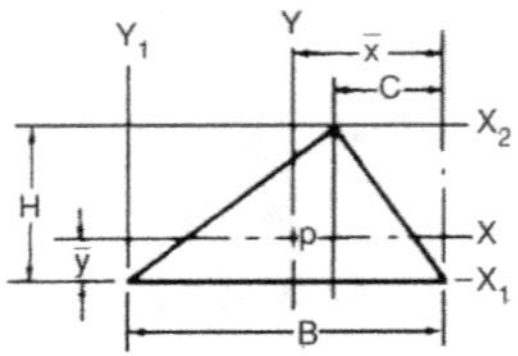 	Area: $A=\frac{1}{2}BH$ Centroid: $\bar{x}=\frac{B+C}{3}$ $\bar{y}=\frac{H}{3}$	$I_x=\frac{BH^3}{36}$ $I_{x_1}=\frac{BH^3}{12}$ $I_{x_2}=\frac{BH^3}{4}$ $I_y=\frac{BH}{36}(B^2+C^2-BC)$ $I_p=\frac{BH}{36}(H^2+B^2+C^2-BC)$	$I_x=\frac{WH^2}{18}$ $I_{x_1}=\frac{WH^2}{6}$ $I_{x_2}=\frac{WH^2}{2}$ $I_y=\frac{W}{18}(B^2+C^2-BC)$ $I_p=\frac{W}{18}(H^2+B^2+C^2-BC)$	$\rho_x=0.236H$ $\rho_{x_1}=0.408H$ $\rho_{x_2}=0.707H$ $\rho_y=0.236\sqrt{B^2+C^2-BC}$ $\rho_p=0.236\sqrt{H^2+B^2+C^2-BC}$
Isosceles trapezoid 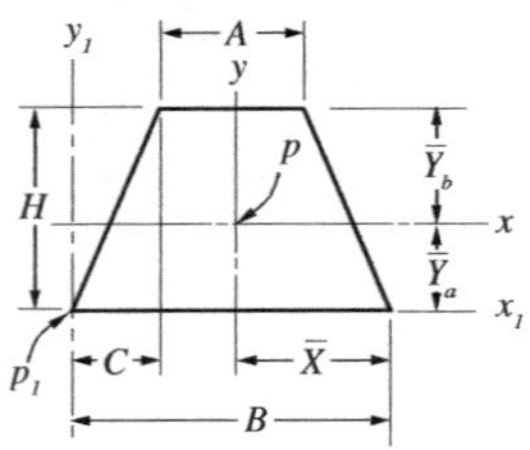 	Area: $\frac{H(A+B)}{2}$ Centroid: $\bar{Y}_a=\frac{H(B+2A)}{3(B+A)}$ $\bar{Y}_b=\frac{H(A+2B)}{3(A+B)}$	$I_x=\frac{H^3(A^2+4AB+B^2)}{36(A+B)}$ $I_{x_1}=\frac{H^3(A+B)(2A+C)}{12(A+C)}$ $I_y=\frac{H(A+B)(A^2+B^2)}{48}$ $I_{y_1}=\frac{H(A+B)(A^2+7B^2)}{48}$ $I_p=I_x+I_y$ $I_{p_1}=I_{x_1}+I_{y_1}$	$I_x=\frac{WH^2}{18}\left[1+\frac{2AB}{(A+B)^2}\right]$ $I_{x_1}=\frac{WH^2(3A+B)}{6(A+B)}$ $I_y=\frac{W}{24}(A^2+B^2)$ $I_{y_1}=\frac{W}{24}(A^2+7B^2)$ $I_p=I_x+I_y$	$\rho_x=H\frac{\sqrt{2(A^2+4AB+B^2)}}{6(A+B)}$ $\rho_y=0.204\sqrt{A^2+B^2}$ $\rho_p=\sqrt{\frac{2I_p}{H(A+B)}}$

(continued)

PLANE AREAS, CONTINUED

Figure	General Properties	Moment of Inertia: Area	Moment of Inertia: Weight	Radius of Gyration, ρ
Regular polygon 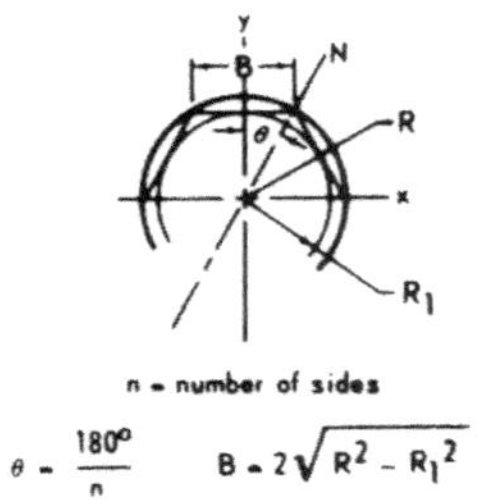	$R=\dfrac{B}{2\sin\theta}$ $\quad R_1=\dfrac{B}{2\tan\theta}$ Area: $\dfrac{nB^2\cot\theta}{4}$ $\dfrac{nR^2\sin 2\theta}{2}$ $nR_1^2\tan\theta$ Centroid: $\bar{x}=\bar{y}=0$	$I_y=I_N=\dfrac{A(6R^2-B^2)}{24}$ $=\dfrac{A(12R_1^2+B^2)}{48}$	$I_y=I_N=\dfrac{W(6R^2-B^2)}{24}$ $=\dfrac{W(12R_1^2+B^2)}{48}$	$\rho_y=\rho_N=\sqrt{\dfrac{6R^2-B^2}{24}}$ $=\sqrt{\dfrac{12R_1^2+B^2}{48}}$
Regular hexagon 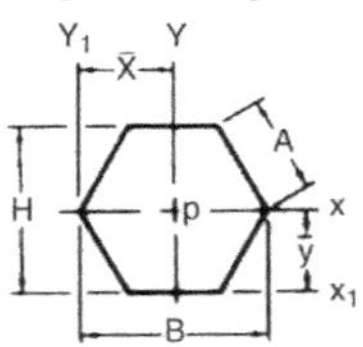	Area: $0.866H^2$ Centroid: $\bar{x}=\dfrac{B}{2}=A$ $\bar{y}=\dfrac{H}{2}$	$I_x=I_y=0.0601H^4$ $I_{x_1}=0.2766H^4$ $I_{y_1}=0.3488H^4$ $I_p=0.1203H^4$	$I_x=I_y=0.0694WH^2$ $=0.0521WB^2$ $I_{x_1}=0.3194WH^2$ $I_p=0.1389WH^2=0.1042WB^2$ $I_{y_1}=0.4028WH^2=0.3021WB^2$	$\rho_x=\rho_y=0.2635H$ $=0.2282B$ $\rho_{x_1}=0.5652H$ $\rho_{y_1}=0.6346H$ $\rho_p=0.3727H$

(continued)

PLANE AREAS, CONTINUED

Figure	General Properties	Moment of Inertia: Area	Moment of Inertia: Weight	Radius of Gyration, ρ
Right angled trapezoid 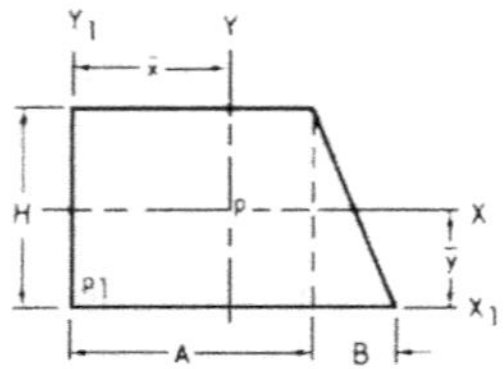	Area: $\frac{H}{2}(2A+B)$ Centroid: $\bar{x}=\frac{3A^2+3AB+B^2}{3(2A+B)}$ $\bar{y}=\frac{H(3A+B)}{3(2A+B)}$	$I_x=\frac{H^3(6A^2+6AB+B^2)}{36(2A+B)}$ $I_{x_1}=\frac{H^3(4A+B)}{12}$ $I_y=I_{y_1}-\frac{H(3A^2+3AB+B^2)^2}{18(2A+B)}$ $I_{y_1}=\frac{H}{12}(2A+B)$ $\times(2A^2+2AB+B^2)$ $I_p=I_x+I_y$ $I_{p_1}=I_{x_1}+I_{y_1}$	$I_x=\frac{WH^2(6A^2+6AB+B^2)}{18(2A+B)^2}$ $I_{x_1}=\frac{WH^2(4A+B)}{6(2A+B)}$ $I_y=I_{y_1}=W\bar{X}^2$ $I_{y_1}=\frac{W}{6}(2A^2+2AB+B^2)$ $I_p=I_x+I_y$ $I_{p_1}=I_{x_1}+I_{y_1}$	$\rho_x=\frac{0.236H}{(2A+B)}\sqrt{6A^2+6AB+B^2}$ $\rho_{x_1}=0.408H\sqrt{\frac{4A+B}{2A+B}}$ $\rho_y=\sqrt{\frac{2I_y}{H(2A+B)}}$ $\rho_{y_1}=\sqrt{\frac{2A^2+2AB+B^2}{6}}$ $\rho_p=\sqrt{\frac{2I_p}{H(2A+B)}}$
Oblique trapezoid 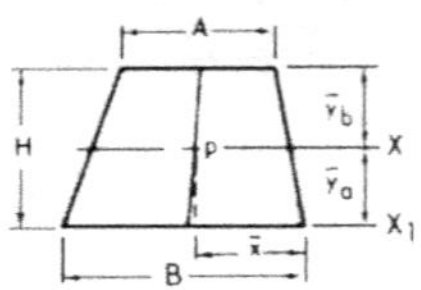	Area: $\frac{1}{2}H(A+B)$ Centroid: $\bar{x}$ is on a line connecting midpoints of sides A and B. $\bar{y}_a=\frac{H(B+2A)}{3(B+A)}$ $\bar{y}_b=\frac{H(A+2B)}{3(A+B)}$	$I_x=\frac{H^3(A^2+4AB+B^2)}{36(A+B)}$ $I_{x_1}=\frac{H^3(B+3A)}{12}$	$I_x=\frac{WH^2}{18}\left(1+\frac{2AB}{(A+B)^2}\right)$ $I_{x_1}=\frac{WH^2}{6}\left(\frac{3A+B}{A+B}\right)$	$\rho_x=H\frac{\sqrt{2(A^2+4AB+B^2)}}{6(A+B)}$ $\rho_{x_1}=H\sqrt{\frac{3A+B}{6A+B}}$

(continued)

PLANE AREAS, CONTINUED

Figure	General Properties	Moment of Inertia: Area	Moment of Inertia: Weight	Radius of Gyration, ρ
Parallelogram 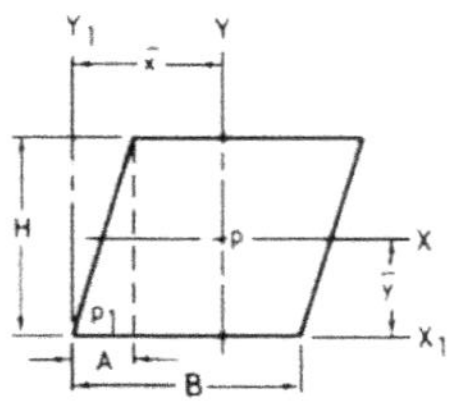	Area: BH Centroid: $\bar{x} = \frac{A+B}{2}$ $\bar{y} = \frac{H}{2}$	$I_x = \frac{BH^3}{12}$ $I_{x_1} = \frac{BH^3}{3}$ $I_y = \frac{BH(A^2+B^2)}{12}$ $I_{y_1} = \frac{BH}{6}(2A^2+2B^2+3AB)$ $I_p = \frac{BH}{12}(A^2+B^2+H^2)$ $I_{p_1} = \frac{BH}{6}(2A^2+2B^2+3AB+2H^2)$	$I_x = \frac{WH^2}{12}$ $I_{x_1} = \frac{WH^2}{3}$ $I_y = \frac{W}{12}(A^2+B^2)$ $I_{y_1} = \frac{W}{6}(2A^2+2B^2+3AB)$ $I_p = \frac{W}{12}(A^2+B^2+H^2)$ $I_{p_1} = \frac{W(2A^2+2B^2+3AB+2H^2)}{6}$	$\rho_x = 0.289H$ $\rho_{x_1} = 0.577H$ $\rho_y = 0.289\sqrt{A^2+B^2}$ $\rho_{y_1} = 0.408\sqrt{2A^2+2B^2+3AB}$ $\rho_p = 0.289\sqrt{A^2+B^2+H^2}$ $\rho_{p_1} = 0.408\sqrt{2A^2+2B^2+3AB+2H^2}$
Regular octagon 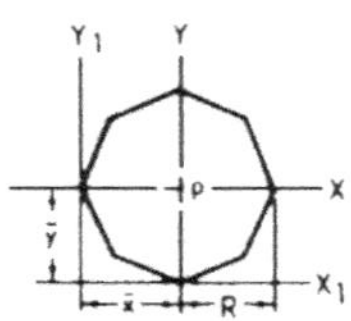	Area: $2.8284R^2$ Centroid: $\bar{x} = \bar{y} = R$	$I_x = I_y = 0.6381R^4$ $I_{x_1} = I_{y_1} = 3.4665R^4$ $I_p = 1.2761R^4$	$I_x = I_y = 0.2256WR^2$ $I_{x_1} = I_{y_1} = 1.2256WR^2$ $I_p = 0.4512WR^2$	$\rho_x = \rho_y = 0.4750R$ $\rho_{x_1} = \rho_{y_1} = 1.1071R$ $\rho_p = 0.672R$

(continued)

PLANE AREAS, CONTINUED

Figure	General Properties	Moment of Inertia: Area	Moment of Inertia: Weight	Radius of Gyration, ρ
Circle 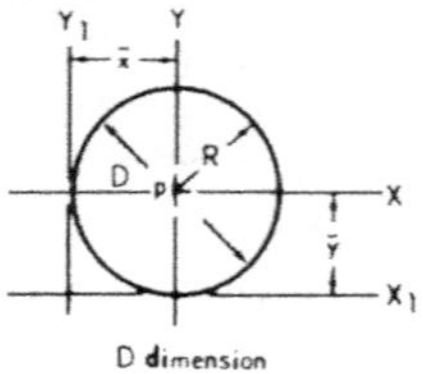	Area: $0.7854D^2$ Centroid: $\bar{x} = \bar{y} = R$	$I_x = I_y = 0.0491D^4$ $I_{x_1} = I_{y_1} = 0.2454D^4$ $I_p = 0.0982D^4$	$I_x = I_y = \frac{WD^2}{16} = \frac{WR^2}{4}$ $I_{x_1} = I_{y_1} = 1.25WR^2$ $I_p = \frac{WD^2}{8} = \frac{WR^2}{2}$	$\rho_x = \rho_y = \frac{D}{4}$ $\rho_{x_1} = \rho_{y_1} = 0.5590D = 1.118R$ $\rho_p = 0.3536D$
Hollow circle 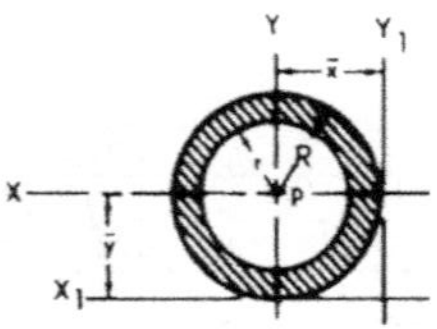	Area: $\pi(R^2 - r^2)$ Centroid: $\bar{x} = \bar{y} = R$	$I_x = I_y = \frac{\pi(R^4 - r^4)}{4}$ $I_p = \frac{\pi(R^4 - r^4)}{2}$ $I_{x_1} = I_{y_1} = \frac{\pi(5R^4 - 4R^2r^2 - r^4)}{4}$	$I_x = I_y = \frac{W(R^2 + r^2)}{4}$ $I_p = \frac{W(R^2 + r^2)}{2}$ $I_{x_1} = I_{y_1} = \frac{W(5R^2 + r^2)}{4}$	$\rho_x = \rho_y = \frac{1}{2}\sqrt{R^2 + r^2}$ $\rho_p = \sqrt{\frac{R^2 + r^2}{2}}$ $\rho_{x_1} = \rho_{y_1} = \frac{1}{2}\sqrt{5R^2 + r^2}$

(continued)

PLANE AREAS, CONTINUED

Figure	General Properties	Moment of Inertia: Area	Moment of Inertia: Weight	Radius of Gyration, ρ
Semicircle 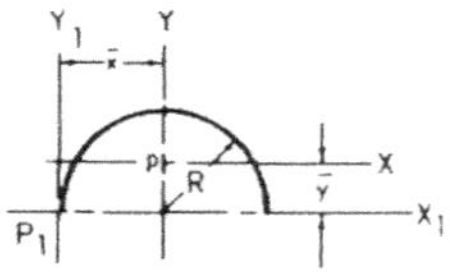 	$D=2R$ Area: $0.3927D^2=1.571R^2$ Centroid: $\bar{x}=R$ $\bar{y}=0.2122D=0.4244R$	$I_x=0.1098R^4$ $I_{x_1}=0.3927R^4$ $I_y=0.3927R^4$ $I_{y_1}=1.9635R^4$ $I_p=0.5025R^4$ $I_{p_1}=2.3562R^4$	$I_x=0.06987WR^2$ $I_{x_1}=0.25WR^2$ $I_y=0.25WR^2$ $I_{y_1}=1.25WR^2$ $I_p=0.3199WR^2$ $I_{p_1}=1.50WR^2$	$\rho_x=0.264R=0.132D$ $\rho_{x_1}=0.5R=0.25D$ $\rho_y=0.5R=0.25D$ $\rho_{y_1}=1.118R=0.559D$ $\rho_p=0.566R=0.2828D$ $\rho_{p_1}=1.225R=0.6124D$
Hollow semicircle 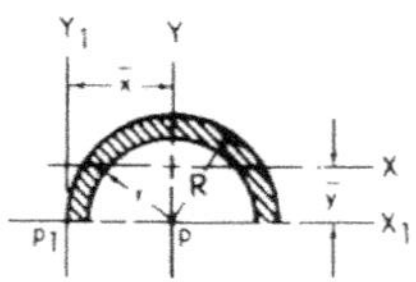 	Area: $\dfrac{\pi(R^2-r^2)}{2}$ Centroid: $\bar{x}=R$ $\bar{y}=0.4244\left(R+\dfrac{r^2}{R+r}\right)$	$I_x=\dfrac{\pi}{8}(R^4-r^4)-\dfrac{\pi(R^2-r^2)}{2}\bar{y}^2$ $I_{x_1}=\dfrac{\pi}{8}(R^4-r^4)$ $I_y=\dfrac{\pi}{8}(R^4-r^4)$ $I_{y_1}=\dfrac{\pi(R^2-r^2)(5R^2+r^2)}{8}$ $I_p=I_x+I_y$ $I_{p_1}=I_{x_1}+I_{y_1}$	$I_x=\dfrac{W(R^2+r^2)}{4}-W\bar{y}^2$ $I_y=\dfrac{W(R^2+r^2)}{4}$ $I_p=I_x+I_y$ $I_{p_1}=\dfrac{W(3R^2+r^2)}{2}$ $I_{x_1}=\dfrac{W(R^2+r^2)}{4}$ $I_{y_1}=\dfrac{W(5R^2+r^2)}{4}$	$\rho_x=\sqrt{\dfrac{2I_x}{\pi(R^2-r^2)}}$ $\rho_{x_1}=\sqrt{\dfrac{R^2+r^2)}{4}}$ $\rho_y=\sqrt{\dfrac{R^2+r^2}{4}}$ $\rho_{y_1}=\sqrt{\dfrac{2I_{y_1}}{\pi(R^2-r^2)}}$ $\rho_p=\sqrt{\dfrac{2I_p}{\pi(R^2-r^2)}}$

(continued)

PLANE AREAS, CONTINUED

Figure	General Properties	Moment of Inertia: Area	Moment of Inertia: Weight	Radius of Gyration, ρ
Ellipse 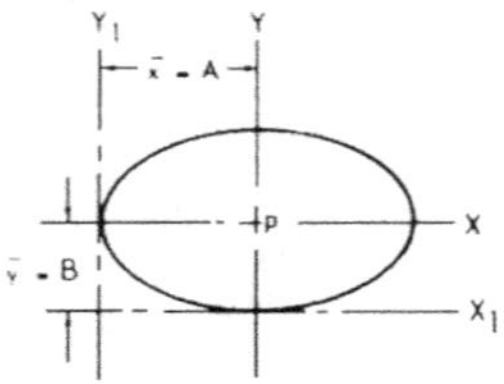	Area: πAB Centroid: $\bar{x} = A$ $\bar{y} = B$	$I_x = \frac{\pi AB^3}{4} = 0.7854AB^3$ $I_{x_1} = 1.25\pi AB^3$ $I_y = \frac{\pi A^3 B}{4} = 0.7854A^3 B$ $I_{y_1} = 1.25\pi A^3 B$ $I_p = \frac{\pi AB(A^2 + B^2)}{4}$	$I_x = \frac{WB^2}{4}$ $I_{x_1} = 1.25WB^2$ $I_y = \frac{WA^2}{4}$ $I_{y_1} = 1.25WA^2$ $I_p = \frac{W(A^2 + B^2)}{4}$	$\rho_x = \frac{B}{2}$ $\rho_{x_1} = 1.118B$ $\rho_y = \frac{A}{2}$ $\rho_{y_1} = 1.118A$ $\rho_p = \sqrt{\frac{A^2 + B^2}{2}}$
Hollow ellipse 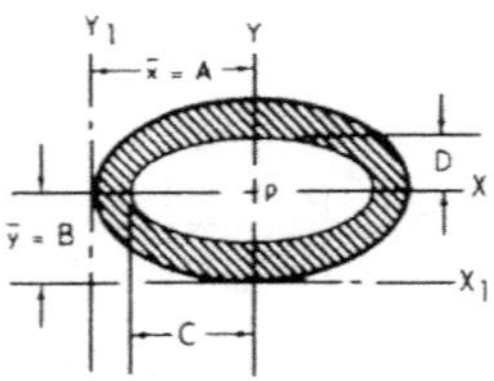	Area: $\pi(AB - CD)$ Centroid: $\bar{x} = A$ $\bar{y} = B$	$I_x = \frac{\pi}{4}(AB^3 - CD^3)$ $I_{x_1} = \frac{\pi}{4}(AB^3 - CD^3)$ $+\pi(AB - CD)(B^2)$ $I_y = \frac{\pi}{4}(A^3B - C^3D)$ $I_{y_1} = \frac{\pi}{4}(A^3B - C^3D)$ $+\pi(AB - CD)(A^2)$ $I_p = I_x + I_y$	$I_x = \frac{W}{4}\left[\frac{(AB^3 - CD^3)}{AB - CD}\right]$ $I_{x_1} = \frac{W}{4}\left[\frac{AB^3 - CD^3}{AB - CD}\right] + WB^2$ $I_y = \frac{W}{4}\left[\frac{A^3B - C^3D}{AB - CD}\right]$ $I_{y_1} = \frac{W}{4}\left[\frac{A^3B - C^3D}{AB - CD}\right] + WA^2$ $I_p = I_x + I_y$	$\rho_x = \sqrt{\frac{AB^3 - CD^3}{4(AB - CD)}}$ $\rho_{x_1} = \sqrt{\frac{I_{x_1}}{\pi(AB - CD)}}$ $\rho_y = \sqrt{\frac{A^3B - C^3D}{4(AB - CD)}}$ $\rho_{y_1} = \sqrt{\frac{I_{y_1}}{\pi(AB - CD)}}$ $\rho_p = \sqrt{\frac{I_p}{\pi(AB - CD)}}$

(continued)

PLANE AREAS, CONTINUED

Figure	General Properties	Moment of Inertia: Area	Moment of Inertia: Weight	Radius of Gyration, ρ
Semiellipse 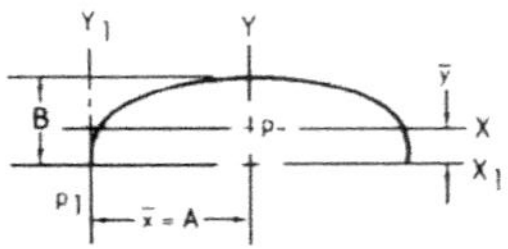	Area: $\frac{\pi AB}{2}$ Centroid: $\bar{x} = A$ $\bar{y} = 0.424B$	$I_x = 0.1098AB^3$ $I_{x_1} = 0.3927AB^3$ $I_y = 0.3927A^3B$ $I_{y_1} = 1.9635A^3B$	$I_x = 0.070WB^2$ $I_{x_1} = 0.25WB^2$ $I_y = 0.25WA^2$ $I_{y_1} = 1.25WA^2$ $I_p = \frac{W(A^2 + 0.280B^2)}{4}$ $I_{p_1} = \frac{W(5A^2 + B^2)}{4}$	$\rho_x = 0.2643B$ $\rho_{x_1} = \frac{B}{2}$ $\rho_y = \frac{A}{2}$ $\rho_{y_1} = 1.118A$ $\rho_p = \sqrt{\frac{A^2}{4} + \frac{B^2}{4} - \frac{16B^2}{9\pi^2}}$ $\rho_{p_1} = \sqrt{\frac{2I_{p_1}}{\pi AB}}$
Hollow semiellipse 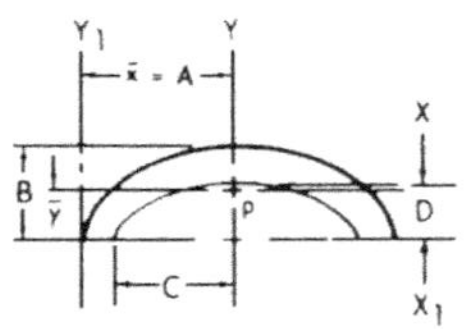	Area: $\frac{\pi(AB - CD)}{2}$ Centroid: $\bar{x} = A$ $\bar{y} = \frac{4}{3\pi}\left[\frac{AB^2 - CD^2}{AB - CD}\right]$	$I_x = \frac{\pi}{8}(AB^3 - CD^3) - \frac{\pi(AB - CD)}{2}\left[\frac{4(AB^2 - CD^2)}{3\pi(AB - CD)}\right]^2$ $I_{x_1} = \frac{\pi}{8}(AB^3 - CD^3)$ $I_y = \frac{\pi}{8}(A^3B - C^3D)$ $I_{y_1} = \frac{\pi}{8}(A^3B - C^3D) + \frac{\pi A^2(AB - CD)}{2}$	$I_x = \frac{W}{4}\left(\frac{AB^3 - CD^3}{AB - CD}\right) - W\left[\left(\frac{4}{3\pi}\right)\left(\frac{AB^2 - CD^2}{AB - CD}\right)\right]^2$ $I_{x_1} = \frac{W}{4}\left(\frac{AB^3 - CD^3}{AB - CD}\right)$ $I_y = \frac{W}{4}\left(\frac{A^3B - C^3D}{AB - CD}\right)$ $I_{y_1} = \frac{W}{4}\left(\frac{A^3B - C^3D}{AB - CD}\right) + WA^2$ $I_p = I_x + I_y$	$\rho_x = \sqrt{\frac{2I_x}{\pi(AB - CD)}}$ $\rho_{x_1} = \sqrt{\frac{AB^3 - CD^3}{4(AB - CD)}}$ $\rho_y = \sqrt{\frac{2I_y}{\pi(AB - CD)}}$ $\rho_{y_1} = \sqrt{\frac{2I_{y_1}}{\pi(AB - CD)}}$ $\rho_p = \sqrt{\frac{2I_p}{\pi(AB - CD)}}$

(continued)

PLANE AREAS, CONTINUED

Figure	General Properties	Moment of Inertia: Area	Moment of Inertia: Weight	Radius of Gyration, ρ
Circular sector 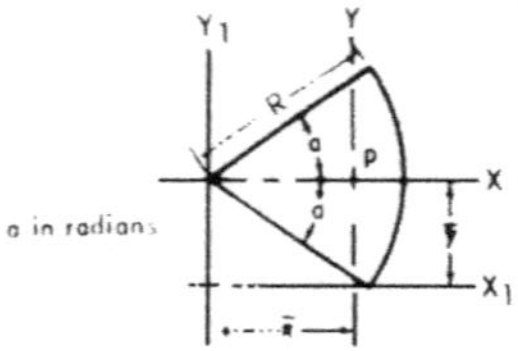 	Area: $R^2 a$ Centroid: $\bar{x} = \frac{2}{3}\left[\frac{R\sin a}{a}\right]$ $\bar{y} = R\sin a$	$I_x = \frac{R^4}{4}(a - \sin a\cos a)$ $I_y = \frac{R^4}{4}\left(a - \frac{16\sin^2 a}{9a} + \frac{\sin 2a}{2}\right)$ $I_{y_1} = \frac{R^4}{4}(a + \sin a\cos a)$ $I_p = \frac{R^4}{4}\left(2a - \frac{16\sin^2 a}{9a}\right)$	$I_x = \frac{WR^2}{4a}(a - \sin a\cos a)$ $I_{x_1} = I_x + WR^2\sin^2 a$ $I_y = \frac{WR^2}{4a}\left(a - \frac{16\sin^2 a}{9a} + \frac{\sin 2a}{2}\right)$ $I_{y_1} = \frac{WR^2}{4a}(a + \sin a\cos a)$ $I_p = I_x + I_y$	$\rho_x = \frac{R}{2}\sqrt{1 - \frac{\sin a\cos a}{a}}$ $\rho_{x_1} = \sqrt{\frac{I_{x_1}}{R^2 a}}$ $\rho_y = \frac{R}{2}\sqrt{1 + \frac{\sin a\cos a}{a} - \frac{16\sin^2 a}{9a^2}}$ $\rho_{y_1} = \frac{R}{2}\sqrt{1 + \frac{\sin a\cos a}{a}}$ $\rho_p = \frac{R}{2}\sqrt{2 - \frac{16\sin^2 a}{9a^2}}$
Hollow circular sector 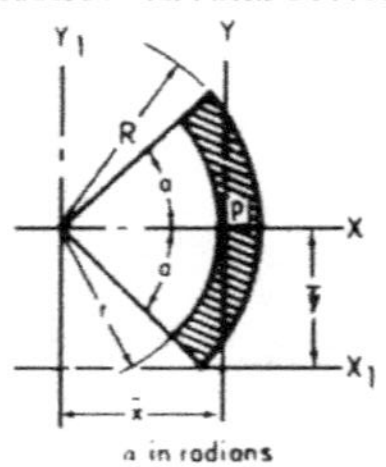 	Area: $(R^2 - r^2)a$ Centroid: $\bar{x} = \frac{2\sin a(R^3 - r^3)}{3a(R^2 - r^2)}$ $\bar{y} = R\sin a$	$I_x = \frac{a}{4}(R^4 - r^4)$ $\times\left(1 - \frac{\sin a\cos a}{a}\right)$ $I_y = I_{y_1} - \frac{1}{a(R_2 - r_2)}$ $\times\left[\frac{2\sin a(R^3 - r^3)}{3}\right]^2$ $I_{y_1} = \frac{a}{4}(R^4 - r^4)$ $\times\left(1 + \frac{\sin a\cos a}{a}\right)$	$I_x = \frac{W}{4}(R^2 + r^2)$ $\times\left(1 - \frac{\sin a\cos a}{a}\right)$ $I_{x_1} = I_x + WR^2\sin^2 a$ $I_y = I_{y_1} - W\left[\frac{2\sin a(R^3 - r^3)}{3a(R^2 - r^2)}\right]^2$ $I_{y_1} = \frac{W}{4}(R^2 + r^2)\left(1 + \frac{\sin a\cos a}{a}\right)$ $I_p = I_x + I_y$	$\rho_x = \sqrt{\frac{R^2 + r^2}{4}\left[1 - \frac{\sin a\cos a}{a}\right]}$ $\rho_{x_1} = \sqrt{\frac{I_{x_1}}{(R^2 - r^2)a}}$ $\rho_y = \sqrt{\frac{I_y}{(R^2 - r^2)a}}$ $\rho_{y_1} = \sqrt{\frac{R^2 + r^2}{4}\left[1 + \frac{\sin a\cos a}{a}\right]}$ $\rho_p = \sqrt{\frac{I_p}{(R^2 - r^2)a}}$

(continued)

PLANE AREAS, CONTINUED

Figure	General Properties	Moment of Inertia: Area	Moment of Inertia: Weight	Radius of Gyration, ρ
Circular segment 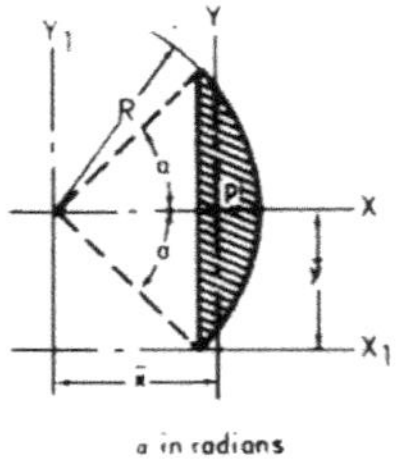	Area: $\frac{R^2}{2}(2a-\sin 2a)$ Centroid: $\bar{x}=\frac{4R\sin^3 a}{3(2a-\sin 2a)}$ $\bar{y}=R\sin a$	$I_x=\frac{AR^2}{4}\times\left[1-\frac{2\sin^3 a\cos a}{3(a-\sin a\cos a)}\right]$ $I_y=I_{y_1}-\frac{4R^6\sin^6 a}{9A}$ $I_{y_1}=\frac{AR^2}{4}\times\left[1+\frac{2\sin^3 a\cos a}{a-\sin a\cos a}\right]$	$I_x=\frac{WR^2}{4}\times\left[1-\frac{2\sin^3 a\cos a}{3(a-\sin a\cos a)}\right]$ $I_{x_1}=I_x+WR^2\sin^2 a$ $I_y=I_{y_1}-W\times\left[\frac{2R\sin^3 a}{3(a-\sin a\cos a)}\right]^2$ $I_{y_1}=\frac{WR^2}{4}\left[1+\frac{2\sin^3 a\cos a}{a-\sin a\cos a}\right]$ $I_p=I_x+I_y$	$\rho_x=\sqrt{\frac{R^2}{4}\left[1-\frac{2\sin^3 a\cos a}{3(a-\sin a\cos a)}\right]}$ $\rho_y=\sqrt{\frac{2I_y}{R^2(2a-\sin 2a)}}$ $\rho_p=\sqrt{\frac{2I_p}{R^2(2a-\sin 2a)}}$
Parabolic segment 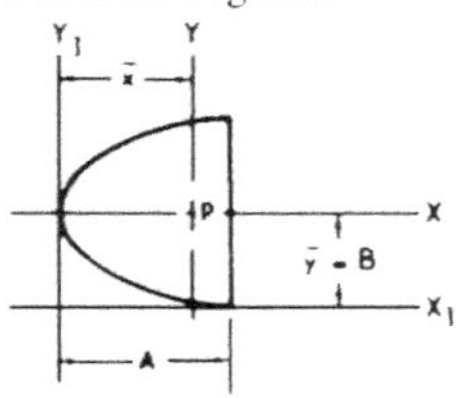	Area: $\frac{4}{3}AB$ Centroid: $\bar{x}=0.6A$ $\bar{y}=B$	$I_x=0.2667AB^3$ $I_{x_1}=1.6AB^3$ $I_y=0.0914A^3B$ $I_{y_1}=0.5714A^3B$ $I_p=I_x+I_y$	$I_x=0.2WB^2$ $I_{x_1}=1.2WB^2$ $I_y=0.0686WA^2$ $I_{y_1}=0.4286WA^2$ $I_p=I_x+I_y$	$\rho_x=0.4472B$ $\rho_{x_1}=1.095B$ $\rho_y=0.2619A$ $\rho_{y_1}=0.6547A$ $\rho_p=\sqrt{\frac{3I_p}{4AB}}$

(continued)

PLANE AREAS, CONTINUED

Figure	General Properties	Moment of Inertia: Area	Moment of Inertia: Weight	Radius of Gyration, ρ
Parabolic half-segment 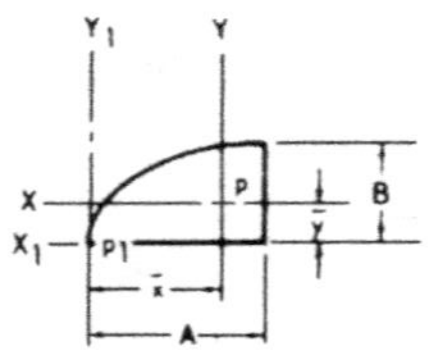	Area: $\frac{2AB}{3}$ Centroid: $\bar{x} = 0.6A$ $\bar{y} = 0.375B$	$I_x = 0.0396AB^3$ $I_{x_1} = 0.1333AB^3$ $I_y = 0.0457A^3B$ $I_{y_1} = 02857A^3B$ $I_p = I_x + I_y$ $I_{p_1} = I_{x_1} + I_{y_1}$	$I_x = 0.0594WB^2$ $I_{x_1} = 0.2WB^2$ $I_y = 0.0686WA^2$ $I_{y_1} = 0.4286WA^2$ $I_p = I_x + I_y$ $I_{p_1} = I_{x_1} + I_{y_1}$	$\rho_x = 0.2437B$ $\rho_{x_1} = 0.4472B$ $\rho_y = 0.2619A$ $\rho_{y_1} = 0.6547A$ $\rho_p = \sqrt{\frac{3Ip}{2AB}}$ $\rho_{p_1} = \sqrt{\frac{3I_{p_1}}{2AB}}$
Nose rib 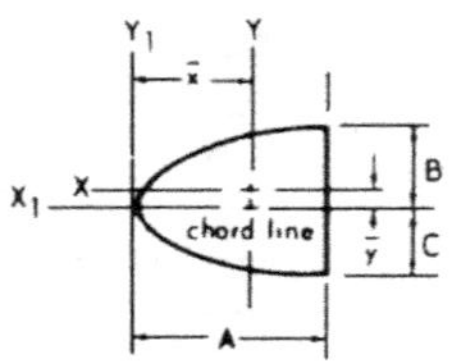 based on parabolic segment	Area: $\frac{2}{3}A(B+C)$ Centroid: $\bar{x} = 0.6A$ $\bar{y} = 0.375(B-C)$	$I_x = \frac{A(B+C)}{480} \times (19B^2 + 26BC + 19C^2)$ $I_y = 0.0457A^3(B+C)$ $I_{y_1} = 0.2857A^3(B+C)$ $I_p = I_x + I_y$ $I_{p_1} = I_{x_1} + I_{y_1}$	$I_x = \frac{W}{320}(19B^2 + 26BC + 19C^2)$ $I_{x_1} = \frac{W}{5}(B^2 - BC + C^2)$ $I_y = 0.0686WA^2$ $I_{y_1} = 0.4286WA^2$ $I_p = I_x + I_y$ $I_{p_1} = I_{x_1} + I_{y_1}$	$\rho_x = \sqrt{\frac{3I_x}{2A(B+C)}}$ $\rho_y = \sqrt{\frac{3I_y}{2A(B+C)}}$

(continued)

PLANE AREAS, CONTINUED

Figure	General Properties	Moment of Inertia: Area	Moment of Inertia: Weight	Radius of Gyration, ρ
Equilateral triangle 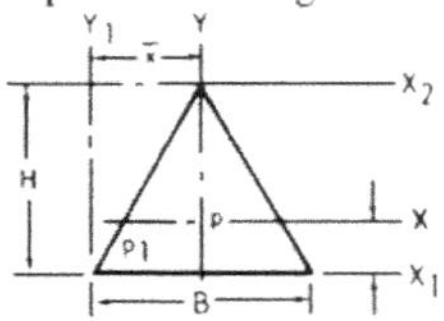	Area: $\frac{BH}{2}$ Centroid: $\bar{x}=\frac{B}{2}$ $\bar{y}=\frac{H}{3}$	$I_x=I_y=\frac{B^3H}{48}$ $I_{x_1}=\frac{B^3H}{16}$ $I_{x_2}=\frac{BH^3}{4}$ $I_{y_1}=\frac{7B^3H}{48}$ $I_p=\frac{B^3H}{24}$ $I_{p_1}=\frac{5B^3H}{24}$	$I_x=I_y=\frac{WB^2}{24}$ $I_{x_1}=\frac{WB^2}{8}$ $I_{x_2}=\frac{WH^2}{2}$ $I_{y_1}=\frac{7WB^2}{24}$ $I_p=\frac{WB^2}{12}$ $I_{p_1}=\frac{5WB^2}{12}$	$\rho_x=0.204B$ $\rho_{x_1}=0.354B$ $\rho_{x_2}=0.707H$ $\rho_y=0.204B$ $\rho_{y_1}=0.540B$ $\rho_p=0.289B$ $\rho_{p_1}=0.645B$
Rhombus 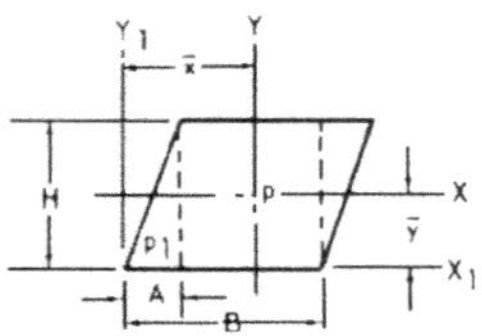	Area: BH Centroid: $\bar{y}=\frac{H}{2}$ $\bar{x}=\frac{A+B}{2}$	$I_x=\frac{BH^3}{12}$ $I_{x_1}=\frac{BH^3}{3}$ $I_y=\frac{BH(A^2+B^2)}{12}$ $I_{y_1}=\frac{BH(2A^2+2B^2+3AB)}{6}$ $I_p=\frac{1}{6}B^3H$ $I_{p_1}=\frac{B^2H(3A+4B)}{6}$	$I_x=\frac{WH^2}{12}$ $I_{x_1}=\frac{WH^2}{3}$ $I_y=\frac{W(A^2+B^2)}{12}$ $I_{y_1}=\frac{W(2A^2+2B^2+3AB)}{6}$ $I_p=\frac{WB^2}{6}$ $I_{p_1}=\frac{WB(3A+4B)}{6}$	$\rho_x=0.289H$ $\rho_{x_1}=0.577H$ $\rho_y=0.289\sqrt{(A^2+B^2)}$ $\rho_{y_1}=0.408\sqrt{2A^2+2B^2+3AB}$ $\rho_p=0.408B$ $\rho_{p_1}=0.408\sqrt{B(3A+4B)}$

(continued)

PLANE AREAS, CONTINUED

Figure	General Properties	Moment of Inertia: Area	Moment of Inertia: Weight	Radius of Gyration, ρ
Obtuse angled triangle	Area:	$I_x = \frac{BH^3}{36}$	$I_x = \frac{WH^2}{18}$	$\rho_x = 0.236H$
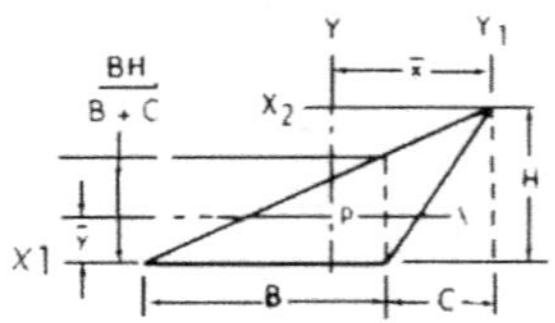	$\frac{BH}{2}$	$I_{x_1} = \frac{BH^3}{12}$	$I_{x_1} = \frac{WH^2}{6}$	$\rho_{x_1} = 0.408H$
	Centroid:	$I_{x_2} = \frac{BH^3}{4}$	$I_{x_2} = \frac{WH^2}{2}$	$\rho_{x_2} = 0.707H$
	$\bar{x} = \frac{B+2C}{3}$	$I_y = \frac{BH}{36}(B^2 + BC + C^2)$	$I_y = \frac{W}{18}(B^2 + BC + C^2)$	$\rho_y = 0.236\sqrt{B^2 + BC + C^2}$
	$\bar{y} = \frac{H}{3}$	$I_{y_1} = \frac{BH}{12}(B^2 + 3BC + 3C^2)$	$I_{y_1} = \frac{W(B^2 + 3BC + 3C^2)}{6}$	$\rho_{y_1} = 0.408\sqrt{B^2 + 3BC + 3C^2}$
		$I_p = \frac{BH}{36}(H^2 + B^2 + BC + C^2)$	$I_p = \frac{W(H^2 + B^2 + BC + C^2)}{18}$	$\rho_p = 0.236\sqrt{H^2 + B^2 + BC + C^2}$

SOLIDS

Figure	General Properties	Weight Moment of Inertia	Radius of Gyration, ρ
Cube 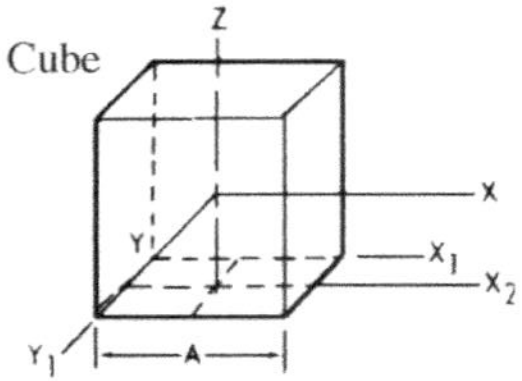	Volume: $V = A^3$ Centroid: $\bar{x} = \bar{y} = \bar{z} = \frac{A}{2}$	$I_x = I_y = I_z = \frac{WA^2}{6}$ $I_{x_1} = I_{y_1} = I_{z_1} = \frac{2WA^2}{3}$ $I_{x_2} = \frac{5WA^2}{12}$	$\rho_x = \rho_y = \rho_z = 0.408A$ $\rho_{x_1} = \rho_{y_1} = \rho_{z_1} = 0.816A$ $\rho_{x_2} = 0.646A$
Rectangular prism 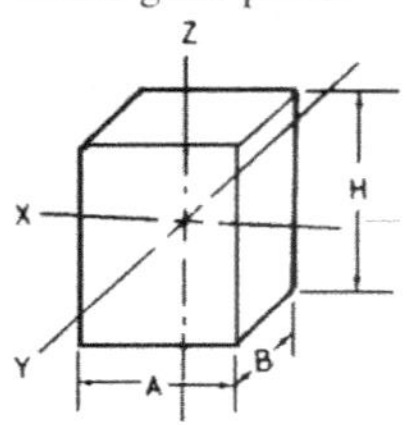	Volume: $V = ABH$ Centroid: $\bar{x} = \frac{A}{2}$ $\bar{y} = \frac{B}{2}$ $\bar{z} = \frac{H}{2}$	$I_x = \frac{W}{12}(B^2 + H^2)$ $I_y = \frac{W}{12}(A^2 + H^2)$ $I_z = \frac{W}{12}(A^2 + B^2)$	$\rho_x = 0.289\sqrt{B^2 + H^2}$ $\rho_y = 0.289\sqrt{A^2 + H^2}$ $\rho_z = 0.289\sqrt{A^2 + B^2}$
Sphere 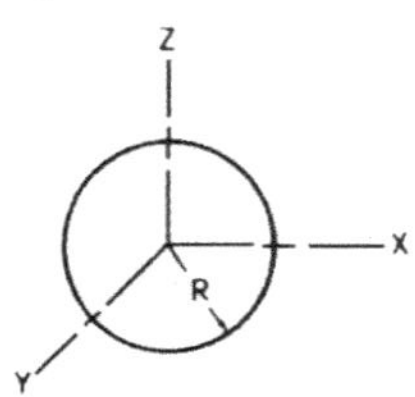	Volume: $\frac{4}{3}\pi R^3$ Surface area: $4\pi R^2$	$I_x = I_y = I_z = \frac{2}{5}WR^2$	$\rho_x = \rho_y = \rho_z = 0.632R$

(continued)

SOLIDS, CONTINUED

Figure	General Properties	Weight Moment of Inertia	Radius of Gyration, ρ
Hollow sphere 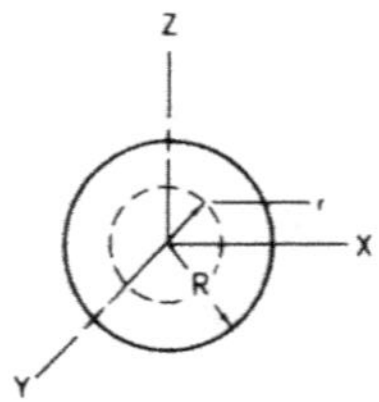	Volume: $V=\frac{4}{3}\pi(R^3-r^3)$	$I_x=I_y=I_z=\frac{2}{5}W\left(\frac{R^5-r^5}{R^3-r^3}\right)$	$\rho_x=\rho_y=\rho_z=0.632\sqrt{\left(\frac{R^5-r^5}{R^3-r^3}\right)}$
Hemisphere 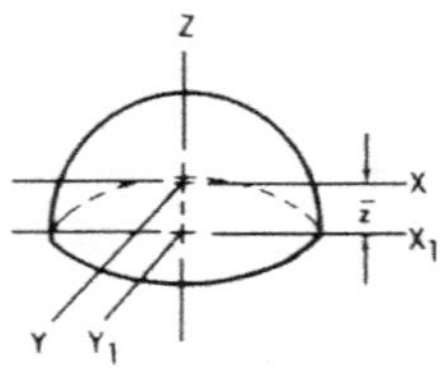	Volume: $V=\frac{2}{3}\pi R^3$ Centroid: $\bar{z}=\frac{3}{8}R$	$I_x=I_y=0.26WR^2$ $I_{x_1}=I_{y_1}=I_z=0.4WR^2$	$\rho_x=\rho_y=0.51R$ $\rho_{x_1}=\rho_{y_1}=\rho_z=0.632R$

(continued)

SOLIDS, CONTINUED

Figure	General Properties	Weight Moment of Inertia	Radius of Gyration, ρ
Right circular cylinder 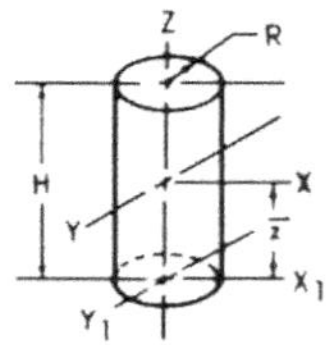	Volume: $V = \pi R^2 H$ Centroid: $\bar{z} = \frac{H}{2}$	$I_x = I_y = \frac{W}{12}(3R^2 + H^2)$ $I_{x_1} = I_{y_1} = \frac{W}{12}(3R^2 + 4H^2)$ $I_z = \frac{WR^2}{2}$	$\rho_x = \rho_y = 0.289\sqrt{3R^2 + H^2}$ $\rho_{x_1} = \rho_{y_1} = 0.289\sqrt{3R^2 + 4H^2}$ $\rho_z = 0.707R$
Hollow right circular cylinder 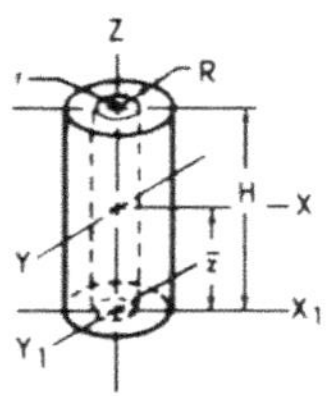	Volume: $V = \pi H(R^2 - r^2)$ Centroid: $\bar{z} = \frac{H}{2}$	$I_x = I_y = \frac{W}{12}[3(R^2 + r^2) + H^2]$ $I_{x_1} = I_{y_1} = W\left[\frac{R^2 + r^2}{4} + \frac{H^2}{3}\right]$ $I_z = \frac{W}{2}(R^2 + r^2)$	$\rho_x = \rho_y = 0.289\sqrt{3(R^2 + r^2) + H^2}$ $\rho_z = 0.707\sqrt{R^2 + r^2}$ $\rho_{x_1} = \rho_{y_1} = \sqrt{\frac{R^2 + r^2}{4} + \frac{H^2}{3}}$

(continued)

SOLIDS, CONTINUED

Figure	General Properties	Weight Moment of Inertia	Radius of Gyration, ρ
Elliptical cylinder 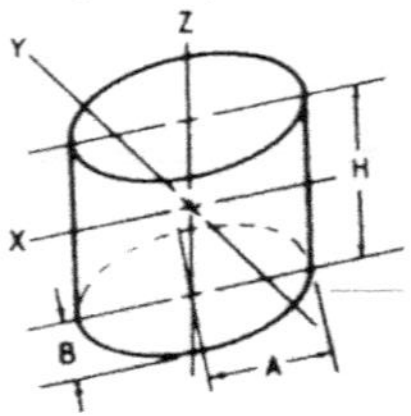 	Volume: $V = \pi ABH$ Centroid: $\bar{z} = \frac{H}{2}$	$I_x = \frac{W}{12}(3B^2 + H^2)$ $I_y = \frac{W}{12}(3A^2 + H^2)$ $I_z = \frac{W}{4}(A^2 + B^2)$	$\rho_x = 0.289\sqrt{3B^2 + H^2}$ $\rho_y = 0.289\sqrt{3A^2 + H^2}$ $\rho_z = \frac{\sqrt{A^2 + B^2}}{2}$
Ellipsoid 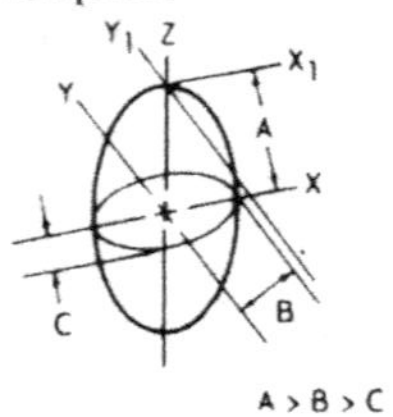 	If $C = B$, then resulting solid is a prolate spheroid (i.e., ellipse rotated about major axis). Volume: $V = \frac{4}{3}\pi ABC$ If $A = B$, then resulting solid is an oblate spheroid (i.e., ellipse rotated about minor axis).	$I_x = \frac{W}{5}(A^2 + C^2)$ $I_{x_1} = \frac{W}{5}(6A^2 + C^2)$ $I_y = \frac{W}{5}(A^2 + B^2)$ $I_{y_1} = \frac{W}{5}(6A^2 + B^2)$ $I_z = \frac{W}{5}(B^2 + C^2)$	$\rho_x = 0.447\sqrt{A^2 + C^2}$ $\rho_{x_1} = 0.447\sqrt{6A^2 + C^2}$ $\rho_y = 0.447\sqrt{A^2 + B^2}$ $\rho_{y_1} = 0.447\sqrt{6A^2 + B^2}$ $\rho_z = 0.447\sqrt{B^2 + C^2}$

(continued)

SOLIDS, CONTINUED

Figure	General Properties	Weight Moment of Inertia	Radius of Gyration, ρ
Paraboloid of revolution 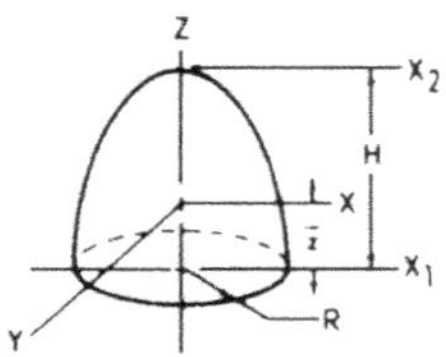 	Volume: $V = \frac{\pi R^2 H}{2}$ Centroid: $\bar{z} = \frac{H}{3}$	$I_x = I_y = \frac{W}{18}(3R^2 + H^2)$ $I_{x_1} = \frac{W}{6}(R^2 + H^2)$ $I_{x_2} = \frac{W}{6}(R^2 + 3H^2)$ $I_z = \frac{WR^2}{3}$	$\rho_x = \rho_y = 0.236\sqrt{3R^2 + H^2}$ $\rho_{x_1} = 0.408\sqrt{R^2 + H^2}$ $\rho_{x_2} = 0.408\sqrt{R^2 + 3H^2}$ $\rho_z = 0.577R$
Solid elliptical hemispheroid 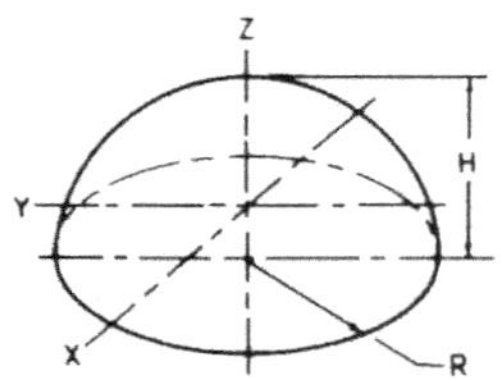 	Volume: $V = 2/3\pi R^2 H$ Centroid: $\bar{z} = 3/8H$	$I_x = I_y = W\left[\frac{R^2}{5} + \frac{19H^2}{320}\right]$ $I_z = 2/5WR^2$	$\rho_x = \rho_y = 0.4475\sqrt{R^2 + 0.297H^2}$ $\rho_z = 0.632R$

(continued)

SOLIDS, CONTINUED

Figure	General Properties	Weight Moment of Inertia	Radius of Gyration, ρ
Right rectangular pyramid 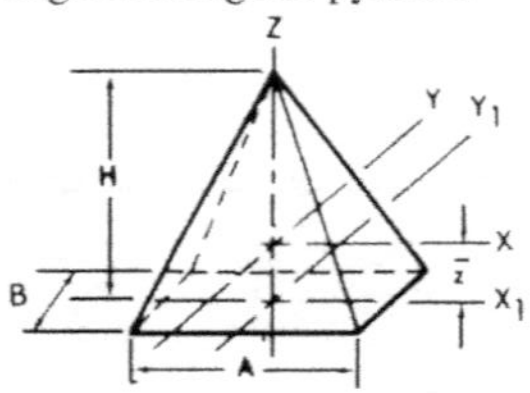	Volume: $V=\frac{ABH}{3}$ Centroid: $\overline{z}=\frac{H}{4}$	$I_x=\frac{W}{20}\left(B^2+\frac{3H^2}{4}\right)$ $I_{x_1}=\frac{W}{20}(B^2+2H^2)$ $I_y=\frac{W}{20}\left(A^2+\frac{3}{4}H^2\right)$ $I_{y_1}=\frac{W}{20}(A^2+2H^2)$ $I_z=\frac{W}{20}(A^2+B^2)$	$\rho_x=0.224\sqrt{B^2+\frac{3H^2}{4}}$ $\rho_y=0.224\sqrt{A^2+\frac{3H^2}{4}}$ $\rho_z=0.224\sqrt{A^2+B^2}$
Right angled wedge 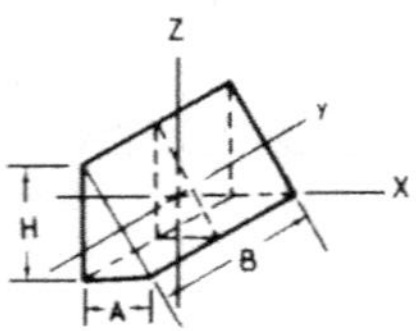	Volume: $V=\frac{ABH}{2}$ Centroid: $\overline{x}=\frac{A}{3}$ $\overline{y}=\frac{B}{2}$ $\overline{z}=\frac{H}{3}$	$I_x=\frac{W}{36}(2H^2+3B^2)$ $I_y=\frac{W}{18}(A^2+H^2)$ $I_z=\frac{W}{36}(2A^2+3B^2)$	$\rho_x=0.167\sqrt{2H^2+3B^2}$ $\rho_y=0.236\sqrt{A^2+H^2}$ $\rho_z=0.167\sqrt{2A^2+3B^2}$

(continued)

SOLIDS, CONTINUED

Figure	General Properties	Weight Moment of Inertia	Radius of Gyration, ρ
Right circular cone 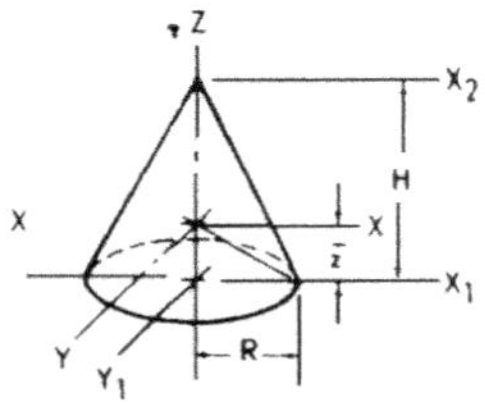	Volume: $V=\frac{\pi R^2 H}{3}$ Centroid: $\bar{z}=\frac{H}{4}$	$I_x=I_y=\frac{3W}{20}\left(R^2+\frac{H^2}{4}\right)$ $I_{x_1}=I_{y_1}=\frac{W}{20}(3R^2+2H^2)$ $I_z=\frac{3W}{10}R^2$ $I_{x_2}=\frac{3W}{20}(R^2+4H^2)$	$\rho_x=\rho_y=0.387\sqrt{R^2+\frac{H^2}{4}}$ $\rho_{x_1}=\rho_{y_1}=0.224\sqrt{3R^2+2H^2}$ $\rho_{x_2}=0.387\sqrt{R^2+4H^2}$ $\rho_z=0.548R$
Frustum of a cone 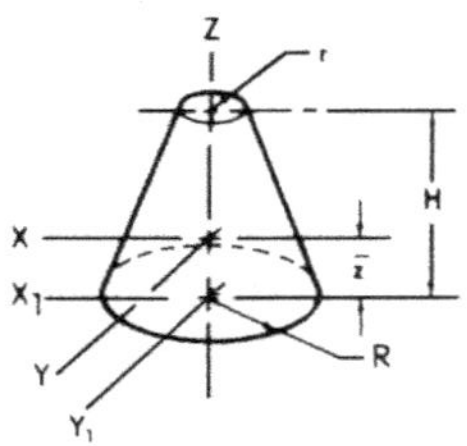	Volume: $V=\frac{\pi H}{3}(R^2+Rr+r^2)$ Centroid: $\bar{z}=\frac{H}{4}\left[\frac{R^2+2Rr+3r^2}{R^2+Rr+r^2}\right]$	$I_{x_1}=I_{y_1}=\frac{WH^2}{10}\left[\frac{R^2+3Rr+6r^2}{R^2+Rr+r^2}\right]+\frac{3W}{10}\left[\frac{R^5-r^5}{R^3-r^3}\right]$ $I_z=\frac{3W}{10}\left[\frac{R^5-r^5}{R^3-r^3}\right]$ $I_x=I_y=I_{x_1}-W(\bar{z})^2$	$\rho_z=0.548\sqrt{\frac{R^5-r^5}{R^3-r^3}}$

(continued)

SOLIDS, CONTINUED

Figure	General Properties	Weight Moment of Inertia	Radius of Gyration, ρ
Spherical sector 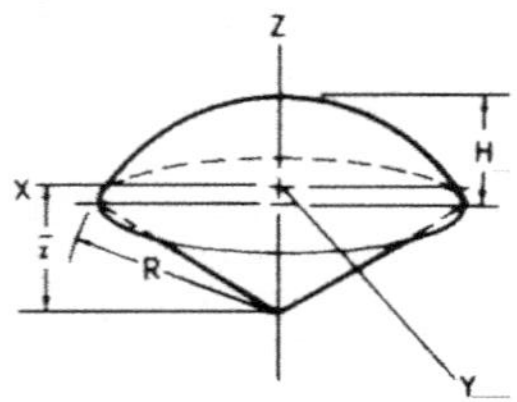	Volume: $V=\frac{2}{3}\pi R^2 H$ Centroid: $\bar{Z}=\frac{3}{8}(2R-H)$	$I_z=\frac{WH}{5}(3R-H)$	$\rho_z=0.447\sqrt{(3R-H)(H)}$
Spherical segment 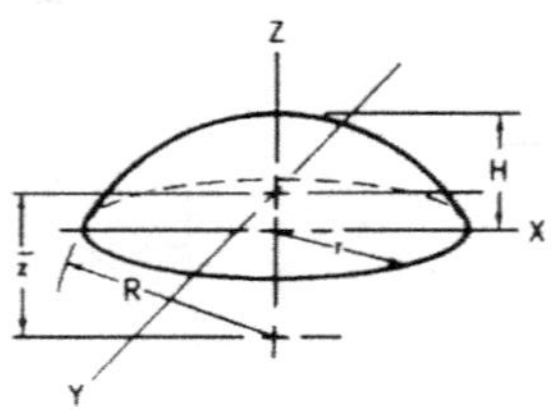	Volume: $V=\frac{\pi H}{6}(3r^2+H^2)$ $V=\frac{\pi H^2}{3}(3R-H)$ Centroid: $\bar{z}=\frac{3}{4}\frac{(2R-H)^2}{(3R-H)}$ Area: $A=2\pi RH$	$I_z=\frac{2HW}{3R-H}\left(R^2-\frac{3}{4}RH+\frac{3}{20}H^2\right)$	$\rho_z=\sqrt{\frac{2H}{3R-H}\left(R^2-\frac{3}{4}RH+\frac{3}{20}H^2\right)}$

(continued)

SOLIDS, CONTINUED

Figure	General Properties	Weight Moment of Inertia	Radius of Gyration, ρ
Elliptic paraboloid 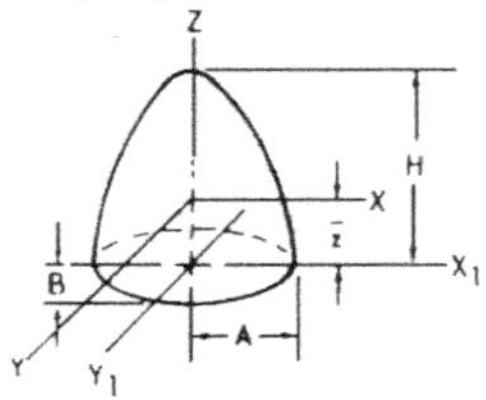	Volume: $V=\frac{\pi ABH}{2}$ Centroid: $\bar{z}=\frac{H}{3}$	$I_x=\frac{W}{18}\left(3B^2+H^2\right)$ $I_{x_1}=\frac{W}{6}\left(B^2+H^2\right)$ $I_y=\frac{W}{18}\left(3A^2+H^2\right)$ $I_{y_1}=\frac{W}{6}\left(A^2+H^2\right)$ $I_z=\frac{W}{6}(A^2+B^2)$	$\rho_x=0.236\sqrt{3B^2+H^2}$ $\rho_y=0.236\sqrt{3A^2+H^2}$ $\rho_z=0.408\sqrt{A^2+B^2}$ $\rho_{x_1}=0.408\sqrt{B^2+H^2}$ $\rho_{y_1}=0.408\sqrt{A^2+H^2}$
Torus 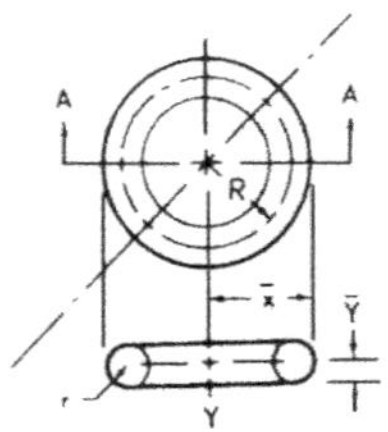	Volume: $V=2\pi^2r^2R$ Centroid: $\bar{x}=\bar{z}=R+r$ $\bar{y}=r$	$I_x=I_z=\frac{W}{8}\left(4R^2+5r^2\right)$ $I_y=\frac{W}{4}\left(4R^2+3r^2\right)$	$\rho_x=\rho_z=0.354\sqrt{4R^2+5r^2}$ $\rho_y=\frac{\sqrt{4R^2+3r^2}}{2}$
Outer half of solid torus 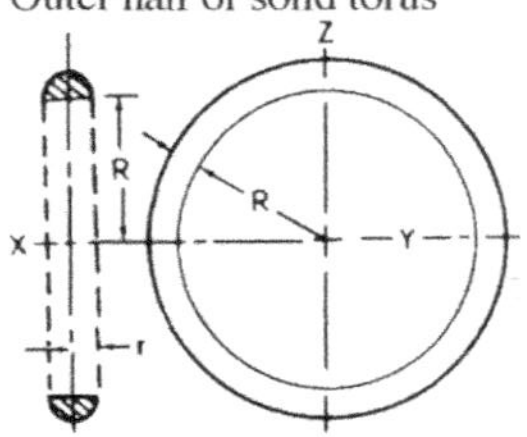	Length: $2\pi R$ Volume: $\frac{4}{3}\pi r^3+\pi^2r^2R$	$I_x=\frac{W}{4.189r+9.870R}[1.6755r^3+7.4022Rr^2$ $+12.5664R^2r+9.869R^3]$ $I_y=I_z=\frac{W}{12}\left[27\pi Rr^2+36R^2r+44r^3\right]$	

(continued)

SOLIDS, CONTINUED

Figure	General Properties	Weight Moment of Inertia

Prismoid

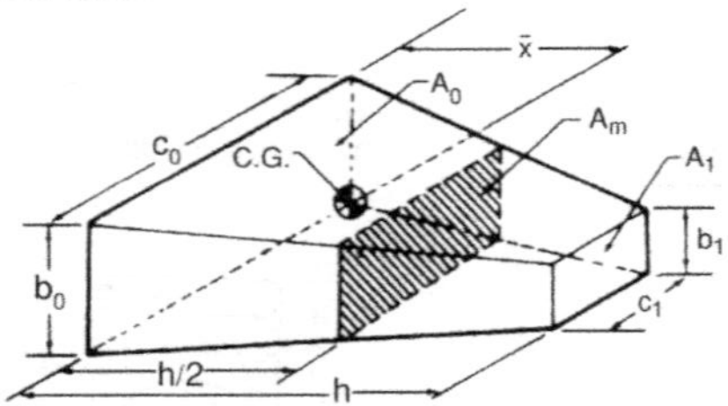

$\overline{x} = A_0$ to C.G.

A_0 = area of base
A_1 = area of top
A_m = area at $h/2$

Note:
A_0, A_1, and A_m must be parallel.

Volume:

$$V = \frac{h}{6}(A_0 + A_1 + 4A_m)$$

Centroid:

$$\overline{X} = \frac{h(A_1 + 2A_m)}{(A_0 + A_1 + 4A_m)}$$

$$I = \frac{\rho h}{20}\left(3I_0 + 16I_m + 3I_1 - \frac{A_1 I_0}{A_0} - \frac{A_0 I_i}{A_i}\right)$$

where: $A_0 = b_0 c_0 \quad A_1 = b_1 c_1$

$$A_m = \left[\frac{b_0 + b_1}{2}\right]\left[\frac{c_0 + c_1}{2}\right]$$

I_0 = area moment of inertia of A_0
I_m = area moment of inertia of A_m
I_1 = area moment of inertia of A_1

Solid ogive

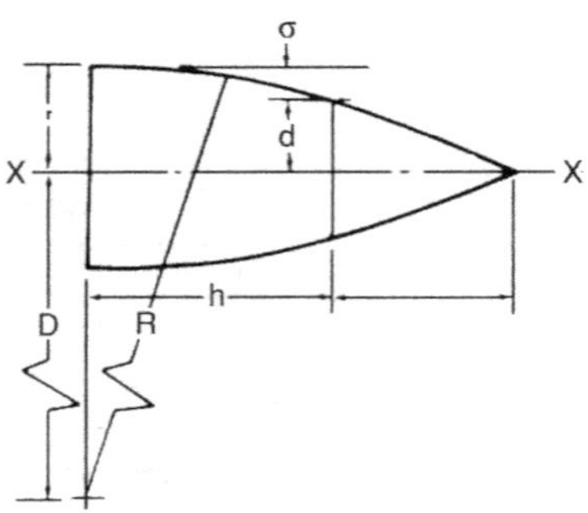

R = ogive radius
d = radius of truncated nose
($d = 0$ for a complete ogive)
$D = R - r$
$\sigma = r - d$

$$V = \frac{\pi h}{9}\left[(3r - a)^2 + 2a^2 - \frac{1.2a^2(R - r)}{R}\right]$$

Complete ogive (1/2 football):

$$V = \pi\left[h\left(R^2 - \frac{h^2}{3}\right) - R^2 D \sin^{-1}\left(\frac{h}{R}\right)\right]$$

Surface area excluding base:

$$A = 2\pi R\left(h - D\sin^{-1}\frac{h}{R}\right)$$

SHELLS

Figure	General Properties	Weight Moment of Inertia	Radius of Gyration, ρ
Lateral cylindrical shell 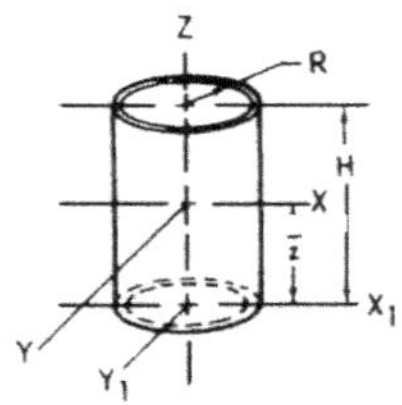	Surface area: $2\pi RH$ Centroid: $\bar{z} = \frac{H}{2}$	$I_x = I_y = \frac{W}{2}\left(R^2 + \frac{H^2}{6}\right)$ $I_z = WR^2$ $I_{x_1} = I_{y_1} = \frac{W}{6}\left(3R^2 + 2H^2\right)$	$\rho_x = \rho_y = 0.707\sqrt{R^2 + \frac{H}{6}}$ $\rho_{x_1} = \rho_{y_1} = 0.408\sqrt{3R^2 + 2H^2}$ $\rho_z = R$
Total cylindrical shell 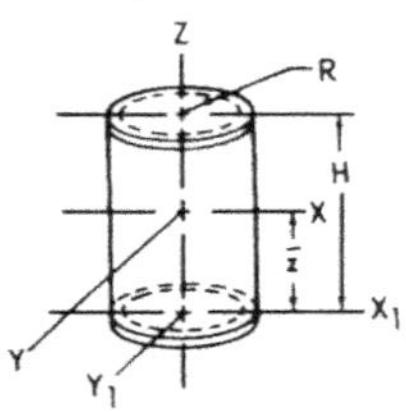	Surface area: $2\pi R(R + H)$ Centroid: $\bar{z} = \frac{H}{2}$	$I_x = I_y = \frac{W}{12(R+H)} \times \left[3R^2(R+2H) + H^2(3R+H)\right]$ $I_{x_1} = I_{y_1} = \frac{W}{12(R+H)} \times \left[3R^2(R+2H) + 2H^2(3R+H)\right]$ $I_z = \frac{WR^2}{2}\left(\frac{R+2H}{R+H}\right)$	$\rho_x = \rho_y = 0.289 \times \sqrt{\frac{3R^2(R+2H) + H^2(3R+H)}{R+H}}$ $\rho_{x_1} = \rho_{y_1} = 0.289 \times \sqrt{\frac{3R^2(R+2H) + 2H^2(3R+2H)}{R+H}}$ $\rho_z = 0.707R\sqrt{\frac{R+2H}{R+H}}$

(continued)

SHELLS, CONTINUED

Figure	General Properties	Weight Moment of Inertia	Radius of Gyration, ρ
Total elliptical shell 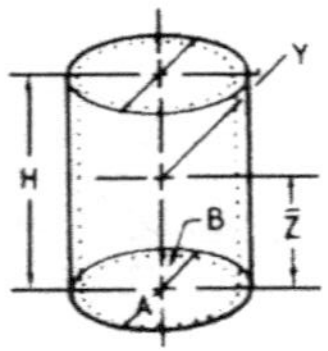	Surface area: $\pi H\left(3A^2+B^2\right)/2A$	$I_x=\frac{WB^2}{4}\left[\frac{7A^2+B^2}{3A^2+B^2}\right]+\frac{WH^2}{12}$ $I_y=\frac{WA^2}{4}\left[\frac{7B^2+A^2}{3A^2+B^2}\right]+\frac{WH^2}{12}$ $I_z=\frac{W}{4}\left[\frac{A^4+14A^2B^2+B^4}{3A^2+B^2}\right]$	
Hollow box 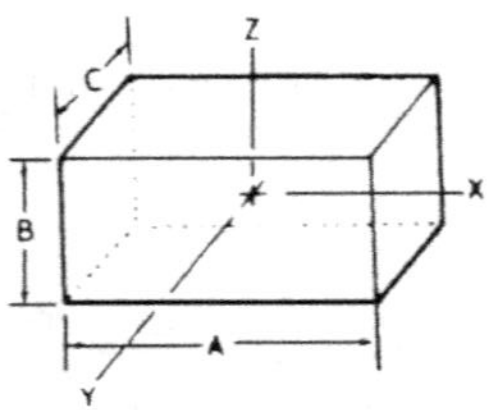	Surface area: $2(AB+BC+AC)$ Surface area (hollow box with open ends): $2C(A+B)$	$I_x=\frac{W}{12}\left(B^2+C^2\right)+\frac{W}{6}\left[\frac{(ABC)(B+C)}{AB+BC+AC}\right]$ $I_y=\frac{W}{12}\left(A^2+B^2\right)+\frac{W}{6}\left[\frac{(ABC)(A+B)}{AB+BC+AC}\right]$ $I_z=\frac{W}{12}\left(A^2+C^2\right)+\frac{W}{6}\left[\frac{(ABC)(A+C)}{AB+BC+AC}\right]$ $I_x=\frac{W}{12}\left(B^2+C^2\right)+\frac{WAB^2}{6(A+B)}$ $I_y=\frac{W}{12}(A+B)^2$ $I_z=\frac{W}{12}\left(A^2+C^2\right)+\frac{WBA^2}{6(A+B)}$	

(continued)

SHELLS, CONTINUED

Figure	General Properties	Weight Moment of Inertia	Radius of Gyration, ρ
Lateral surface of a circular cone 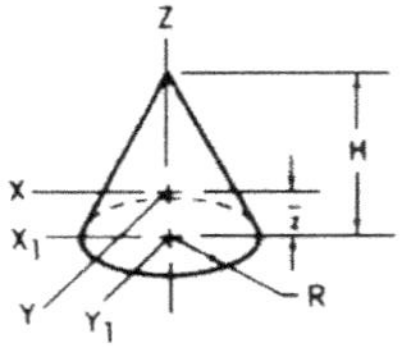	Surface area: $\pi R\sqrt{R^2+H^2}$ Centroid: $\bar{z}=\frac{H}{3}$	$I_x=I_y=\frac{W}{4}\left(R^2+\frac{2}{9}H^2\right)$ $I_z=\frac{WR^2}{2}$ $I_{x_1}=I_{y_1}=\frac{W}{12}\left(3R^2+2H^2\right)$	$\rho_x=\rho_y=\frac{\sqrt{9R^2+2H^2}}{6}$ $\rho_{x_1}=\rho_{y_1}=0.289\sqrt{3R^2+2H^2}$ $\rho_z=0.707R$
Lateral surface of frustum of a circular cone 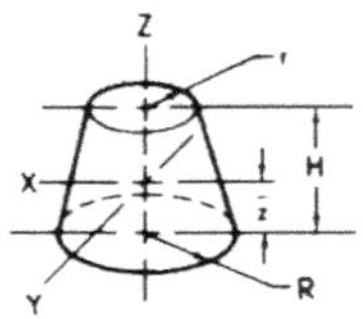	Surface area: $\pi(R+r)\sqrt{H^2+(R-r)^2}$ Centroid: $\bar{z}=\frac{H}{3}\left(\frac{2r+R}{r+R}\right)$	$I_x=I_y=\frac{W}{4}\left(R^2+r^2\right)+\frac{WH^2}{18}\left[1+\frac{2Rr}{(R+r)^2}\right]$ $I_z=\frac{W}{2}\left(R^2+r^2\right)$	$\rho_x=\rho_y=\sqrt{\frac{\left(R^2+r^2\right)}{4}+\frac{H^2}{18}\left[1+\frac{2Rr}{(R+r)^2}\right]}$ $\rho_z=0.707\sqrt{R^2+r^2}$

(continued)

SHELLS, CONTINUED

Figure	General Properties	Weight Moment of Inertia	Radius of Gyration, ρ
Spherical shell 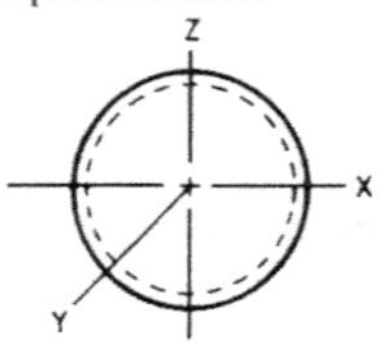	Surface area: $4\pi R^2$	$I_x = I_y = I_z = \frac{2}{3} WR^2$	$\rho_x = \rho_y = \rho_z = 0.816R$
Hemispherical shell 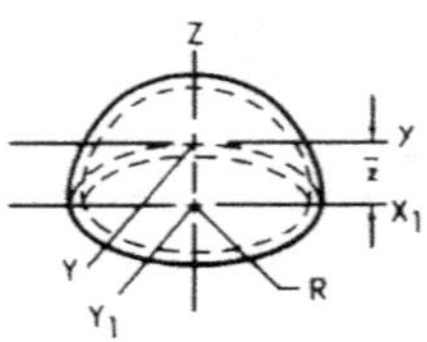	Surface area: $2\pi R^2$ Centroid: $\bar{z} = \frac{R}{2}$	$I_x = I_y = \frac{5}{12} WR^2$ $I_z = I_{x_1} = I_{y_1} = \frac{2}{3} WR^2$	$\rho_x = \rho_y = 0.646R$ $\rho_z = \rho_{x_1} = \rho_{y_1} = 0.816R$
Elliptical hemispheroidal shell 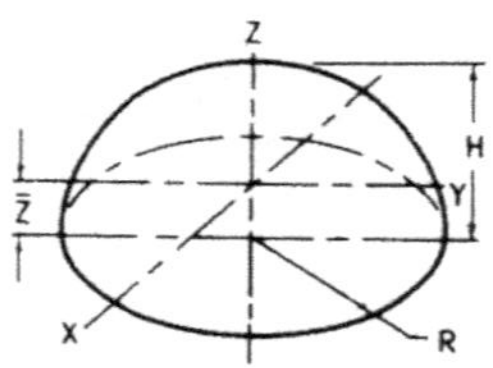 2H is minor axis	Eccentricity: $E = \frac{\sqrt{R^2 - H^2}}{R}$ Surface area: $\pi\left[R^2 + \frac{H^2}{2E} \text{LOG}_e \frac{1+E}{1-E}\right]$ Centroid: $\bar{Z} = \frac{2\pi H\left(R^3 - H^3\right)}{3E^2 R\left(\text{Surface Area}\right)}$	Notes: Surface area formula is for $\frac{1}{2}$ of an oblate spheroid, i.e., an ellipse rotated about its minor axis 2H. (Looks like M & M candy.) Surface area for a prolate spheroid, i.e., ellipse rotated about its major axis 2R. (Looks like a watermelon.) $2\pi(H^2 + \frac{RH}{E} \sin^{-1} E)$	

(continued)

SHELLS, CONTINUED

<table>
<tr><th>Figure</th><th>General Properties</th><th>Weight Moment of Inertia</th></tr>
<tr><td>Paraboloid of a revolution shell
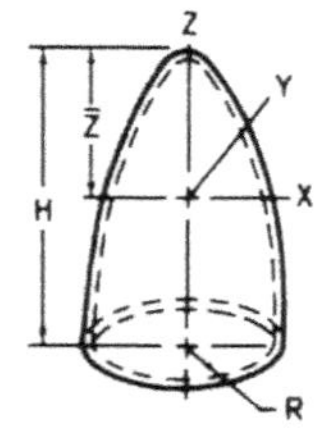
</td><td>Surface area:
$\frac{\pi R}{6H^2}\left[P-R^3\right]$
Centroid:
$\bar{Z}=\frac{P\left(6H^2-R^2\right)+R^5}{10H\left(P-R^3\right)}$
$P=\left(4H^2+R^2\right)^{\frac{3}{2}}$</td><td>$I_x=I_y=\frac{W}{28H^2}\left[\frac{P\left(12H^4+6R^2H^2-R^4\right)+R^7}{P-R^3}\right]$
$-\frac{W}{100H^2}\left[\frac{P\left(6H^2-R^2\right)+R^5}{P-R^3}\right]^2$
$I_Z=\frac{WR^2}{10H^2}\left[\frac{P\left(6H^2-R^2\right)+R^5}{P-R^3}\right]$
or
$I_Z=\frac{WR^2\bar{Z}}{H}$</td></tr>
<tr><td>Sector of a hollow torus
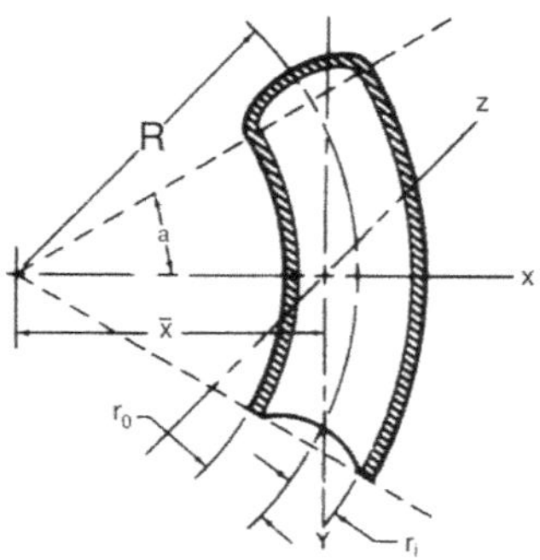
</td><td>$A=r_0^2+r_i^2$
$K=\frac{2\sin^2 a}{a}$
$\left[2R^2+A+\frac{A^2}{BR^2}\right]$
Length:
$2aR$
Centroid:
$\bar{X}=\frac{\sin a}{a}\left[R+\frac{A}{4R}\right]$
Volume:
$2\pi Ra\left(r_0^2-r_i^2\right)$</td><td>$I_x=\frac{W}{16a}\left[4R^2(2a-\sin 2a)+A(10a-3\sin 2a)\right]$
$I_y=\frac{W}{16a}\left[4R^2(2a+\sin 2a)+A(10a-3\sin 2a)-4K\right]$
$I_z=\frac{W}{4a}\left[4R^2a+3Aa-K\right]$</td></tr>
</table>

SECTION 5

STRUCTURAL DESIGN

BEAM FORMULAS

Bending moment, vertical shear, and deflection of beams of uniform cross section under various conditions of loading

NOMENCLATURE

E	= modulus of elasticity, N/m^2 = Pa (lb/in.^{2})
I	= area moment of inertia, m^4(in.^{4})
l	= length of beam, m (in.)
M	= maximum bending moment, N-m (lb-in.)
M_x	= bending moment at any section, N-m (lb-in.)
P	= concentrated loads, N (lb)
R_1, R_2	= reactions, N (lb)
V	= maximum vertical shear, N (lb)
V_x	= vertical shear at any section, N (lb)
w	= uniformly distributed load per unit length, N/m (lb/in.)
W	= total load on beam, N (lb)
x	= distance from support to any section, m (in.)
y	= deflection, m (in.)

SIMPLE BEAM—UNIFORMLY DISTRIBUTED LOAD

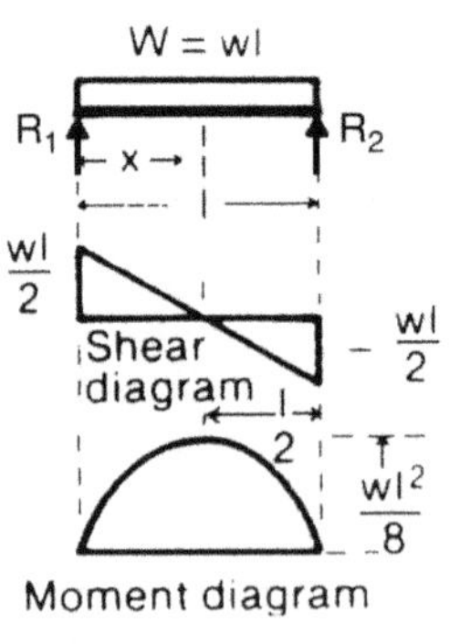

$$R_1 = R_2 = \frac{wl}{2}$$

$$V_x = \frac{wl}{2} - wx$$

$$V_{max} = \pm\frac{wl}{2} \quad \left(\text{when } \begin{matrix} x=0 \\ x=l \end{matrix}\right)$$

$$M_x = \frac{wlx}{2} - \frac{wx^2}{2}$$

$$M_{max} = \frac{wl^2}{8} \quad \left(\text{when } x = \frac{l}{2}\right)$$

$$y_x = \frac{wx}{24EI}(x^3 - 2lx^2 + l^3)$$

$$y_{max} = \frac{5wl^4}{384EI} \quad \text{(at center of span)}$$

Source: The beam formulas appearing on pages 5-2–5-11 are from *Handbook of Engineering Fundamentals*, 3rd Edition, pages 518–520, by O. W. Eshbach and M. Souders.

BEAM FORMULAS, CONTINUED

SIMPLE BEAM—CONCENTRATED LOAD AT ANY POINT

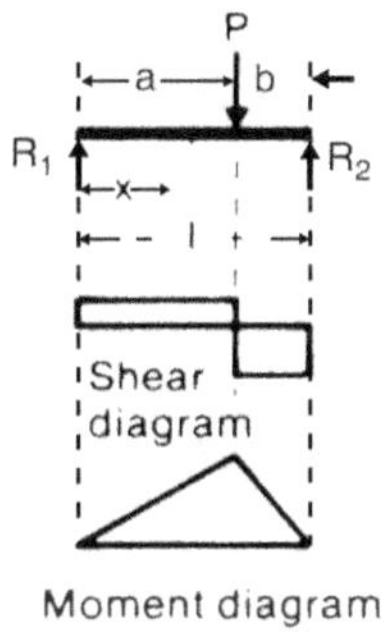

$$R_1 = \frac{Pb}{l}$$

$$R_2 = \frac{Pa}{l}$$

$$V_x = R_1 \quad (\text{when } x < a)$$

$$= -R_2 \quad (\text{when } x > a)$$

$$M_x = \frac{Pbx}{l} \quad (\text{when } x < a)$$

$$= \frac{Pbx}{l} - P(x-a) \quad (\text{when } x > a)$$

$$M_{max} = \frac{Pab}{l} \quad (\text{at point of load})$$

$$y_x = \frac{Pbx}{6EIl}(l^2 - b^2 - x^2) \quad (\text{when } x < a)$$

$$y_x = \frac{Pa(l-x)}{6EIl}(2lx - x^2 - a^2) \quad (\text{when } x > a)$$

$$y_{max} = \frac{Pab(a+2b)\sqrt{3a(a+2b)}}{27EIl} \left(\text{when } x = \sqrt{\frac{a(a+2b)}{3}} \text{ when } a > b\right)$$

SIMPLE BEAM—CONCENTRATED LOAD AT CENTER

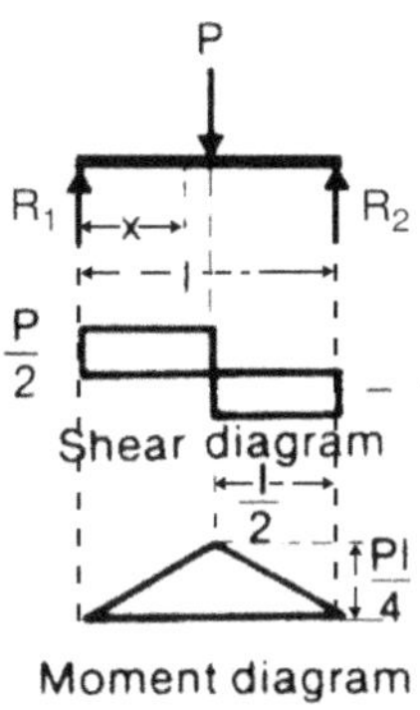

$$R_1 = R_2 = \frac{P}{2}$$

$$V_x = V_{max} = \pm\frac{P}{2}$$

$$M_x = \frac{Px}{2} \left(\text{when } x < \frac{l}{2}\right)$$

$$M_x = \frac{P(l-x)}{2} \left(\text{when } x > \frac{l}{2}\right)$$

$$M_{max} = \frac{Pl}{4} \left(\text{when } x = \frac{l}{2}\right)$$

$$y_x = \frac{Px}{48EI}\left(3l^2 - 4x^2\right) \left(\text{when } x < \frac{l}{2}\right)$$

$$y_{max} = \frac{Pl^3}{48EI} \quad (\text{at center of span})$$

BEAM FORMULAS, CONTINUED

SIMPLE BEAM—TWO EQUAL CONCENTRATED LOADS AT EQUAL DISTANCES FROM SUPPORTS

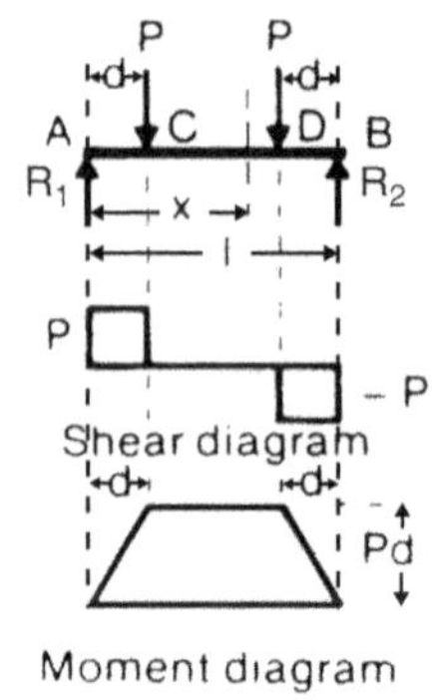

$$R_1 = R_2 = P$$

$$V_x = P \quad \text{for } AC$$

$$= 0 \quad \text{for } CD$$

$$= -P \quad \text{for } DB$$

$$V_{max} = \pm P$$

$$M_x = Px \quad \text{for } AC$$

$$= Pd \quad \text{for } CD$$

$$= P(l - x) \quad \text{for } DB$$

$$M_{max} = Pd$$

$$y_x = \frac{Px}{6EI}(3ld - 3d^2 - x^2) \quad \text{for } AC$$

$$y_x = \frac{Pd}{6EI}(3lx - 3x^2 - d^2) \quad \text{for } CD$$

$$y_{max} = \frac{Pd}{24EI}(3l^2 - 4d^2) \quad \text{(at center of span)}$$

SIMPLE BEAM—LOAD INCREASING LINEARLY FROM SUPPORTS TO CENTER OF SPAN

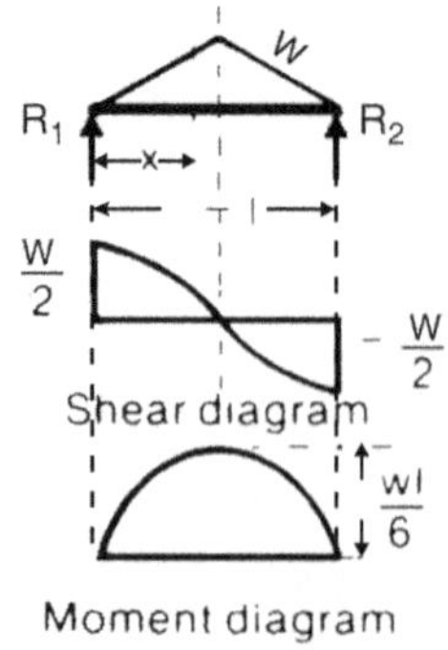

$$R_1 = R_2 = \frac{W}{2}$$

$$V_x = W\left(\frac{1}{2} - \frac{2x^2}{l^2}\right) \quad \left(\text{when } x < \frac{l}{2}\right)$$

$$V_{max} = \pm\frac{W}{2} \text{ (at supports)}$$

$$M_x = Wx\left(\frac{1}{2} - \frac{2x^2}{3l^2}\right)$$

$$M_{max} = \frac{Wl}{6} \quad \text{(at center of span)}$$

$$y_x = \frac{Wx}{480EIl^2}\left(5l^2 - 4x^2\right)^2$$

$$y_{max} = \frac{Wl^3}{60EI} \quad \text{(at center of span)}$$

BEAM FORMULAS, CONTINUED

CANTILEVER BEAM—LOAD CONCENTRATED AT FREE END

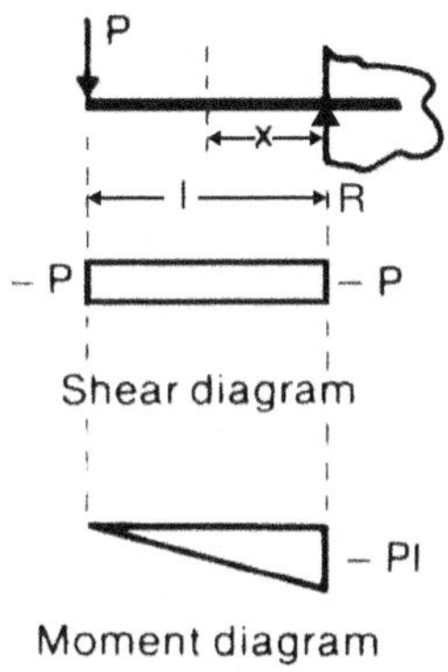

$$R = P$$

$$V_x = V_{max} = -P$$

$$M_x = -P(l - x)$$

$$M_{max} = -Pl \quad (\text{when } x = 0)$$

$$y_x = \frac{P}{6EI}(3lx^2 - x^3)$$

$$y_{max} = \frac{Pl^3}{3EI}$$

SIMPLE BEAM—LOAD INCREASING LINEARLY FROM CENTER TO SUPPORTS

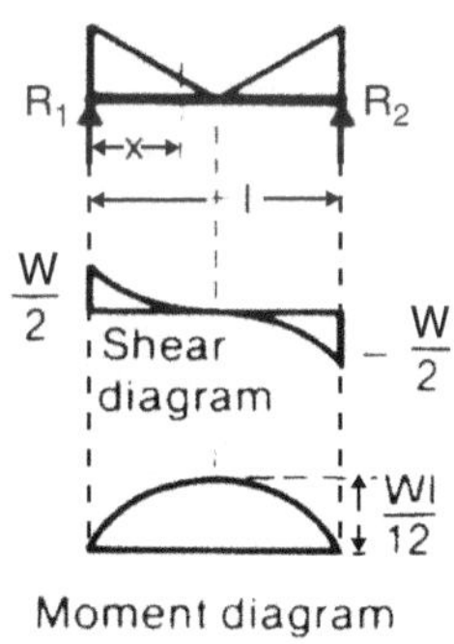

$$R_1 = R_2 = \frac{W}{2}$$

$$V_x = W\left(\frac{1}{2} - \frac{2x}{l} + \frac{2x^2}{l^2}\right) \quad \left(\text{when } x < \frac{l}{2}\right)$$

$$V_{max} = \pm\frac{W}{2}$$

$$M_x = Wx\left(\frac{1}{2} - \frac{x}{l} + \frac{2x^2}{3l^2}\right) \quad \left(\text{when } x < \frac{l}{2}\right)$$

$$M_{max} = \frac{Wl}{12} \quad (\text{at center of span})$$

$$y_{max} = \frac{3Wl^3}{320\,EI} \quad (\text{at center of span})$$

BEAM FORMULAS, CONTINUED

CANTILEVER BEAM—UNIFORMLY DISTRIBUTED LOAD

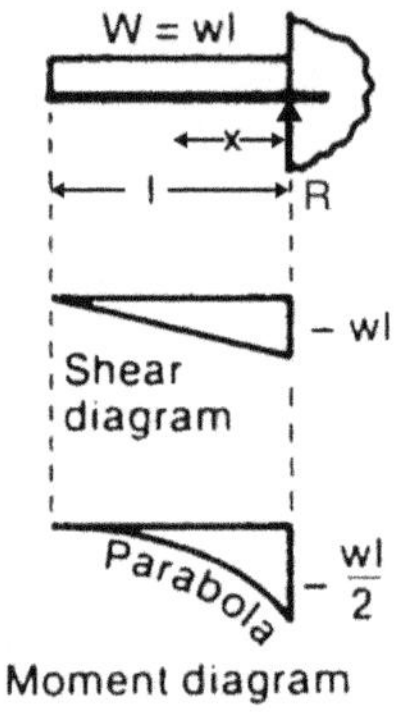

$$R = W = wl$$

$$V_x = -w(l - x)$$

$$V_{max} = -wl \quad (\text{when } x = 0)$$

$$M_x = -w(l - x)\frac{l - x}{2}$$

$$M_{max} = -\frac{wl^2}{2} \quad (\text{when } x = 0)$$

$$y_x = \frac{wx^2}{24EI}(x^2 - 4lx + 6l^2)$$

$$y_{max} = \frac{wl^4}{8EI}$$

SIMPLE BEAM—LOAD INCREASING LINEARLY FROM ONE SUPPORT TO THE OTHER

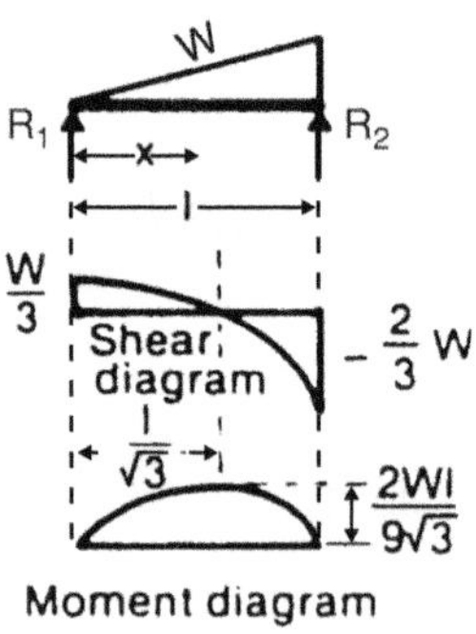

$$R_1 = \frac{W}{3} \quad R_2 = \frac{2}{3}W$$

$$V_x = W\left(\frac{1}{3} - \frac{x^2}{l^2}\right)$$

$$V_{max} = -\frac{2}{3}W \quad (\text{when } x = l)$$

$$M_x = \frac{Wx}{3}\left(1 - \frac{x^2}{l^2}\right)$$

$$M_{max} = \frac{2}{9\sqrt{3}}Wl \quad \left(\text{when } x = \frac{l}{\sqrt{3}}\right)$$

$$y_x = \frac{Wx}{180EIl^2}\left(3x^4 - 10l^2x^2 + 7l^4\right)$$

$$y_{max} = \frac{0.01304}{EI}Wl^3 \quad (\text{when } x = 0.5193l)$$

BEAM FORMULAS, CONTINUED

CANTILEVER BEAM—LOAD INCREASING LINEARLY FROM FREE END TO SUPPORT

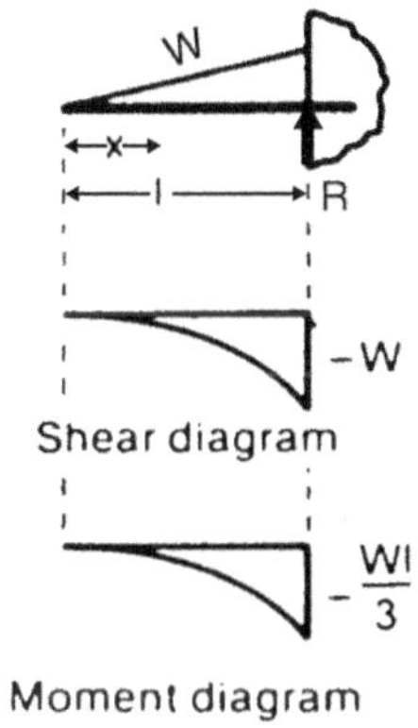

$$R = W$$

$$V_x = -W\frac{(l-x)^2}{l^2}$$

$$V_{max} = -W \quad (\text{when } x = 0)$$

$$M_x = -\frac{W}{3}\frac{(l-x)^3}{l^2}$$

$$M_{max} = -\frac{Wl}{3} \quad (\text{when } x = 0)$$

$$y_x = \frac{W}{60EIl^2}\left(x^5 - 5l^4x + 4l^5\right)$$

$$y_{max} = \frac{Wl^3}{15EI}$$

FIXED BEAM—CONCENTRATED LOAD AT CENTER OF SPAN

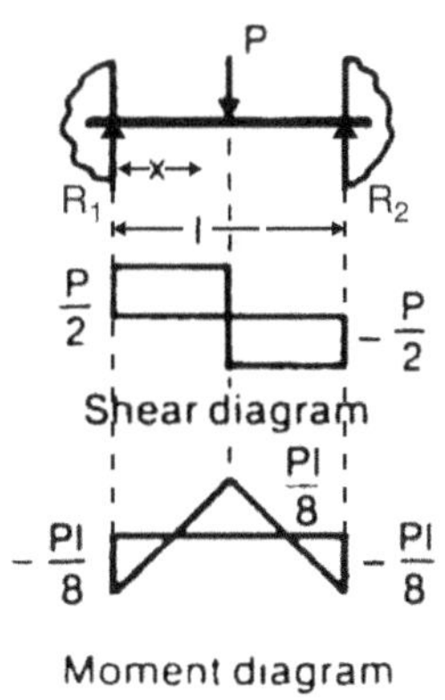

$$R_1 = R_2 = \frac{P}{2}$$

$$V_x = V_{max} = \pm\frac{P}{2}$$

$$M_x = P\left(\frac{x}{2} - \frac{l}{8}\right) \quad \left(\text{when } x < \frac{l}{2}\right)$$

$$M_x = -\frac{Pl}{8} \quad \left(\text{when } \begin{matrix} x = 0 \\ x = l \end{matrix}\right)$$

$$M_{max} = +\frac{Pl}{8} \quad (\text{at center of span})$$

$$y_x = \frac{Px^2}{48EI}(3l - 4x) \quad \left(\text{when } x < \frac{l}{2}\right)$$

$$y_{max} = \frac{Pl^3}{192EI}$$

Beam Formulas, continued

Fixed Beam—Uniformly Distributed Load

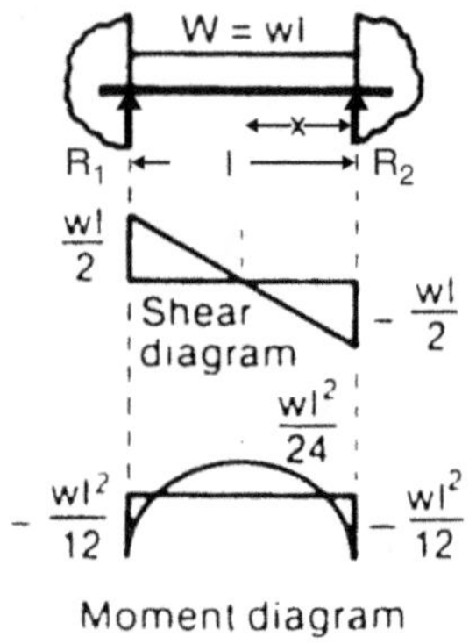

$$R_1 = R_2 = \frac{wl}{2} = \frac{W}{2}$$

$$V_x = \frac{wl}{2} - wx$$

$$V_{max} = \pm\frac{wl}{2} \quad \text{(at ends)}$$

$$M_x = \frac{wl^2}{2}\left(\frac{1}{6} - \frac{x}{l} + \frac{x^2}{l^2}\right)$$

$$M_{max} = -\frac{wl^2}{12} \quad \left(\text{when } \begin{matrix} x = 0 \\ x = l \end{matrix}\right)$$

$$M_{max} = \frac{wl^2}{24} \quad \left(\text{when } x = \frac{l}{2}\right)$$

$$y_x = \frac{wx^2}{24EI}(l-x)^2$$

$$y_{max} = \frac{wl^4}{384EI}$$

Simple Beam—Uniformly Distributed Load over Part of Beam

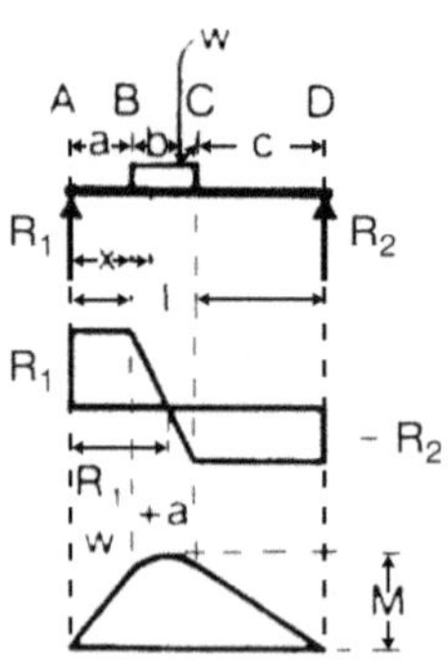

$$R_1 = \frac{wb(2c+b)}{2l}$$

$$R_2 = \frac{wb(2a+b)}{2l}$$

$$V_x = \frac{wb(2c+b)}{2l} - w(x-a)$$

$$V_{max} = R_1 \quad (\text{when } a < c)$$

$$= R_2 \quad (\text{when } a > c)$$

$$M_x = \frac{wbx(2c+b)}{2l} \qquad \text{for } AB$$

$$= R_1x - \frac{w(x-a)^2}{2} \qquad \text{for } BC$$

$$= R_2(l-x) \qquad \text{for } CD$$

$$M_{max} = \frac{wb(2c+b)[4al + b(2c+b)]}{8l^2}$$

BEAM FORMULAS, CONTINUED

BEAM SUPPORTED AT ONE END, FIXED AT OTHER—CONCENTRATED LOAD AT ANY POINT

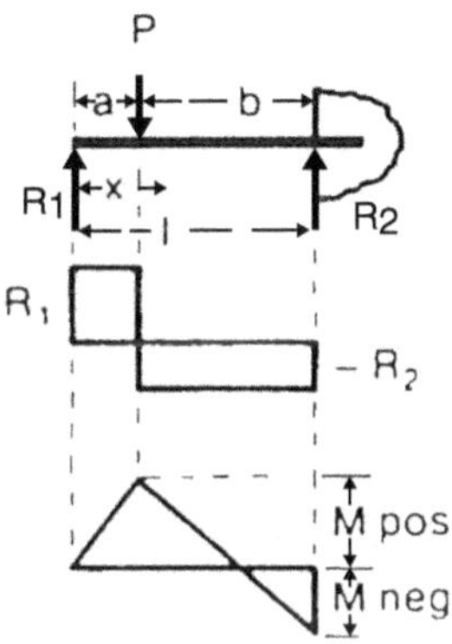

$$R_1 = \frac{Pb^2(2l+a)}{2l^3}$$

$$R_2 = P - R_1$$

$$V_x = R_1 \quad \text{(when } x < a)$$

$$= R_2 \quad \text{(when } x > a)$$

$$M_x = \frac{Pb^2x(2l+a)}{2l^3} \quad \text{(when } x < a)$$

$$= R_1x - P(x-a) \quad \text{(when } x > a)$$

$$M_{\text{max, positive}} = \frac{Pab^2(2l+a)}{2l^3} \quad \text{(when } x = a)$$

$$M_{\text{max, negative}} = -\frac{Pab(l+a)}{2l^2} \quad \text{(when } x = l)$$

$$y_x = \frac{Pb^2x}{12EIl^3}(3al^2 - 2lx - ax^2) \quad \text{(when } x < a)$$

$$y_x = \frac{Pa}{12EIl^3}(l-x)^2(3l^2x - a^2x - 2a^2l) \quad \text{(when } x > a)$$

Beam Formulas, continued

Fixed Beam—Concentrated Load at Any Point

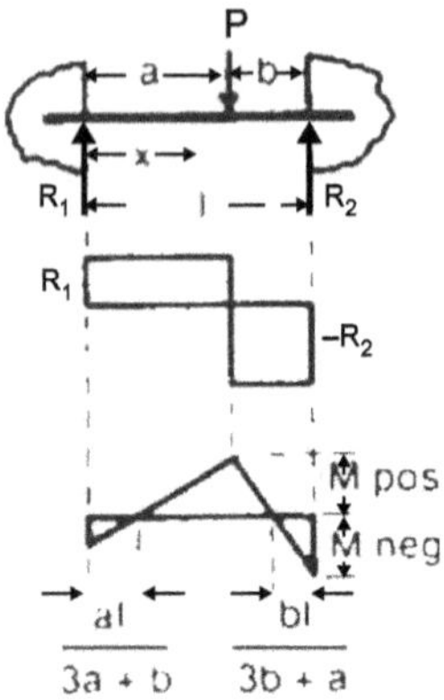

$$R_1 = \frac{Pb^2(l+2a)}{l^3}$$

$$R_2 = \frac{Pa^2(l+2b)}{l^3}$$

$$V_x = R_1 \quad (\text{when } x < a)$$

$$= R_2 \quad (\text{when } x > a)$$

$$M_x = R_1 x - \frac{Pab^2}{l^2} \quad (\text{when } x < a)$$

$$= R_2(l-x) - \frac{Pa^2b}{l^2} \quad (\text{when } x > a)$$

$$M_{\text{positive}} = \frac{2Pa^2b^2}{l^3}$$

$$M_{\text{negative}_1} = -\frac{Pab^2}{l^2}$$

$$M_{\text{negative}_2} = -\frac{Pa^2b}{l^2}$$

$$y_x = \frac{Pb^2x^2}{6EIl^3}(3al - 3ax - bx) \quad (\text{when } x < a)$$

$$y_x = \frac{Pa^2(l-x)^2}{6EIl^3}(ax + 3bx - al) \quad (\text{when } x > a)$$

$$y_{\max} = -\frac{2Pa^3b^2}{3EI(3a+b)^2} \quad \left(\text{when } a > \text{b at } x = \frac{2al}{l+2a}\right)$$

BEAM FORMULAS, CONTINUED

BEAM SUPPORTED AT ONE END, FIXED AT OTHER—UNIFORMLY DISTRIBUTED LOAD

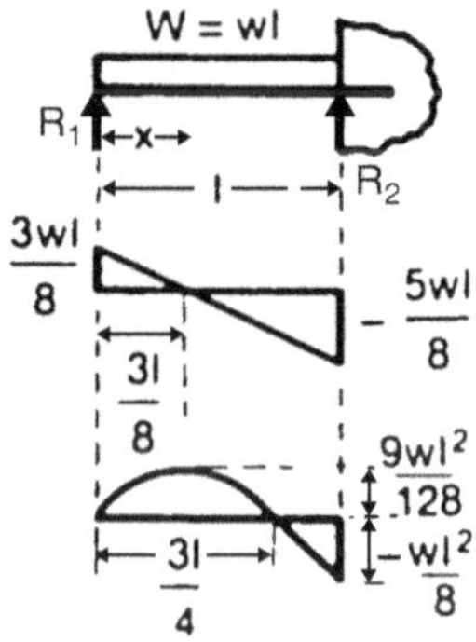

$$R_1 = \frac{3wl}{8}$$

$$R_2 = \frac{5wl}{8}$$

$$V_x = \frac{3wl}{8} - wx$$

$$V = \frac{3wl}{8} \quad \text{(at left support)}$$

$$= -\frac{5wl}{8} \quad \text{(at right support)}$$

$$M_x = wx\left(\frac{3l}{8} - \frac{x}{2}\right)$$

$$M_{\text{max, positive}} = \frac{9wl^2}{128} \quad \left(\text{when } x = \frac{3}{8}l\right)$$

$$M_{\text{max, negative}} = \frac{wl^2}{8} \quad (\text{when } x = l)$$

$$y_x = \frac{wx}{48EI}(l^3 - 3lx^2 + 2x^3)$$

$$y_{\max} = \frac{wl^4}{185EI} \quad (\text{at } 0.4215l \text{ from } R_1)$$

TORSION FORMULAS—SOLID AND TUBULAR SECTIONS

f_s = shear stress, N/m^2 ($lb/in.^2$) (formulas for f_s apply for stress not exceeding the shear yield strength)
F_{su} = ultimate shear strength, N/m^2 ($lb/in.^2$)
G = shear modulus of elasticity, N/m^2 ($lb/in.^2$)
J = torsion constant,* m^4 ($in.^4$)
l = length of section along torsion axis, m (in.)
T = torque, N-m (in.-lb)
T_u = approximate ultimate torque, N-m (in.-lb)
θ = angular deflection

$$\theta = \frac{Tl}{GJ}$$

*The torsion constant J is a measure of the stiffness of a member in pure twisting.

TORSION FORMULAS—SOLID AND TUBULAR SECTIONS, CONTINUED

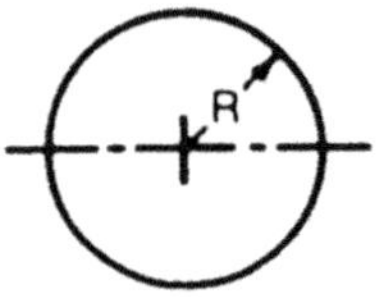

$$\text{Maximum } f_s = \frac{2T}{\pi R^3} \quad \text{at boundary}$$

$$T_u = \frac{2\pi R^3 F_{su}}{3}$$

$$J = \frac{\pi R^4}{2}$$

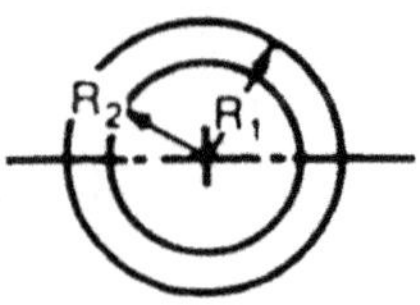

$$\text{Maximum } f_s = \frac{2R_1 T}{\pi\left(R_1^4 - R_2^4\right)} \quad \text{at boundary}$$

$$T_u = \frac{2\pi\left(R_1^3 - R_2^3\right)F_{su}}{3}$$

$$J = \frac{\pi}{2}\left(R_1^4 - R_2^4\right)$$

Thin-walled sections should be checked for buckling.

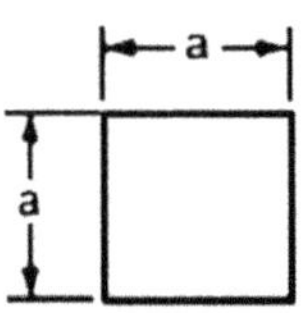

$$\text{Maximum } f_s = \frac{4.8T}{a^3} \quad \text{at midpoints of sides}$$

$$T_u = \frac{a^3 F_{su}}{3}$$

$$J = 0.141a^4$$

(continued)

Note: Formulas for C_w are not given in this table because, for these cross sections, C_w is negligibly small in comparison to J.

TORSION FORMULAS—SOLID AND TUBULAR SECTIONS, CONTINUED

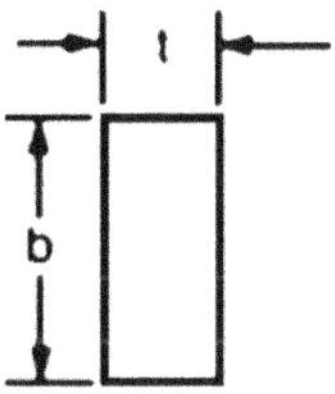

Maximum $f_s = \frac{3T}{bt^2}\left(1 + 0.6\frac{t}{b}\right)$ at midpoints of long sides

(approximately $3T/bt^2$ for narrow rectangles)

$$T_u = \frac{bt^2 F_{su}}{2}\left(1 - \frac{t}{3b}\right)$$

$$J = \frac{bt^3}{3}\left[1 - 0.63\frac{t}{b} + 0.052\left(\frac{t}{b}\right)^5\right]$$

(approximately $bt^3/3$ for narrow rectangles)

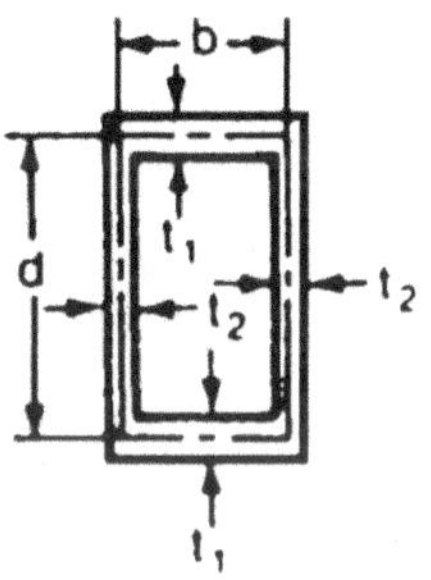

For sides with thickness t_1, average $f_s = \frac{T}{2t_1bd}$

For sides with thickness t_2, average $f_s = \frac{T}{2t_2bd}$

$$T_u = 2bdt_{min}F_{su}$$

($t_{min} = t_1$ or t_2, whichever is smaller)

$$J = \frac{2b^2d^2}{\left(\frac{b}{t_1}\right) + \left(\frac{d}{t_2}\right)}$$

Thin-walled sections should be checked for buckling.

Note: Formulas for C_w are not given in this table because, for these cross sections, C_w is negligibly small in comparison to J.

TORSION FORMULAS—THIN-WALLED OPEN SECTIONS

C_w = torsion warping constant, mm^6 (in.^{6})
J = torsion constant,* mm^4 (in.^{4})

*The torsion constant J is a measure of the stiffness of a member in pure twisting. The torsion warping constant C_w is a measure of the resistance to rotation that arises because of restraint of warping of the cross section.

TORSION FORMULAS—THIN-WALLED OPEN SECTIONS, CONTINUED

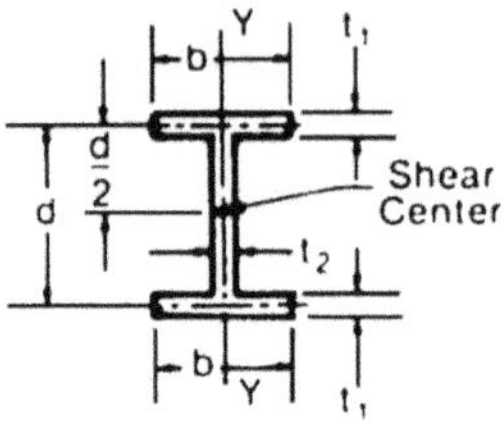

$$J=\frac{1}{3}\left(2bt_1^3+dt_2^3\right)$$

$$C_w=\frac{d^2b^3t_1}{24}$$

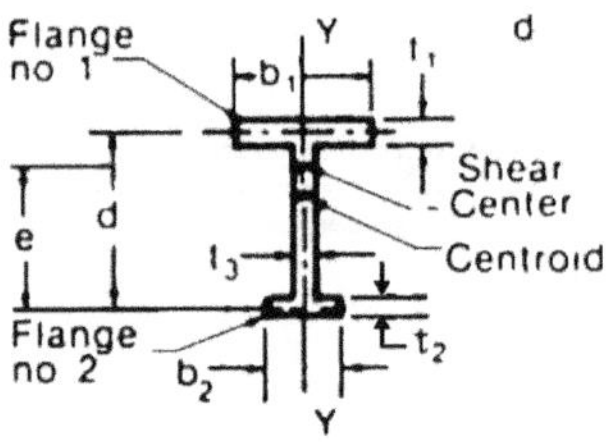

$$t_1\neq t_2$$

$$J=\frac{1}{3}\left(b_1t_1^3+b_2t_2^3+dt_3^3\right)$$

$$C_w=\frac{d^2t_1t_2b_1^3b_2^3}{12(t_1b_1^3+t_2b_2^3)}$$

$$e=\frac{t_1b_1^3d}{t_1b_1^3+t_2b_2^3}$$

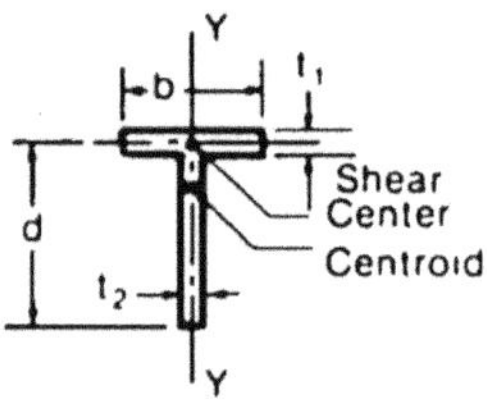

$$J=\frac{1}{3}\left(bt_1^3+dt_2^3\right)$$

$$C_w=\frac{b^3t_1^3}{4}+d^3t_2^3$$

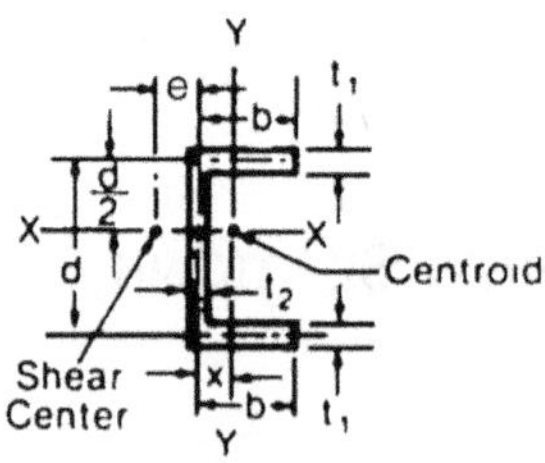

$$J=\frac{1}{3}\left(2bt_1^3+dt_2^3\right)$$

$$C_w=\left(\frac{d^2b^3t_1}{12}\right)\left(\frac{2d+3b}{d+6b}\right)$$

$$e=\frac{3b^2}{d+6b}$$

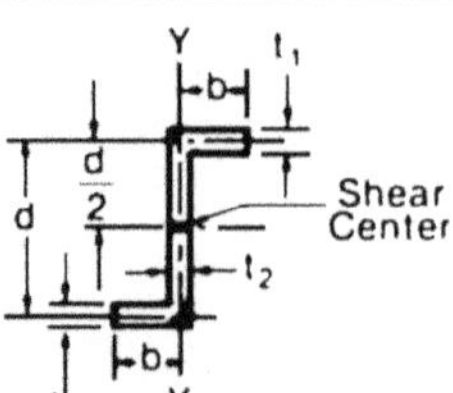

$$J=\frac{1}{3}\left(2bt_1^3+dt_2^3\right)$$

$$C_w=\left(\frac{t_1d^2b^3}{12}\right)\left(\frac{b+2d}{2b+d}\right)$$

POSITION OF FLEXURAL CENTER Q FOR DIFFERENT SECTIONS

Form of Section	Position of Q
Any narrow section symmetrical about the x axis. Centroid at $x = 0$, $y = 0$ 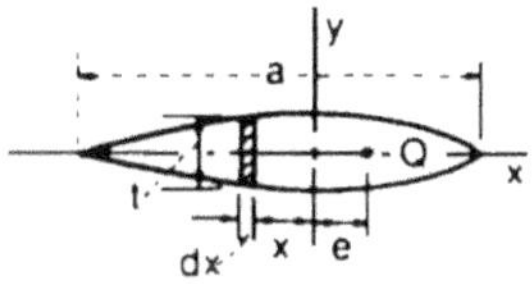	$e = \frac{1+3\nu\int xt^3dx}{1+\nu\int t^3dx}$ For narrow triangle (with $\nu = 0.25$), $e = 0.187a$ For any equilateral triangle, $e = 0$
Sector of thin circular tube 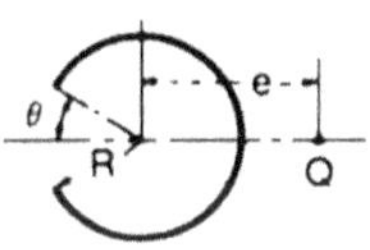	$e = \frac{2R}{(\pi-\theta)+\sin\theta\cos\theta}[(\pi-\theta)\cos\theta+\sin\theta]$ For complete tube split along element ($\theta = 0$), $e = 2R$
Semicircular area 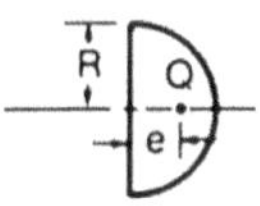	$e = \left(\frac{8}{15\pi}\frac{3+4\nu}{1+\nu}\right)R$ (Q is to right of centroid) For sector of solid or hollow circular area
Angle 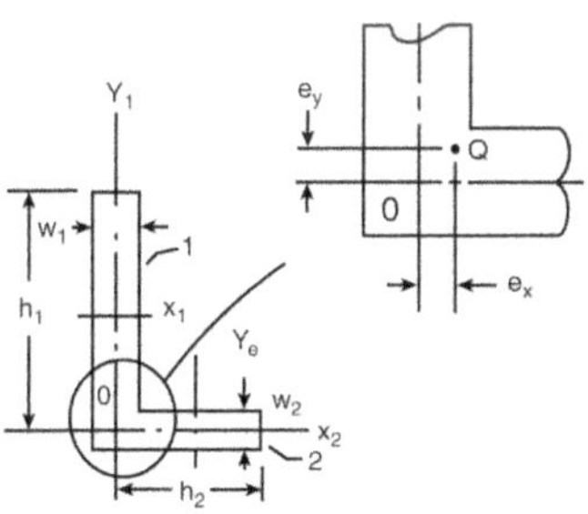	Leg 1 = rectangle w_1h_1; leg 2 = rectangle w_2h_2 I_1 = moment of inertia of leg 1 about Y_1 (central axis) I_2 = moment of inertia of leg 2 about Y_2 (central axis) $e_x = \frac{1}{2}h_2\left(\frac{I_2}{I_1+I_2}\right)$ (For e_x, use x_1 and x_2 central axes.) $e_y = \frac{1}{2}h_1\left(\frac{I_1}{I_1+I_2}\right)$ If w_1 and w_2 are small, $e_x = e_y = 0$ (practically) and Q is at 0.

(continued)

Source: The Position of Flexural Center Q for Different Sections table appearing on pages 5-15–5-17 is from *Formulas for Stress and Strain,* 4th Edition, pages 142 and 143, by R. J. Roark. Copyright ©1965, McGraw–Hill, New York. Reproduced with permission of The McGraw–Hill Companies.

Position of Flexural Center Q for Different Sections, continued

Form of Section	Position of Q
Channel 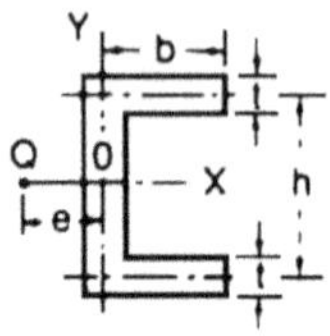	$$e = h\left(\frac{H_{xy}}{I_x}\right)$$ where H_{xy} = product of inertia of the half section (above X) with respect to axes X and Y and I_x = moment of inertia of whole section with respect to X axis. If t is uniform, $$e = \frac{b^2h^2t}{4I_x}$$
Tee 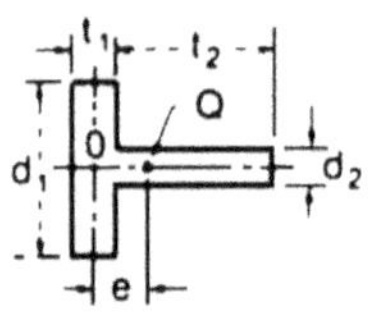	$$e = \frac{1}{2}(t_1+t_2)\left(\frac{1}{1+\left(d_1^3t_1/d_2^3t_2\right)}\right)$$ For a T-beam of ordinary proportions, Q may be assumed to be at 0.
I with unequal flanges and thin web 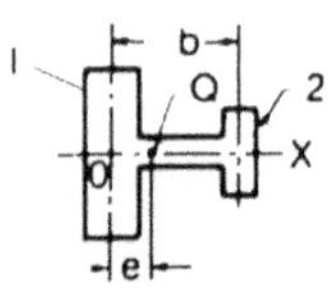	$$e = b\left(\frac{I_2}{I_1+I_2}\right)$$ where I_1 and I_2, respectively, denote moments of inertia about X axis of flange 1 and flange 2
Beam composed of n elements, of any form, connected or separate, with common neutral axis (e.g., multiple-spar airplane wing) 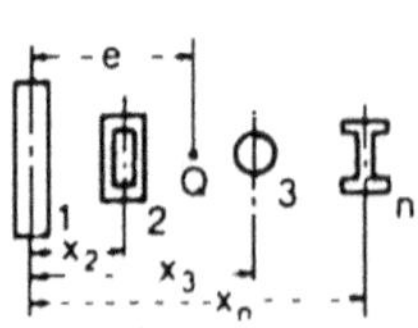	$$e = \frac{E_2I_2x_2 + E_3I_3x_3 + \cdots + E_nI_nx_n}{E_1I_1 + E_2I_2 + E_3I_3 + \cdots + E_nI_n}$$ where I_1, I_2, etc., are moments of inertia of the several elements about the X axis (i.e., Q is at the centroid of the products EI for the several elements)

(continued)

POSITION OF FLEXURAL CENTER Q FOR DIFFERENT SECTIONS, CONTINUED

Form of Section	Position of Q

Lipped channel (t small)

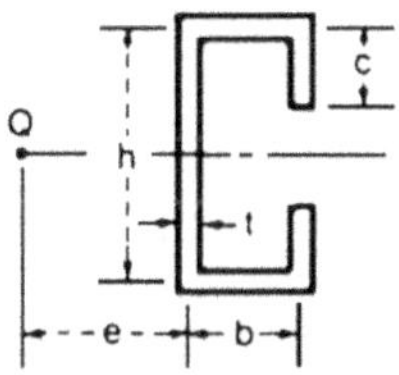

Values of e/h

c/h \ b/h	1.0	0.8	0.6	0.4	0.2
0	0.430	0.330	0.236	0.141	0.055
0.1	0.477	0.380	0.280	0.183	0.087
0.2	0.530	0.425	0.325	0.222	0.115
0.3	0.575	0.470	0.365	0.258	0.138
0.4	0.610	0.503	0.394	0.280	0.155
0.5	0.621	0.517	0.405	0.290	0.161

Hat section (t small)

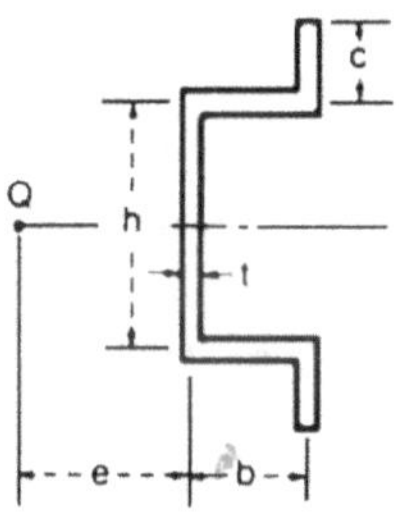

Values of e/h

c/h \ b/h	1.0	0.8	0.6	0.4	0.2
0	0.430	0.330	0.236	0.141	0.055
0.1	0.464	0.367	0.270	0.173	0.080
0.2	0.474	0.377	0.280	0.182	0.090
0.3	0.453	0.358	0.265	0.172	0.085
0.4	0.410	0.320	0.235	0.150	0.072
0.5	0.355	0.275	0.196	0.123	0.056
0.6	0.300	0.225	0.155	0.095	0.040

D section (A = enclosed area)

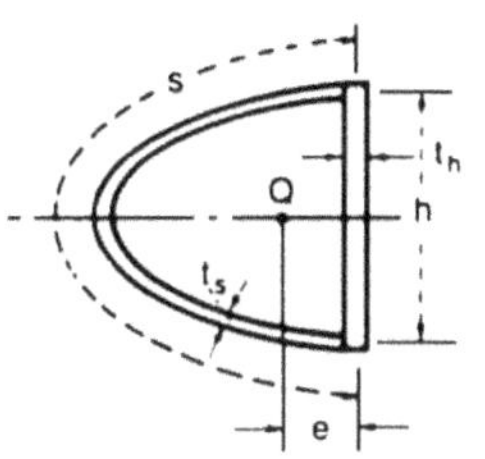

Values of e/h

t_1/t_s \ S/h	1	1.5	2	3	4	5	6	7
0.5	—	—	1.0	0.800	0.665	0.570	0.500	0.445
0.6	—	—	0.910	0.712	0.588	0.498	0.434	0.386
0.7	—	0.980	0.831	0.641	0.525	0.443	0.384	0.338
0.8	—	0.910	0.770	0.590	0.475	0.400	0.345	0.305
0.9	—	0.850	0.710	0.540	0.430	0.360	0.310	0.275
1.0	1.0	0.800	0.662	0.500	0.400	0.330	0.285	0.250
1.2	0.905	0.715	0.525	0.380	0.304	0.285	0.244	0.215
1.6	0.765	0.588	0.475	0.345	0.270	0.221	0.190	0.165
2.0	0.660	0.497	0.400	0.285	0.220	0.181	0.155	0.135
3.0	0.500	0.364	0.285	0.200	0.155	0.125	0.106	0.091

BUCKLING AND STRESS

SHEAR-BUCKLING CURVES FOR SHEET MATERIALS

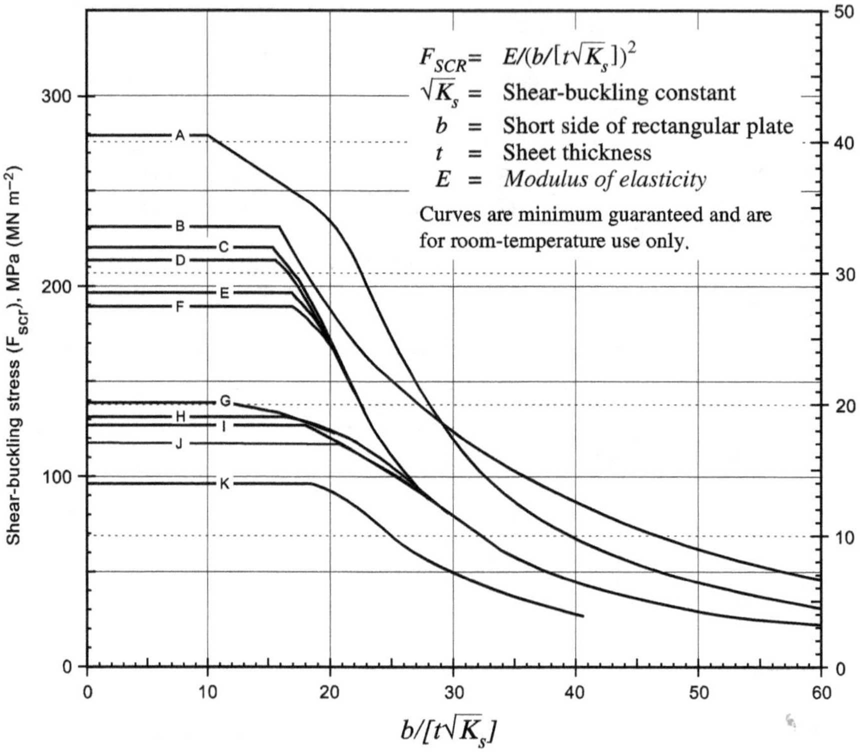

A	AMS 4901 Ti		
B	18-8 Type 301	1/2 Hard	
C	7075.T6 ALC	$1.0 \le t \le 6.3$ mm	($0.040 \le t \le 0.249$ in.)
D	7075.T6 ALC	$0.3 \le t \le 1.0$ mm	($0.012 \le t \le 0.039$ in.)
E	2024.T81 ALC	$t \ge 1.6$ mm	($t \ge 0.063$ in.)
F	2024.T81 ALC	$t \le 1.6$ mm	($t \le 0.063$ in.)
G	6061.T6	$t \ge 6.3$ mm	($t \ge 0.250$ in.)
H	2024.T3 ALC	$1.6 \le t \le 6.3$ mm	($0.063 \le t \le 0.249$ in.)
I	2024.T3 ALC	$0.25 \le t \le 1.6$ mm	($0.010 \le t \le 0.062$ in.)
J	2024.T42 ALC	$t < 1.6$ mm	($t < 0.063$ in.)
K	AZ31B-H24	$0.4 \le t \le 6.3$ mm	($0.016 \le t \le 0.250$ in.)

Source: Vought Aerospace Corporation handbook. Copyright © Northrop Grumman, Los Angeles. Reproduced with permission of Northrop Grumman.

BUCKLING AND STRESS, CONTINUED

ALLOWABLE SHEAR FLOW—7075-T6 CLAD WEB

Allowable Shear Flow—ISO AlZn6MgCu (7075-T6) Clad Web

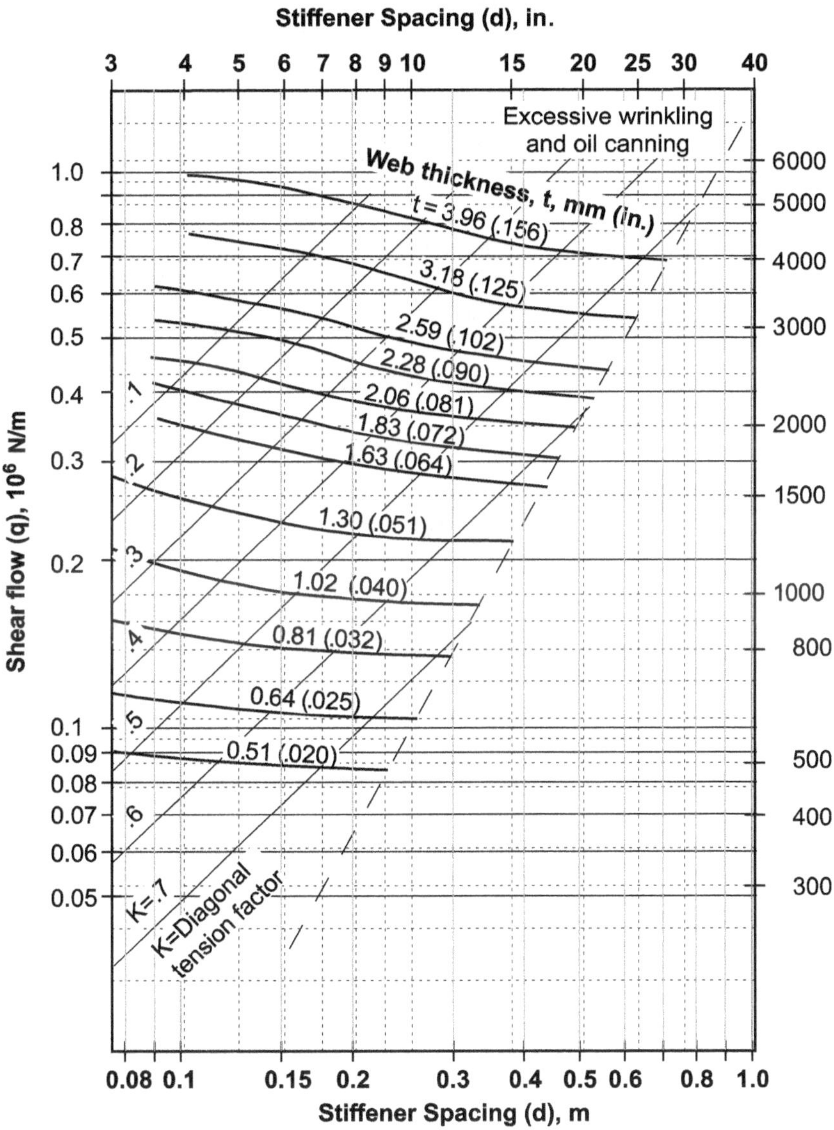

BUCKLING AND STRESS, CONTINUED

SHEAR-BUCKLING CONSTANT $\left(\sqrt{K_s}\right)$

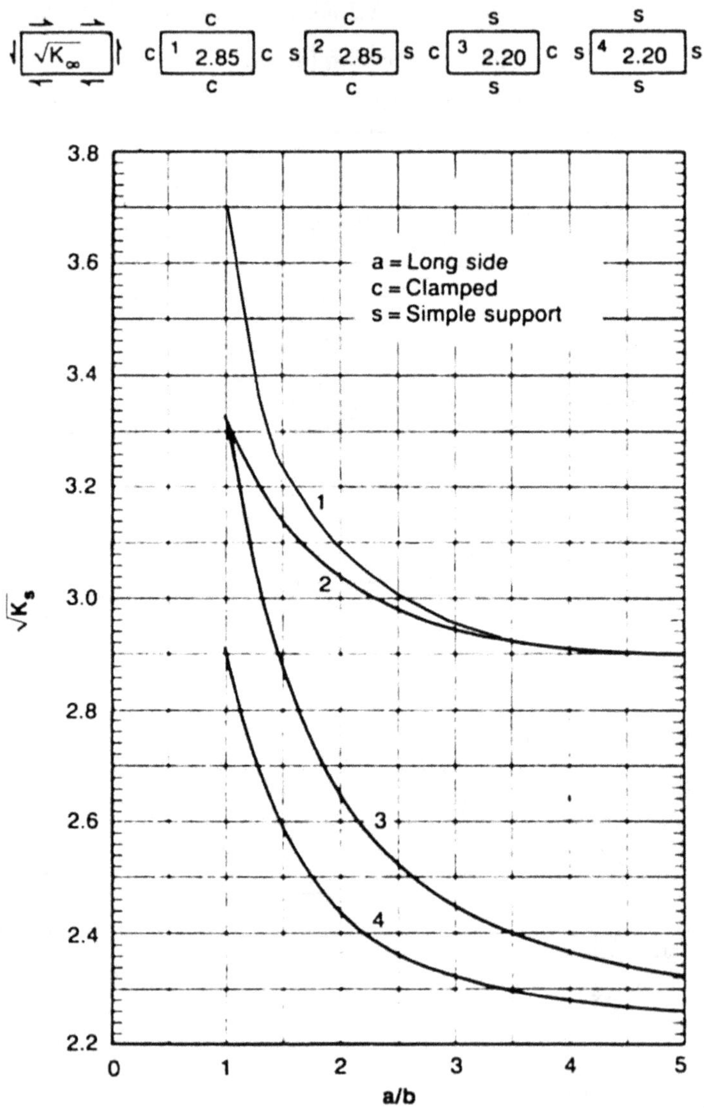

BUCKLING AND STRESS, CONTINUED

BEAM DIAGONAL TENSION NOMENCLATURE

A = cross-sectional area, m^2 (in.^{2})
d = spacing of uprights, m (in.)
q = shear flow (shear force per unit length), N/m (lb/in.)
t = thickness, m (in.) (when used without subscript, signifies thickness of web)
k = diagonal-tension factor
h = depth of beam, m (in.)
h_U = length of upright (measured between centroids of upright-to-flange rivet patterns), m (in.)
σ = normal stress, N/m^2 = Pa (lb/in.^{2})
τ = shear stress, N/m^2 = Pa (lb/in.^{2})
ωd = flange flexibility factor
α = angle between neutral axis of beams and direction of diagonal tension, deg

Subscripts:

DT = diagonal tension
IDT = incomplete diagonal tension
PDT = pure diagonal tension
U = upright
e = effective

MAXIMUM STRESS TO AVERAGE STRESS IN WEB STIFFENER

For curved webs: for rings, read abscissa as d/h; for stringers, read abscissa as h/d.

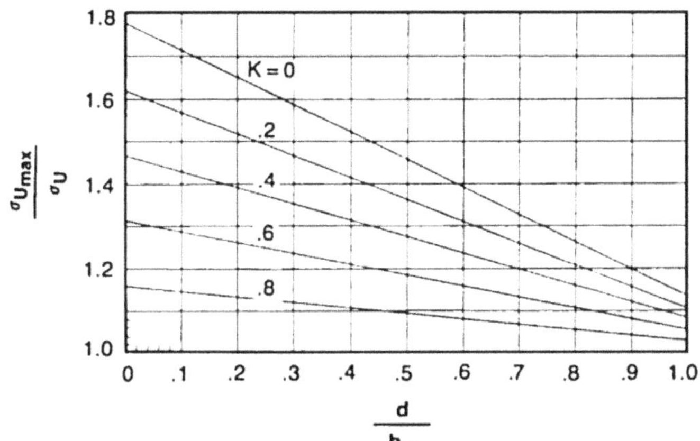

Source: The diagonal tension material appearing on pages 5-21 through 5-24 is from *Summary of Diagonal Tension, Part I,* NACA-TN-2661.

BUCKLING AND STRESS, CONTINUED

ANGLE FACTOR C_1

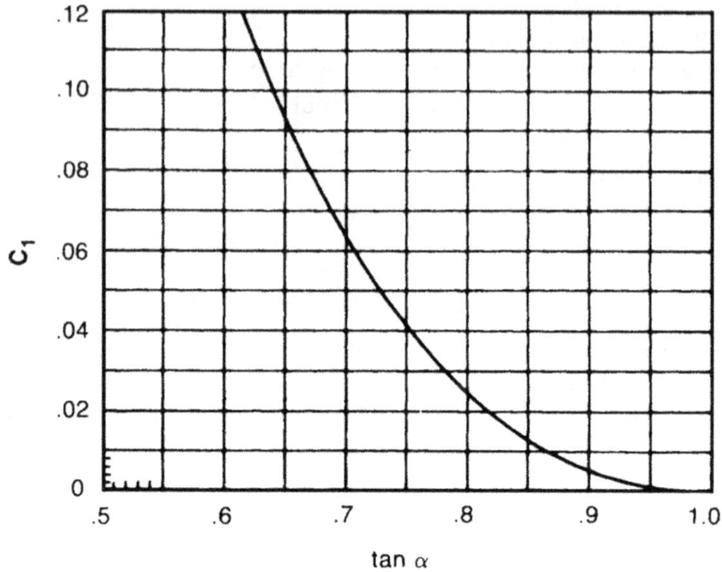

STRESS-CONCENTRATION FACTORS C_2 *AND* C_3

$$\left(\omega d = 0.7d\sqrt[4]{\frac{t}{(I_c + I_T)h_e}}\right)$$

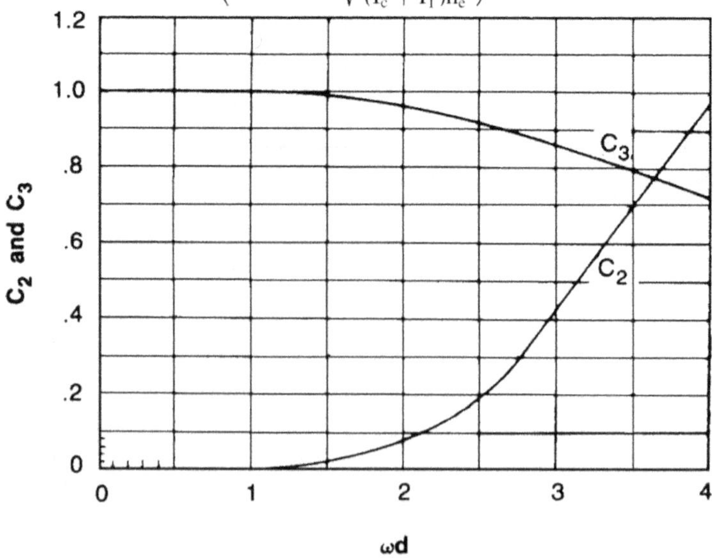

I_C = I of compression flange.

I_T = I of tension flange.

Buckling and Stress, continued

Diagonal-Tension Analysis Chart

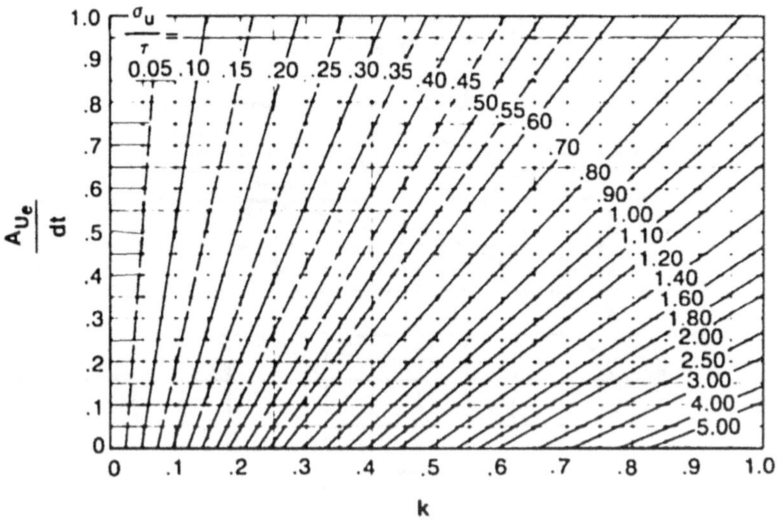

Angle of Diagonal Tension (Incomplete)

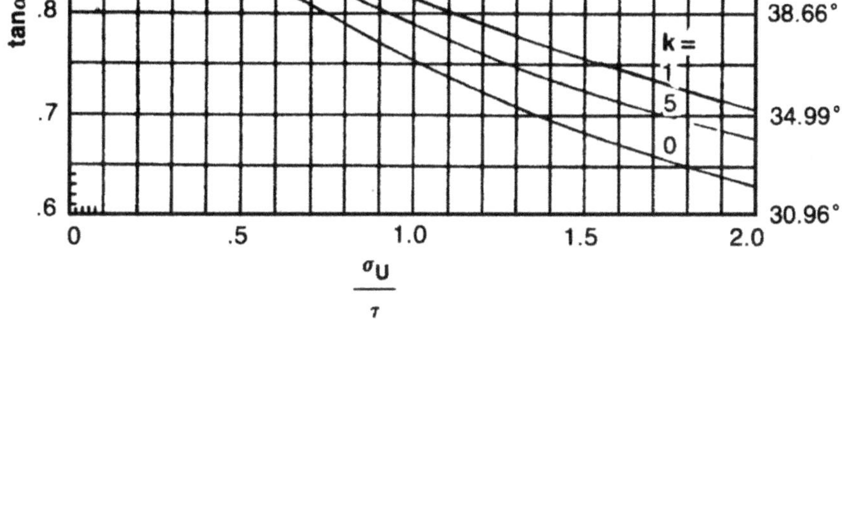

BUCKLING AND STRESS, CONTINUED

ANGLE OF DIAGONAL TENSION (PURE)

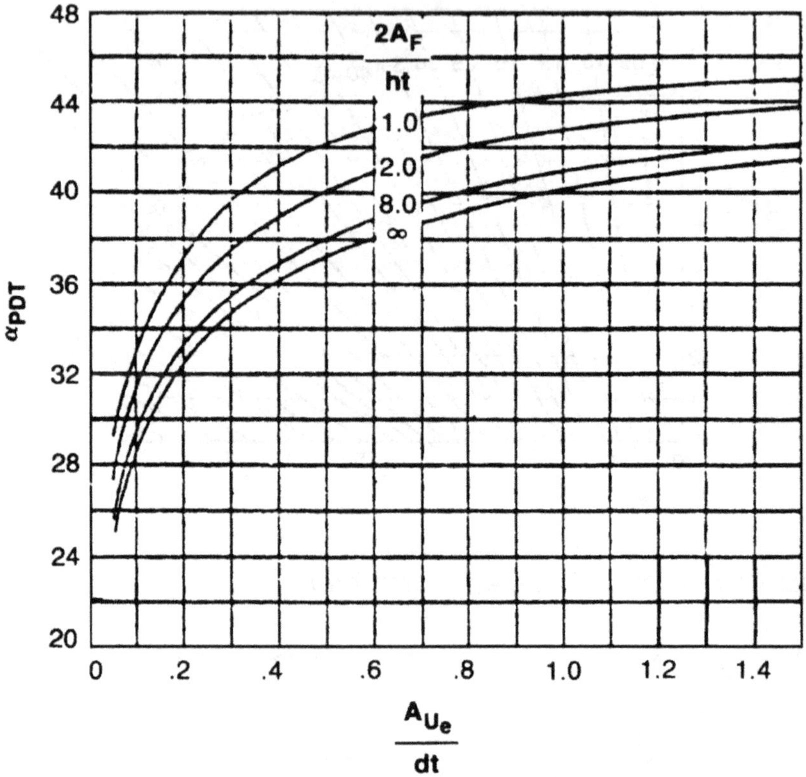

COLUMNS

INTERACTION OF COLUMN FAILURE WITH LOCAL FAILURE (CRIPPLING)

The method of analysis of columns subject to local failure can be summarized as follows:

a) Sections having four corners	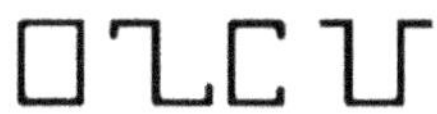
b) Sections having two corners, attached to sheets along both flanges	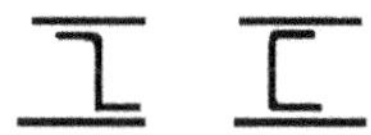
c) Sections having only two corners but restrained against column failure about axis thru corners	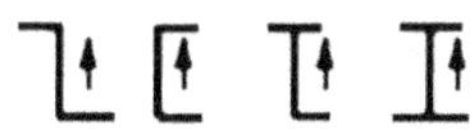 (Arrow represents direction of column failure)
d) Sections having three corners, attached to a sheet along one uplipped flange	
For local failing stress (upper limit of column curve) use crippling stress F_{cc} (see note)	
e) Sections having only two corners (with no restraint in any direction)	 (Arrow represents direction of column failure)
f) Sections having only two corners, attached to sheet	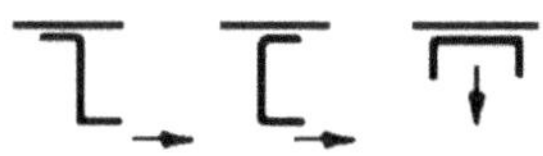 (Arrow represents direction of column failure)
For local failing stress (upper limit of column curve) use local buckling stress $F_{c_{Cr}}$ (see note)	
Column curve	For columns that fail by Euler buckling, select the proper column curve and determine the allowable ultimate stress F_C.

Note: The calculations for F_{cc} and $F_{c_{cr}}$ should be made by reference to sources not included in this handbook.

COLUMNS, CONTINUED

COMPARISON OF DIFFERENT COLUMN CURVES

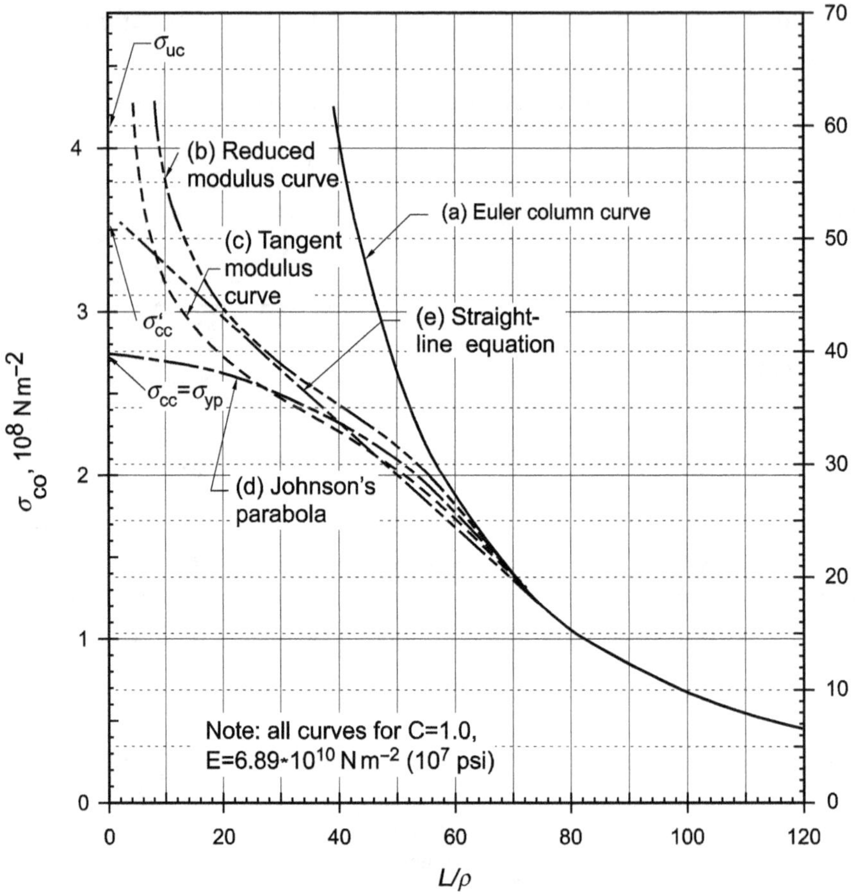

EULER COLUMN FORMULA

$$P_E = \frac{C\pi^2 EI}{L^2}$$

where P_E = critical column load and C = end fixity coefficient (pin end = 1.0, restrained = 4.0). The critical column stress σ_E is

$$\sigma_E = \frac{C\pi^2 E}{(L/\rho)^2}$$

where ρ = radius of gyration = $\sqrt{I/A}$.

COLUMNS, CONTINUED

COMPARISON OF DIFFERENT COLUMN CURVES, CONTINUED

REDUCED MODULUS CURVE

The L/ρ corresponding to the critical stress σ_{sc} in the short column range is

$$\left(\frac{L}{\rho}\right)_{E_r} = \pi\sqrt{\frac{E_r}{\sigma_{sc}}}$$

where

$$E_r = \left(\sqrt{\frac{4EE_t}{\sqrt{E}+\sqrt{E_t}}}\right)^2 \quad E_r = \sqrt{EE_t} \quad E_t = \text{tangent modulus}$$

TANGENT MODULUS CURVE

$$\left(\frac{L}{\rho}\right)_{E_t} = \pi\sqrt{\frac{E_t}{\sigma_{sc}}}$$

JOHNSON PARABOLIC FORMULA

The Johnson equation that gives the critical short column stress σ_{sc} is

$$\sigma_{sc} = \sigma_{cc} - \frac{\sigma_{cc}^2 (L/\rho)^2}{4C\pi^2 E}$$

STRAIGHT LINE EQUATION

$$\sigma_{sc} = \sigma'_{cc}\frac{1-k(L/\rho)}{\sqrt{C}}$$

where σ_{cc}, k, and $\sqrt{C}$ are chosen to give best agreement with experimental data.

COLUMNS, CONTINUED

COLUMN CURVES FOR ALUMINUM ALLOYS—BASED ON TANGENT MODULUS

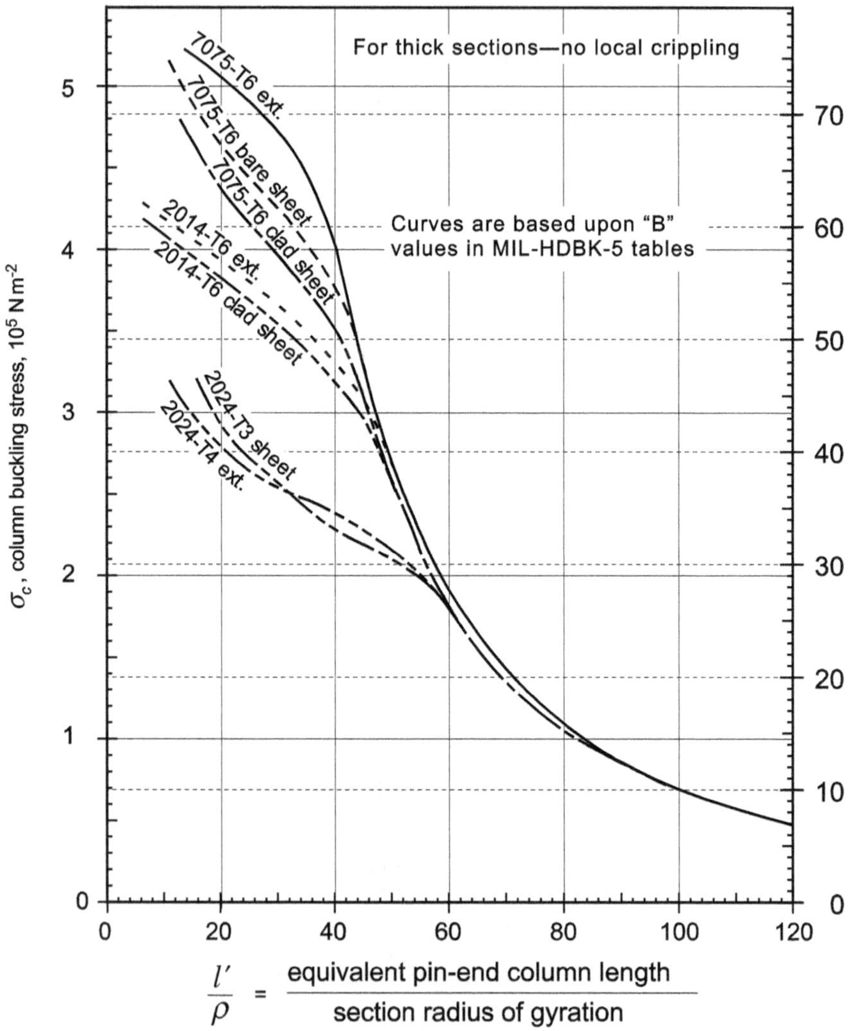

where

$l' = l/\sqrt{C} =$ equivalent pin-end column length

$l =$ total column length

$C =$ end fixity coefficient

$\rho =$ section radius of gyration

Source: Northrop Grumman structures manual. Copyright © Northrop Grumman Corporation, Los Angeles. Reproduced with permission of Northrop Grumman.

COLUMNS, CONTINUED

END FIXITY COEFFICIENTS

Column Shape and End Conditions		End Fixity Coefficient
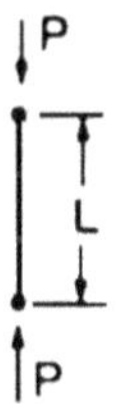	Uniform column, axially loaded, pinned ends	$C = 1$ $\frac{1}{\sqrt{C}} = 1$
	Uniform column, axially loaded, fixed ends	$C = 4$ $\frac{1}{\sqrt{C}} = 0.5$
	Uniform column, axially loaded, one end fixed, one end pinned	$C = 2.05$ $\frac{1}{\sqrt{C}} = 0.70$
	Uniform column, axially loaded, one end fixed, one end free	$C = 0.25$ $\frac{1}{\sqrt{C}} = 2$

(continued)

COLUMNS, CONTINUED

END FIXITY COEFFICIENTS, CONTINUED

Column Shape and End Conditions		End Fixity Coefficient
	Uniform column, distributed axial load, one end fixed, one end free	$C = 0.794$ $\frac{1}{\sqrt{C}} = 1.12$
	Uniform column, distributed axial load, pinned ends	$C = 1.87$ $\frac{1}{\sqrt{C}} = 0.732$
	Uniform column, distributed axial load, fixed ends	$C = 7.5$ $\frac{1}{\sqrt{C}} = 0.365$
	Uniform column, distributed axial load, one end fixed, one end pinned	$C = 3.55$ (approx.) $\frac{1}{\sqrt{C}} = 0.530$

COLUMNS, CONTINUED

COLUMN STRESS FOR ALUMINUM ALLOY COLUMNS

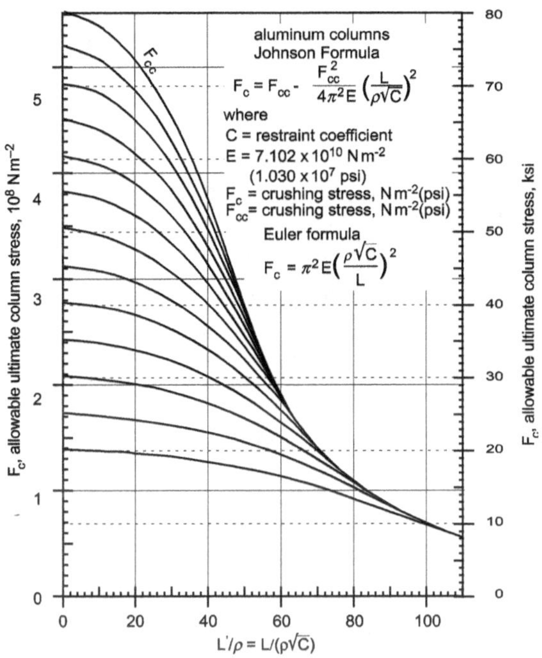

COLUMN STRESS FOR MAGNESIUM ALLOY COLUMNS

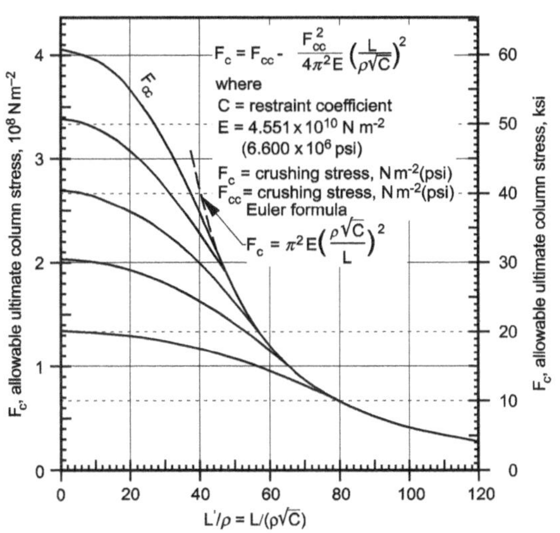

COLUMNS, CONTINUED

COLUMN STRESS FOR STEEL COLUMNS

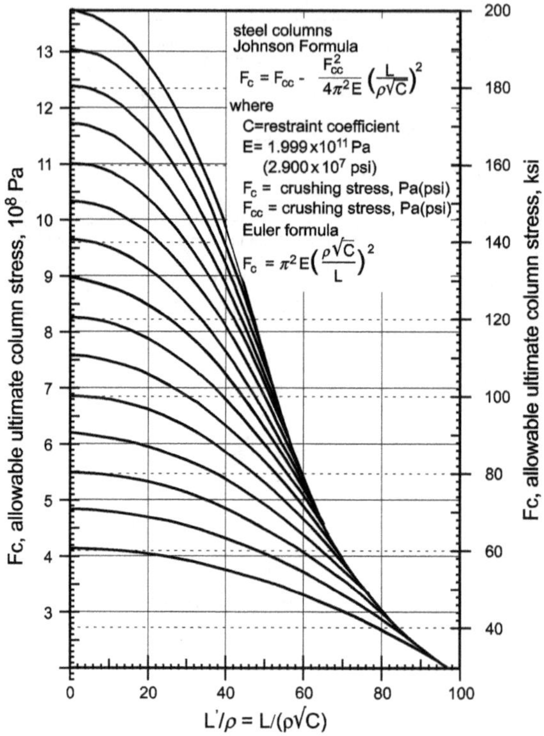

COLUMN STRESS FOR TITANIUM ALLOY COLUMNS

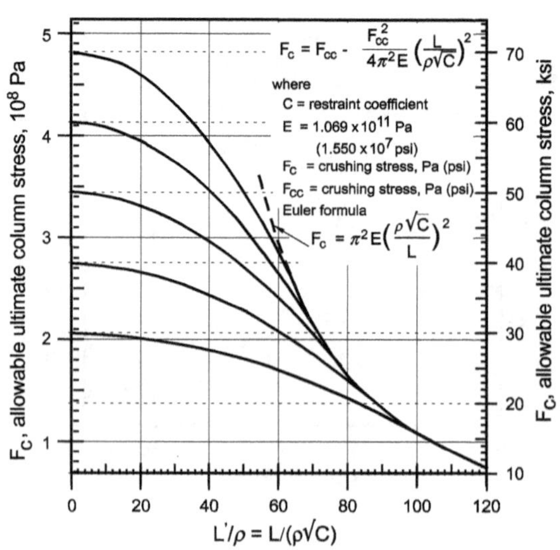

COLUMNS, CONTINUED

BEAM-COLUMN CURVES FOR CENTRALLY LOADED COLUMNS WITH END COUPLES

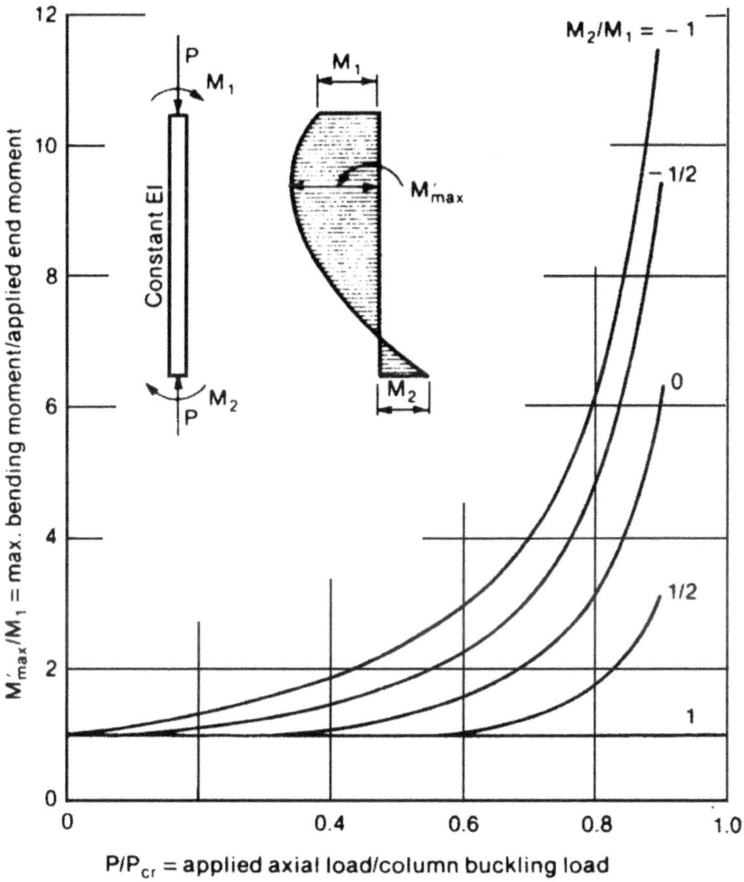

P = applied axial load

P_{cr} = $F_c \times A$ = column buckling load

F_c = column buckling stress

M_1, M_2 = applied end moment

M'_{max} = max. total bending moment including secondaries

Source: Westergaard, H. M., "Buckling of Elastic Structure," *Transactions of the ASCE*, 1922, p. 594.

PLATES

BUCKLING STRESS COEFFICIENTS FOR FLAT PLATES IN SHEAR

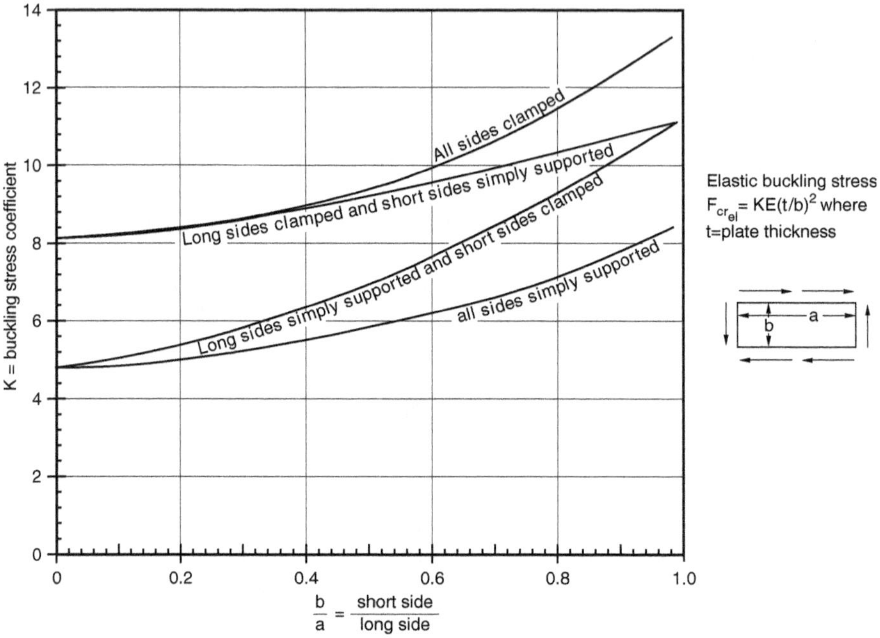

Sources: TN 1222, 1559; Timoshenko, "Theory of Elastic Stability"; RAS Data Sheets; "Stressed Skin Structures."

BUCKLING STRESS COEFFICIENTS FOR FLAT RECTANGULAR PLATES LOADED IN COMPRESSION ON THE LONG SIDE

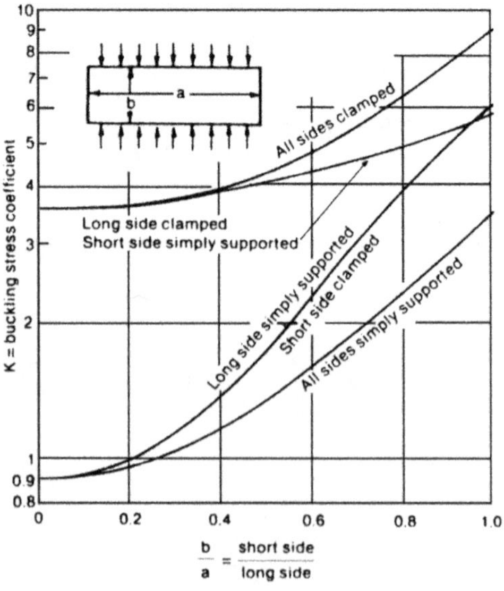

PLATES, CONTINUED

BUCKLING STRESS COEFFICIENTS FOR FLAT RECTANGULAR PLATES LOADED IN COMPRESSION ON THE SHORT SIDE

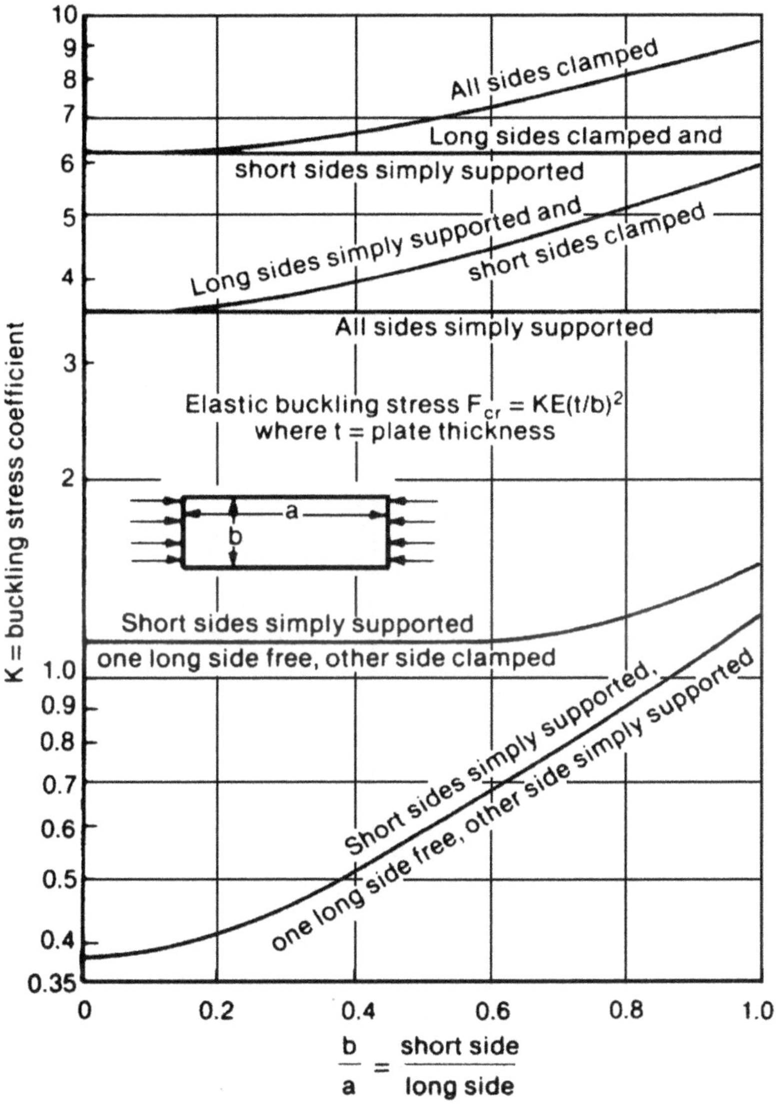

Sources: Timoshenko, "Theory of Elastic Stability"; RAS Data Sheets; "Stressed Skin Structures."

PLATES, CONTINUED

STRESS CONCENTRATION FACTORS FOR AN AXIALLY LOADED PLATE WITH AN ECCENTRIC CIRCULAR HOLE

Note: These factors apply to stresses in the elastic region

$f_{max} = K \times f_{net}$ = maximum stress (at point "n")

$f_{net} = P/(d_1 + d_2)(t)$ = average net section stress

t = plate thickness

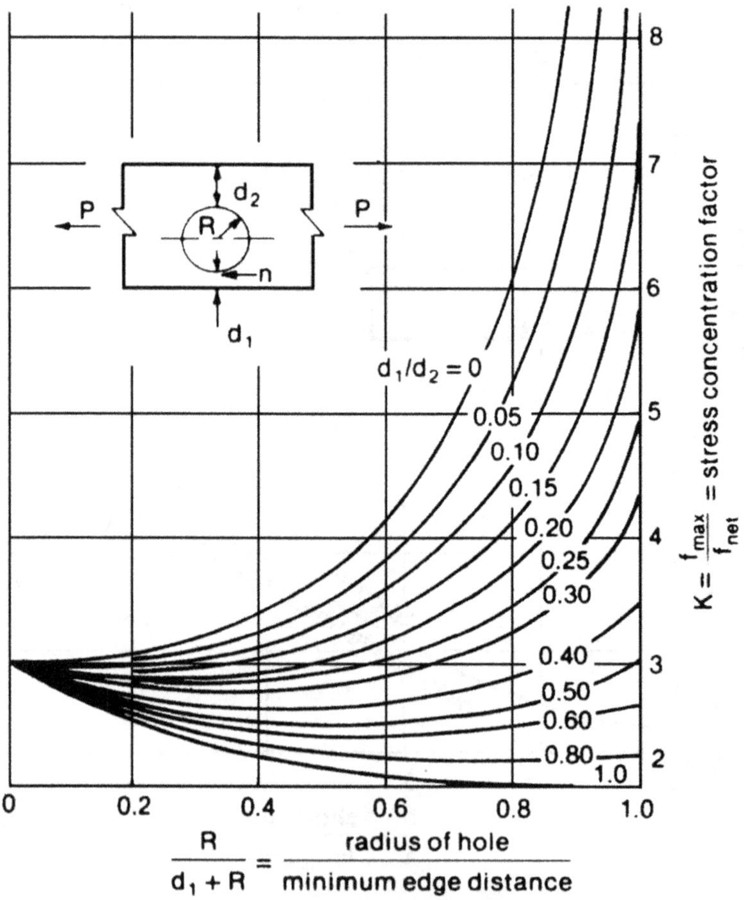

Source: Aeronautical Research Inst. of Sweden, Rept. No. 36.

PLATES, CONTINUED

STRESS CONCENTRATION FACTORS FOR A FLAT PLATE WITH A SHOULDER FILLET UNDER AXIAL LOAD

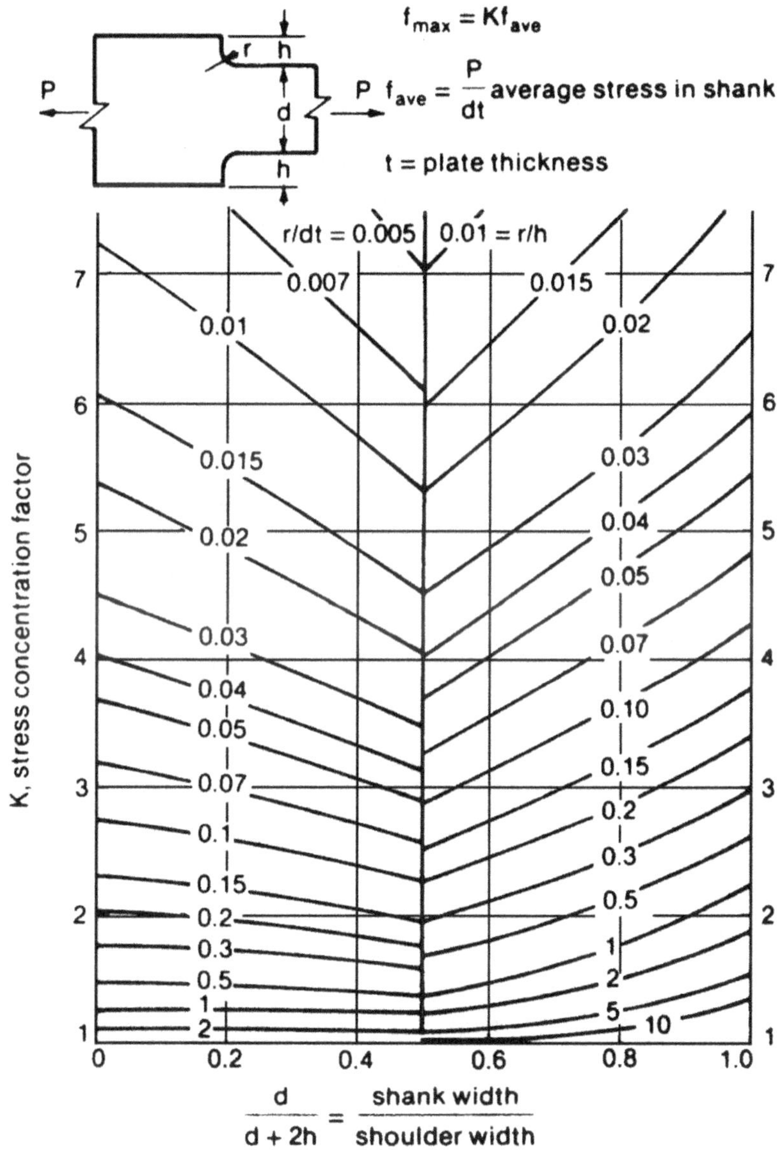

Sources: Peterson, R. E., "Stress Concentration Factors," Fig. 57; TN 2442; Grumman Aeronautical and Engineering Company Data.

RIVET STRENGTHS

STATIC JOINT STRENGTH OF 100° FLUSH HEAD ALUMINUM ALLOY SOLID RIVETS IN MACHINE—COUNTERSUNK ALUMINUM ALLOY SHEET

Sheet material	3.1354T3, 3.1354T351, 3.1364T3											
Rivet material	3.1124T31							3.1324T31				
Rivet diameter, mm (nominal hole diameter, mm)	1.6 (1.65)	2.0 (2.05)	2.5 (2.55)	3.0 (3.05)	3.5 (3.55)	4.0 (4.05)	5.0 (5.05)	3.0 (3.05)	3.5 (3.55)	4.0 (4.05)	5.0 (5.05)	6.0 (6.05)
Sheet thickness, mm	Ultimate Strength, N											
0.8		640										
1.0			980			(a)					(a)	
1.2			1040	1410				1560				
1.4				1470	1910			1740	2120			
1.5				1500	1950	2450		1780	2270	2600		
1.6	400	680			1980	2490		1820	2350	2770		
1.8			1060			2570		1890	2430	3040		
2.0				1510		2650	3870		2520	3140	4310	
2.5					2050		4130	1900			4880	6460
3.0						2670	4140		2570	3350	5190	7020
4.0											5200	7450
Rivet ultimate shear strength, N	440	680	1060	1510	2050	2670	4140	1900	2570	3350	5200	7470

[a]Countersinking in the unpopulated portion of the table is not allowed due to creating a knife-edge condition.

Sources: DIN 65495-101 and DIN 65495-102.

RIVET STRENGTHS, CONTINUED

STATIC JOINT STRENGTH OF 100° FLUSH HEAD ALUMINUM ALLOY SOLID RIVETS IN MACHINE—COUNTERSUNK ALUMINUM ALLOY SHEET

Sheet material	Clad 2024-T42								
Rivet type	MS20426AD (2117-T3) (F_{su} = 30 ksi)				MS20426D (2017-T3) (F_{su} = 38 ksi)			MS20426DD (2024-T31) (F_{su} = 41 ksi)	
Rivet diameter, in. (Nominal hole diameter, in.)	3/32 (0.096)	1/8 (0.1285)	5/32 (0.159)	3/16 (0.191)	5/32 (0.159)	3/16 (0.191)	1/4 (0.257)	3/16 (0.191)	1/4 (0.257)
	Ultimate Strength,[a] lb								
Sheet thickness, in.									
0.032	178[c]	—	—	—	—	—	—	—	—
0.040	193	309[c]	—	—	—	—	—	—	—
0.050	206	340	479[c]	—	580[b,c]	—	—	—	—
0.063	216	363	523	705[c]	657[b]	859[b,c]	—	886[b]	—
0.071	—	373	542	739	690	917[b]	—	942	—
0.080	—	—	560	769	720	969[b]	—	992	—
0.090	—	—	575	795	746	1015	1552[b,c]	1035	1647[b,c]
0.100	—	—	—	818	—	1054	1640[b]	1073	1738[b]
0.125	—	—	—	853	—	1090	1773	1131	1877
0.160	—	—	—	—	—	—	1891	—	2000
0.190	—	—	—	—	—	—	1970	—	2084
Rivet shear strength[d]	217	388	596	862	755	1090	1970	1175	2125

[a]Test data from which the yield and ultimate strength listed were derived can be found in "Report on Flush Riveted Joint Strength" (Airworthiness Requirements Committee, A/C Industries Association of America, Inc., Airworthiness Project 12 [Revised May 25, 1948]).
[b]Yield value is less than 2/3 of the indicated ultimate strength value.
[c]Values above line are for knife-edge condition, and the use of fasteners in this condition is undesirable. The use of knife-edge condition in design of military aircraft requires specific approval of the procuring activity.
[d]River shear strength is documented in MS20426.

Source: MMPDS-05 (Metallic Materials Properties Development and Standardization).

FASTENER STRENGTHS

Thread		Minimum Ultimate Strength, N: 1100 MPa Tensile Strength		1250 MPa Tensile Strength		1550 MPa Tensile Strength		1800 MPa Tensile Strength			
Dia mm	Pitch mm	Flush & Protruding Head	Double[b] Shear	Flush & Protruding Head	Double[b] Shear	Flush & Protruding Head	Double[b] Shear	Flush & Protruding Head	Double[b] Shear	Tensile[a] Stress Area mm^2	Shear Area mm^2
3	0.50	5534	9331	6289	10,603	7798	13,148	9056	15,268	5.03	7.07
4	0.70	9657	16,588	10,973	18,850	13,607	23,373	15,802	27,143	8.78	12.57
5	0.80	15,601	25,918	17,728	29,452	21,983	36,521	25,529	42,412	14.2	19.6
6	1.00	22,136	37,322	25,154	42,412	31,191	52,590	36,222	61,073	20.1	28.3
8	1.00	43,084	66,350	48,959	75,398	60,709	93,494	70,501	108,573	39.2	50.3
10	1.25	67,318	103,673	76,498	117,810	94,858	146,084	110,157	169,646	61.2	78.5
12	1.25	101,279	149,288	115,090	169,646	142,711	210,361	165,729	244,290	92.1	113
14	1.50	137,000	203,198	155,682	230,907	193,046	286,325	224,182	332,506	125	154
16	1.50	183,973	265,402	209,060	301,593	259,235	373,975	301,047	434,294	167	201
18	1.50	237,858	335,899	270,293	381,704	335,163	473,312	389,222	549,653	216	254
20	1.50	298,654	414,690	339,379	471,239	420,830	584,336	488,706	678,584	272	314
22	1.50	366,361	501,775	416,320	570,199	516,236	707,047	599,500	821,087	333	380
24	2.00	422,858	597,154	480,521	678,584	595,845	841,444	691,950	977,161	384	452
30	2.00	683,321	933,053	776,501	1,060,288	962,861	1,314,757	1,118,162	1,526,814	621	707
36	2.00	1,005,987	1,343,596	1,143,168	1,526,814	1,417,528	1,893,249	1,646,161	2,198,612	915	1018
42	2.00	1,390,857	1,828,784	1,580,520	2,078,164	1,959,845	2,576,923	2,275,948	2,992,555	1264	1385

[a]Tensile stress area based on ASME B13.1M-2005.
[b]Shear strength is estimated at 60% of the tensile strength.

FASTENER STRENGTHS, CONTINUED

	Minimum Ultimate Strength, lbs									
	160 ksi Tensile Strength		180 ksi Tensile Strength		220 ksi Tensile Strength		260 ksi Tensile Strength			
Thread Size (UNJF-3A)	Flush & Protruding Head	Double[b] Shear	Flush & Protruding Head	Double[b] Shear	Flush & Protruding Head	Double[b] Shear	Flush & Protruding Head	Double[b] Shear	Tensile[a] Stress Area in.^{2}	Shear Area in.^{2}
0.1900-32	3199	5444	3599	6124	4399	7485	5198	8846	0.0200	0.0284
0.2500-28	5820	9425	6547	10,603	8002	12,959	9457	15,315	0.0364	0.0491
0.3125-24	9291	14,726	10,452	16,567	12,775	20,249	15,097	23,930	0.0581	0.0767
0.3750-24	14,053	21,206	15,809	23,856	19,322	29,158	22,835	34,459	0.0878	0.1104
0.4375-20	18,995	28,863	21,369	32,471	26,118	39,687	30,866	46,903	0.1187	0.1503
0.5000-20	25,593	37,699	28,792	42,412	35,190	51,836	41,588	61,261	0.1600	0.1963
0.5625-18	32,477	47,713	36,537	53,677	44,656	65,605	52,775	77,534	0.2030	0.2485
0.6250-18	40,953	58,905	46,073	66,268	56,311	80,994	66,549	95,720	0.2560	0.3068
0.7500-16	59,674	84,823	67,133	95,426	82,052	116,632	96,970	137,837	0.3730	0.4418
0.8750-14	81,516	115,454	91,705	129,885	112,084	158,749	132,463	187,612	0.5095	0.6013
1.0000-12	106,087	150,796	119,348	169,646	145,869	207,345	172,391	245,044	0.6630	0.7854
1.1250-12	136,916	190,852	154,030	214,708	188,259	262,421	222,488	310,134	0.8557	0.9940
1.2500-12	171,671	235,619	193,130	265,072	236,048	323,977	278,966	382,882	1.073	1.227
1.3750-12	210,354	285,100	236,648	320,737	289,237	392,012	341,825	463,287	1.315	1.485
1.5000-12	252,964	339,292	284,584	381,704	347,825	466,527	411,066	551,350	1.581	1.767

[a]Tensile stress area based on ASME B1.1-2005.
[b]Shear strength is estimated at 60% of the tensile strength.

BOLT AND SCREW STRENGTHS

SUMMARY OF FASTENER MATERIALS

Material	Surface Treatment	Useful Design Temperature Limit, °C (°F)	Ultimate Tensile Strength at Room Temperature, MPa (ksi)	Comments
Carbon steel	Zinc plate	–54 to 121 (–65 to 250)	380 (55) and up	—
Alloy steels	Cadmium plate, nickel plate, zinc plate, or chromium plate	–54 (–65) to limiting temperature of plating	Up to 2068 (300)	Some can be used at 482°C (900°F)
A-286 stainless	Passivated per MIL-S-5002	–253 to 649 (–423 to 1200)	Up to 1517 (220)	—
17-4PH stainless	None	–184 to 316 (–300 to 600)	Up to 1517 (220)	—
17-7PH stainless	Passivated	–129 to 316 (–200 to 600)	Up to 1517 (220)	—
300 series stainless	Furnace oxidized	–253 to 427 (–423 to 800)	483 to 965 (70 to 140)	Oxidation reduces galling
410, 416, and 430 stainless	Passivated	–157 to 649 (–250 to 1200)	Up to 1241 (180)	324 MPa at 649°C (47 ksi at 1200°F); will corrode slightly
U-212 stainless	Cleaned and passivated per MIL-S-5002	649 (1200)	1276 (185)	965 MPa at 649°C (140 ksi at 1200°F)
Inconel 718 stainless	Passivated per QQ-P-35 or cadmium plated	–253 to 482 (–423 to 900) or cadmium plate limit	Up to 1517 (220)	—
Inconel X-750 stainless	None	–196 to 649 (–320 to 1200)	Up to 1241 (180)	938 MPa at 649°C (136 ksi at 1200°F)
Waspalloy stainless	None	–253 to 871 (–423 to 1600)	1034 (150)	—
Titanium	None	–212 to 260 (–350 to 500)	Up to 1103 (160)	—

Source: "Fastener Design Manual," by R. T. Barrett. NASA Reference Publication 1228, Lewis Research Center, Cleveland, OH, 1990.

VIBRATION

NATURAL FREQUENCIES

MASS-SPRING SYSTEMS IN TRANSLATION (RIGID MASS AND MASSLESS SPRING)

k = spring stiffness, N/m (lb/in.)
m = mass, kg (lb-s^2/in.)
ω_n = angular natural frequency, rad/s

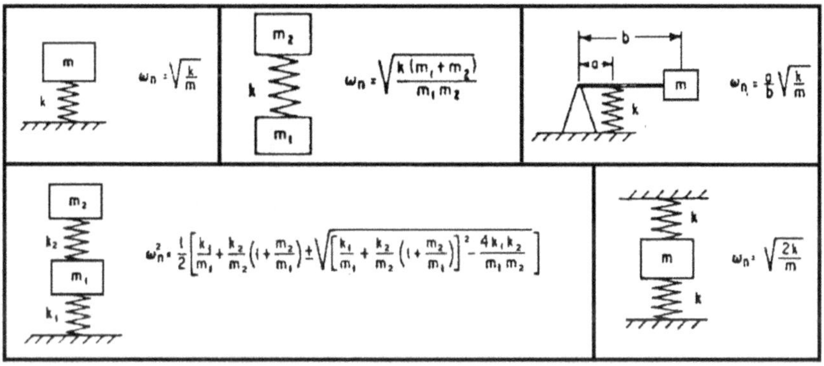

SPRINGS IN COMBINATION

k_r = resultant stiffness of combination

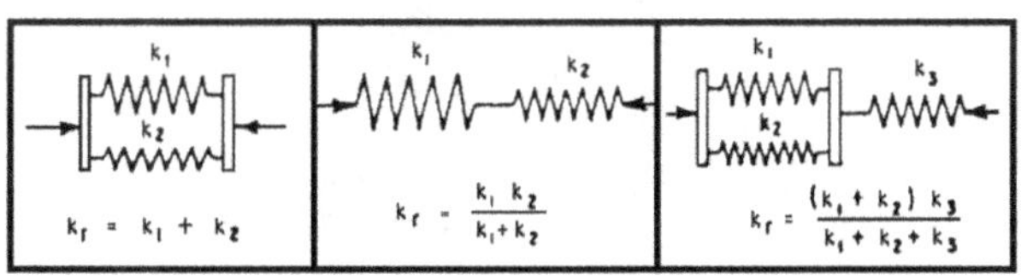

CHARACTERISTICS OF HARMONIC MOTION

A body that experiences simple harmonic motion follow a displacement pattern defined by

$$x = x_0 \sin(2\pi f t) = x_0 \sin \omega t$$

where f is the frequency of the simple harmonic motion, $\omega = 2\pi f$ is the corresponding angular frequency, and x_0 is the amplitude of the displacement.

Source: The material on pages 5-43–5-49 is from *Shock and Vibration Handbook* pages 1-9, 1-11, 1-13, 1-14, 2-7, 2-9, 7-30, and 7-32, by C. M. Harris and C. E. Crede. Copyright © 1976, McGraw–Hill, New York. Reproduced with permission of The McGraw–Hill Companies.

CHARACTERISTICS OF HARMONIC MOTION, CONTINUED

Relation of frequency to the amplitudes of displacement, velocity, and acceleration in harmonic motion.

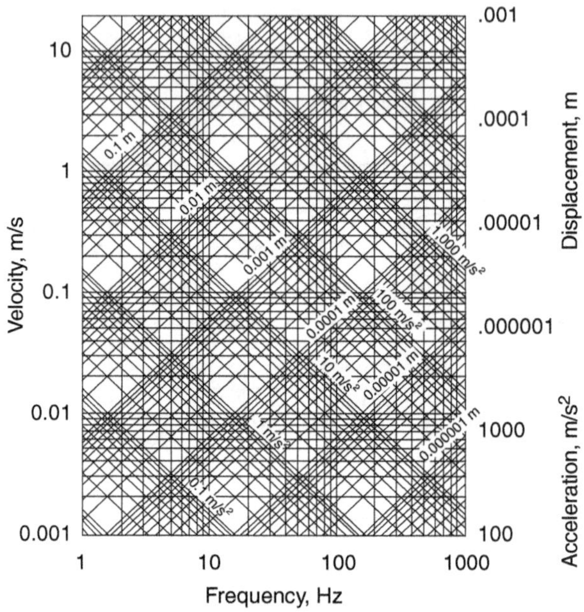

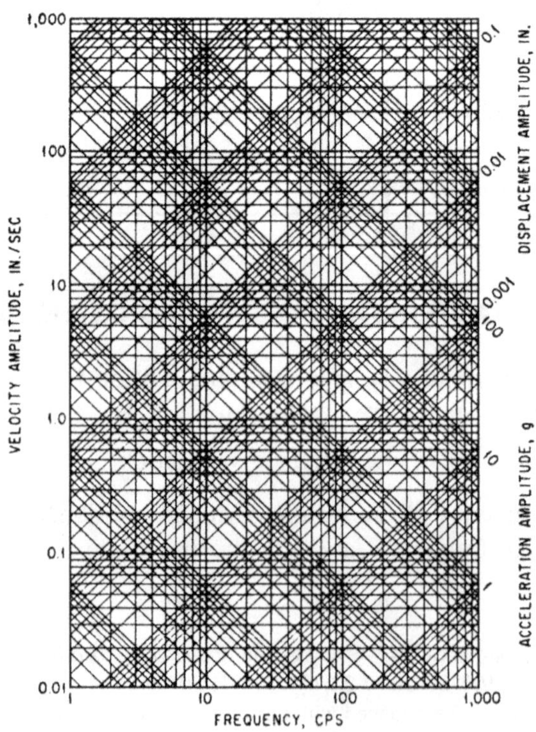

(g = 386 in. s^{-2})

FORCED VIBRATION

SINGLE DEGREE OF FREEDOM

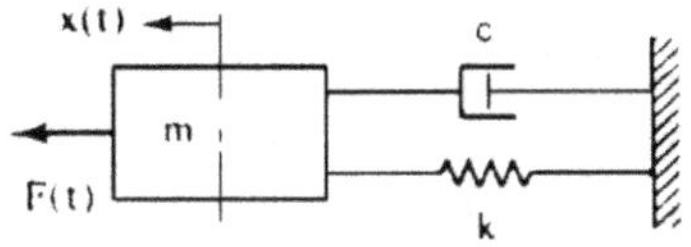

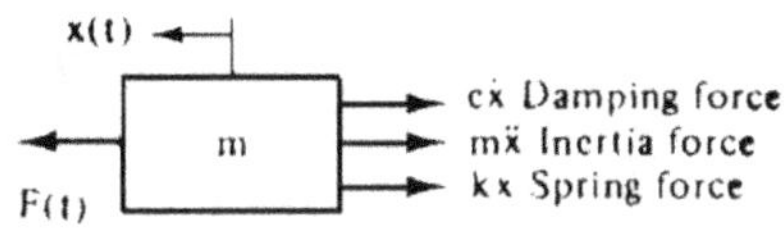

The equation of motion of the system is

$$m\ddot{x} + c\dot{x} + kx = F_0 \sin \omega t$$

where

m = mass
c = viscous damping constant
k = spring stiffness
$F(t)$ = $F_0 \sin \omega t$ = sinusoidal external force
ω = forcing frequency

The steady-state solution is

$$x = a \sin(\omega t - \theta)$$

where

$$a = \frac{a_s}{\sqrt{\left(1 - \omega^2/\omega_n^2\right)^2 + \left[2\xi\left(\omega/\omega_n\right)\right]^2}}$$

$$\tan\theta = \frac{2\xi\left(\omega/\omega_n\right)}{1 - \left(\omega/\omega_n\right)^2}$$

a_s = F_0/k = maximum static-load displacement
θ = phase angle
ξ = c/c_c = fraction of critical damping
c_c = $2\sqrt{mk} = 2m\omega_n$ = critical damping
ω_n = $\sqrt{k/m}$, natural frequency, rad/s

FORCED VIBRATION, CONTINUED

SINGLE DEGREE OF FREEDOM, CONTINUED

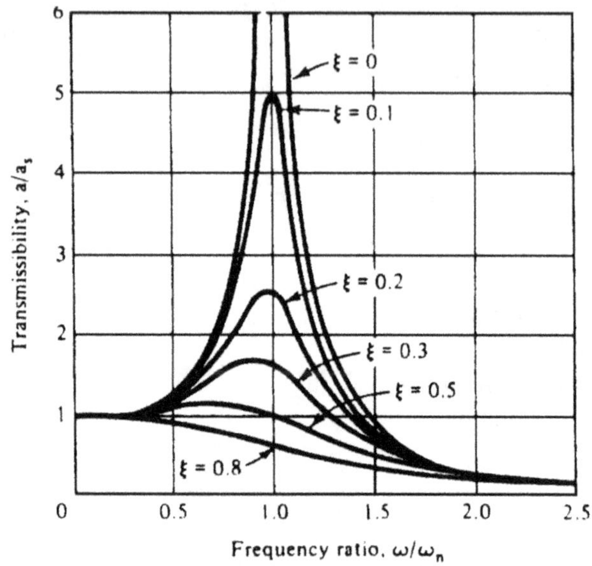

Forced response of a single-degree-of-freedom system.

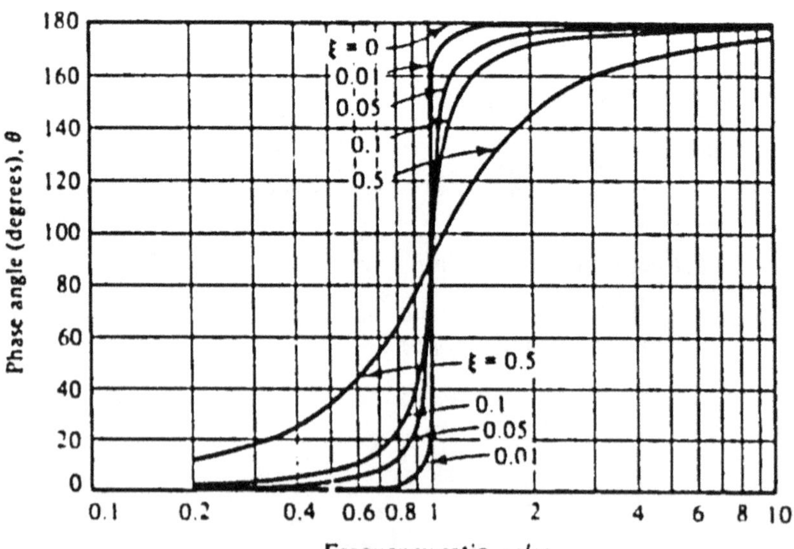

Forced response of a single-degree-of-freedom system (continued).

Beam Natural Frequency

Massive Springs (Beams) with Concentrated Mass Loads

m = mass of load, kg (lb-s^2/in.)
$m_s(m_b)$ = mass of spring (beam), kg (lb-s^2/in.)
k = stiffness of spring, N/m (lb/in.)
l = length of beam, m (in.)
I = area moment of inertia of beam cross section, m^4 (in.^{4})
E = Young's modulus, N/m^2 = Pa (lb/in.^{2})
ω_n = angular natural frequency, rad/s

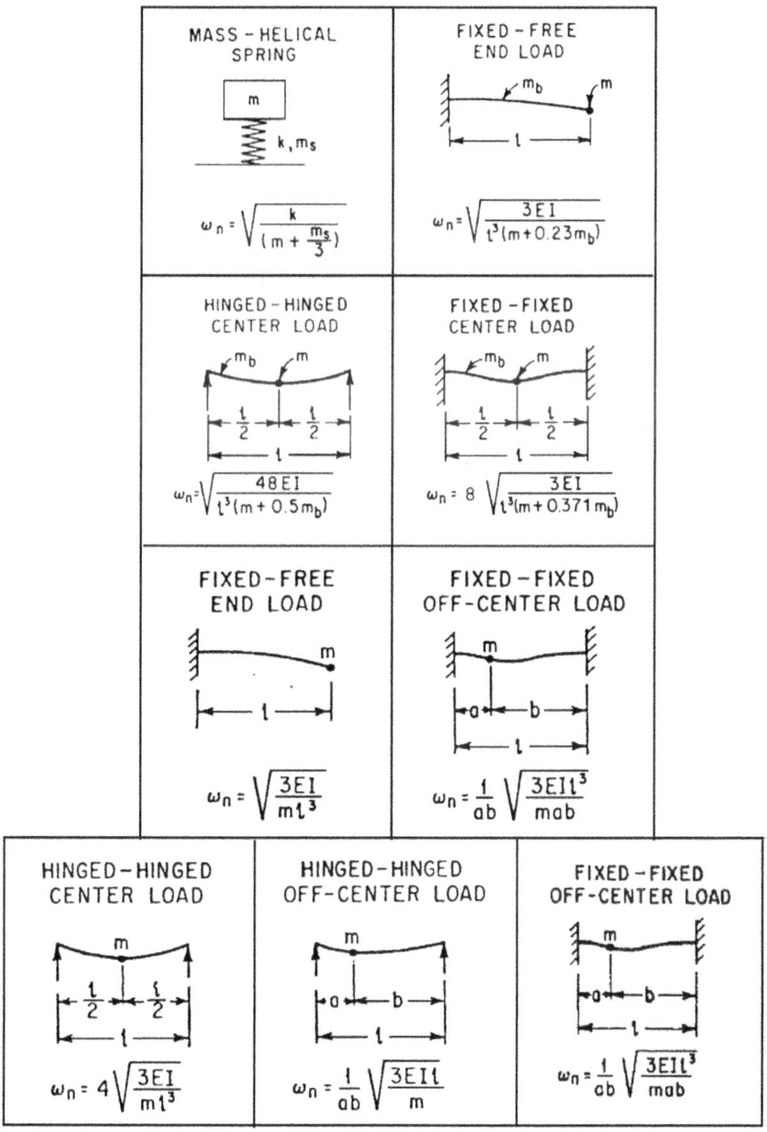

BEAM NATURAL FREQUENCY, CONTINUED

MASSIVE SPRINGS (BEAMS) WITH CONCENTRATED MASS LOADS, CONTINUED

ω_n = angular natural frequency, $A\sqrt{EI/\mu l^4}$ rad/s
E = Young's modulus, N/m^2 = Pa (lb/in.^{2})
I = area moment of inertia of beam cross section, m^4 (in.^{4})
l = length of beam, m (in.)
μ = mass per unit length of beam, kg/m (lb-s^2/in.^{2})
A = coefficient from the following table

Nodes are indicated in the following table as a proportion of length l measured from left end.

	First Mode	Second Mode
Fixed–free (cantilever)	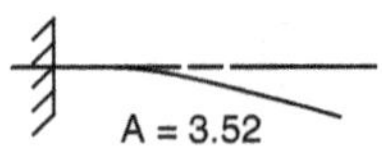	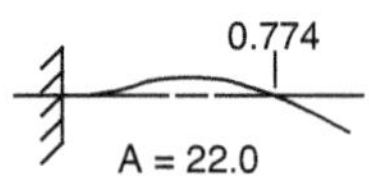
Hinged–hinged (simple)	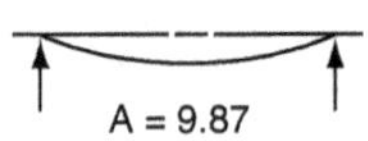	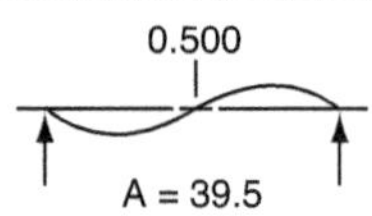
Fixed–fixed (built-in)	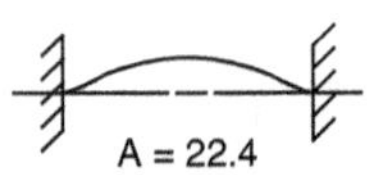	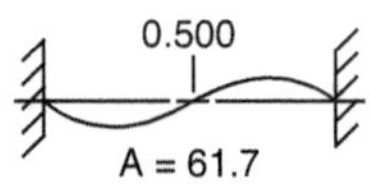
Free–free	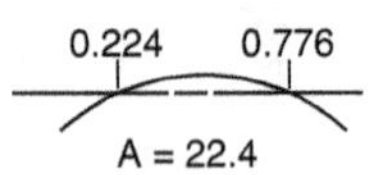	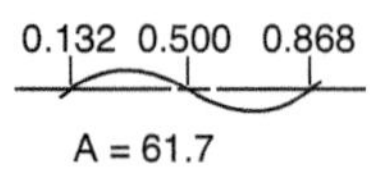
Fixed–hinged	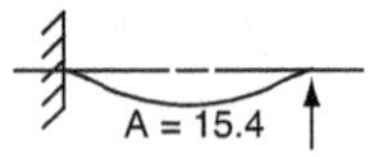	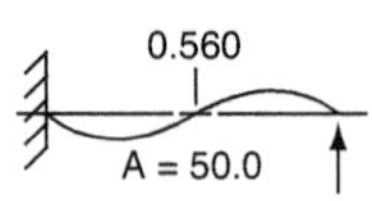
Hinged–free	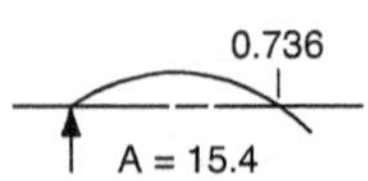	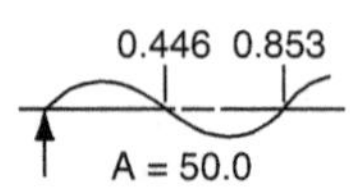

PLATE NATURAL FREQUENCY

$\omega_n = 2\pi f_n$
$D = Eh^3/12(1 - \mu^2)$
γ = weight density, N/m^3 (lb/in.^{3})
h = plate thickness, m (in.)
s = denotes simply supported edge
c = denotes built-in or clamped edge
a = length of plate, m (in.)
b = width of plate, m (in.)
g = acceleration of gravity = 9.81m/s^2 (32.17 ft/s^2)

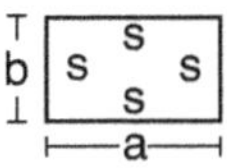	b/a	1.0	1.5	2.0	2.5	3.0	∞
	$\omega_n/\sqrt{Dg/\gamma h a^4}$	19.74	14.26	12.34	11.45	10.97	9.87
	b/a	1.0	1.5	2.0	2.5	3.0	∞
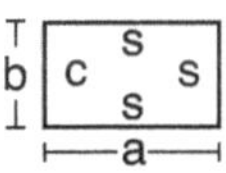	$\omega_n/\sqrt{Dg/\gamma h a^4}$	23.65	18.90	17.33	16.63	16.26	15.43
	a/b	1.0	1.5	2.0	2.5	3.0	∞
	$\omega_n/\sqrt{Dg/\gamma h a^4}$	23.65	15.57	12.92	11.75	11.14	9.87
	b/a	1.0	1.5	2.0	2.5	3.0	∞
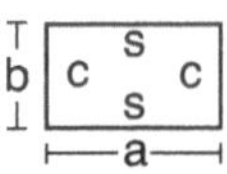	$\omega_n/\sqrt{Dg/\gamma h a^4}$	28.95	25.05	23.82	23.27	22.99	22.37
	a/b	1.0	1.5	2.0	2.5	3.0	∞
	$\omega_n/\sqrt{Dg/\gamma h a^4}$	28.95	17.37	13.69	12.13	11.36	9.87
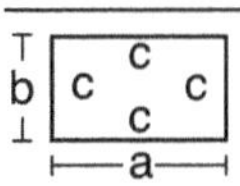	b/a	1.0	1.5	2.0	2.5	3.0	∞
	$\omega_n/\sqrt{Dg/\gamma h a^4}$	35.98	27.00	24.57	23.77	23.19	22.37

SECTION 6

MECHANICAL DESIGN

SPRINGS

SPRING NOMENCLATURE

A = cross section area of wire, m^2 ($in.^2$)
C = spring index = D/d
CL = compressed length, m (in.)
D or D_m = mean coil diameter, m (in.)
d = diameter of wire or side of square, m (in.)
E = elastic modulus (Young's modulus), N/m^2 = Pa (psi)
f or f_{bt} = stress, tensile or compressive, N/m^2 = Pa (psi)
FL = free length, unloaded spring, m (in.)
G = shear modulus of elasticity in torsion, $E/(2[1+\mu])$, N/m^2 = Pa (psi)
I = area moment of inertia, m^4 ($in.^4$)
in. = inch
IT = initial tension
J = torsional constant, m^4 ($in.^4$)
k = spring rate, P/s, N/m (lb/in.)
L = active length subject to deflection, m (in.)
l = length, m (in.)
lb = pound
m = meter
N = Newtons
n = number of active coils
OD = outside diameter, m (in.)
P = load, N (lb)
p = pitch, m (in.)
Pa = Pascal
P_1 = applied load (also P_2, etc.), N (lb)
psi = pounds per square inch
r = radius of wire, m (in.)
SH = solid height of compressed spring, m (in.)
s = deflection, m (in.)
T = torque, N-m (in.-lb)
t = leaf spring thickness, m (in.)
TC = total number of coils
U = elastic energy (strain energy), N-m (in.-lb)
w = leaf spring width, m (in.)
μ = Poisson's ratio
π = pi, 3.1416
τ = stress, torsional, N/m^2 = Pa (psi)
τ_{it} = stress, torsional, due to initial tension, N/m^2 = Pa (psi)

SPRINGS, CONTINUED

Maximum shear stress in wire, τ:

$$\tau = \frac{Tr}{J} + \frac{P}{A}$$

where

$$A = \frac{\pi d^2}{4} \quad T = \frac{PD}{2} \quad r = \frac{d}{2} \quad J = \frac{\pi d^4}{32}$$

$$\tau = \frac{(PD/2)(d/2)}{\pi d^4/32} + \frac{P}{\pi d^2/4}$$

Let $C = D/d$ and $K_s = 1 + 1/2C$, then

$$\tau = \frac{8PD}{\pi d^3}\left(1 + \frac{1}{2C}\right) = K_s \frac{8PD}{\pi d^3}$$

Spring deflection:

$$U = \frac{T^2 l}{2GJ} = \frac{4P^2 D^3 n}{d^4 G}$$

$$s = \frac{\partial U}{\partial P} = \frac{8PD^3 n}{d^4 G} \text{ (Castigliano's Theorem)}$$

HELICAL SPRINGS

Lateral buckling of compression springs

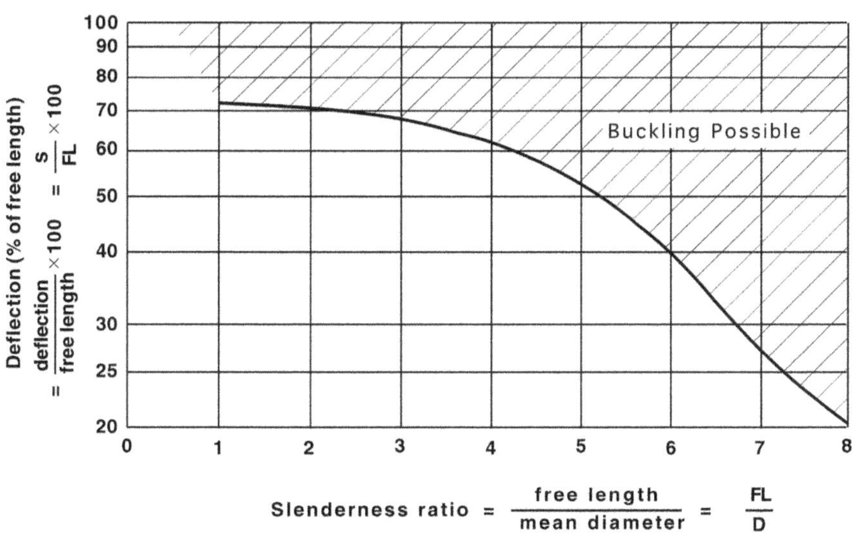

SPRINGS, CONTINUED

HELICAL SPRINGS, CONTINUED

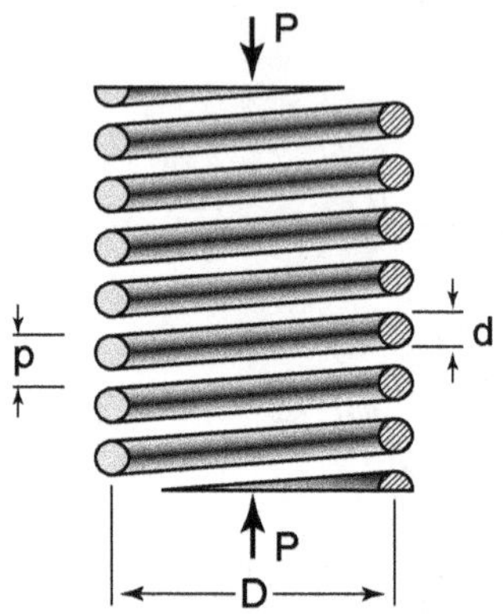

Helical spring

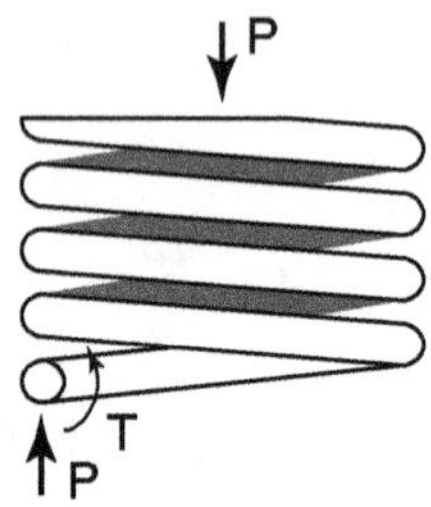

Free-body diagram

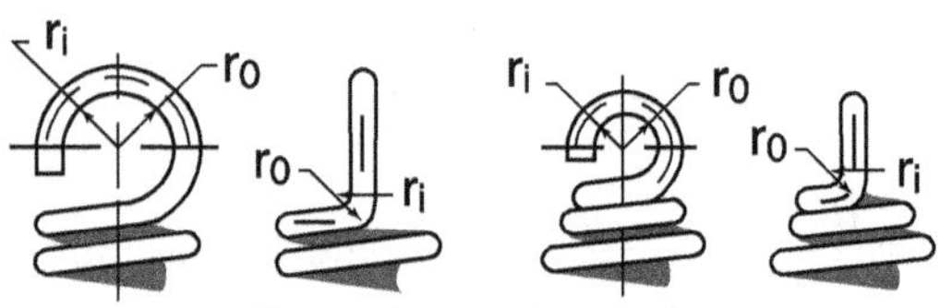

Extension spring ends

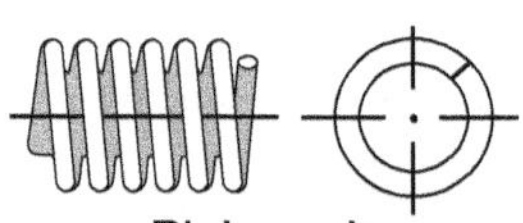

Plain end,
right hand

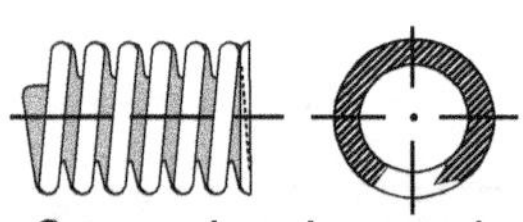

Squared and ground
end, left hand

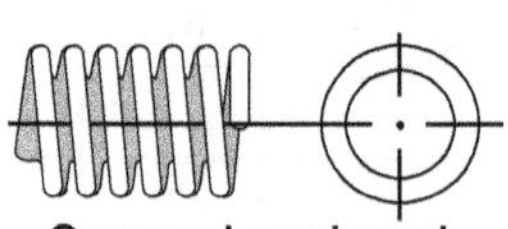

Squared or closed
end, right hand

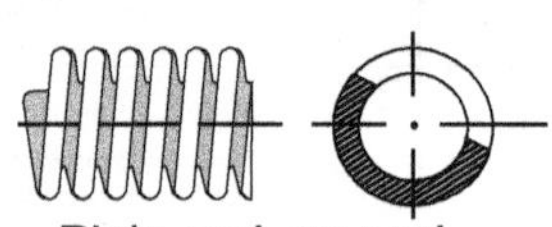

Plain end, ground,
left hand

Compression spring ends

Source: *Mechanical Engineering Design*, 3rd Ed., pages 296, 301, and 302, by Joseph E. Shigley. Copyright © 1977, McGraw–Hill, New York. Reproduced with permission of The McGraw–Hill Companies.

SPRINGS, CONTINUED

PREFERRED SIZES FOR SPRING MATERIALS—WIRE, STRIP, AND BARS

		Spring Steels				Corrosion Resisting				Copper and Nickel Alloys			
Music Wire		High Carbon and Alloy Steels		Valve Spring Quality Steels		18-8 Chrome Nickel Austenitic 300 Series		Straight Chrome Martensitic 400 Series		Spring Quality Brass Phosphor Bronze Beryllium Cop.		K Monel and Inconel X-750	
mm	(in.)	mm	(in.)	mm	(in.)	mm	(in.)	mm	(in.)	mm	(in.)	mm	(in.)
0.10	0.004	0.81	0.032	2.34	0.092	0.10	0.004	0.81	0.032	0.25	0.010	3.18	0.125
0.15	0.006	0.89	0.035	2.67	0.105	0.15	0.006	0.89	0.035	0.30	0.012	3.96	0.156
0.20	0.008	1.04	0.041	3.18	0.125	0.20	0.008	1.04	0.041	0.36	0.014	4.11	0.162
0.25	0.010	1.19	0.047	3.43	0.135	0.25	0.010	1.19	0.047	0.41	0.016	4.78	0.188
0.30	0.012	1.37	0.054	3.76	0.148	0.30	0.012	1.37	0.054	0.46	0.018	6.35	0.250
0.36	0.014	1.60	0.063	3.96	0.156	0.36	0.014	1.60	0.063	0.51	0.020	7.95	0.313
0.41	0.016	1.83	0.072	4.11	0.162	0.51	0.020	1.83	0.072	0.64	0.025	9.53	0.375
0.46	0.018	2.03	0.080	4.50	0.177	0.66	0.026	2.03	0.080	0.81	0.032	12.07	0.475
0.51	0.020	2.34	0.092	4.78	0.188	0.81	0.032	2.34	0.092	0.91	0.036	12.70	0.500
0.56	0.022	2.67	0.105	4.88	0.192	1.07	0.042	2.67	0.105	1.02	0.040	14.30	0.563
0.61	0.024	3.18	0.125	5.26	0.207	1.22	0.048	3.18	0.125	1.14	0.045	17.48	0.688
0.66	0.026	3.43	0.135	5.54	0.218	1.37	0.054	3.43	0.135	1.30	0.051	19.05	0.750
0.71	0.028	3.76	0.148	5.72	0.225	1.60	0.063	3.76	0.148	1.45	0.057	22.23	0.875
0.81	0.032	3.96	0.156	6.20	0.244	1.83	0.072	3.96	0.156	1.63	0.064	25.40	1.000
1.07	0.042	4.11	0.162	6.35	0.250	2.03	0.080	4.11	0.162	1.83	0.072	28.58	1.125
1.22	0.048	4.50	0.177			2.34	0.092	4.50	0.177	2.06	0.081	31.75	1.250
1.60	0.063	4.78	0.188			2.67	0.105	4.78	0.188	2.31	0.091	34.93	1.375
1.83	0.072	4.88	0.192			3.05	0.120	4.88	0.192	2.59	0.102	38.10	1.500
2.03	0.080	5.26	0.207			3.18	0.125	5.26	0.207	2.90	0.114	41.28	1.625
2.29	0.090	5.54	0.218			3.43	0.135	5.54	0.218	3.18	0.125	44.45	1.750
2.72	0.107	5.72	0.225			3.76	0.148	5.72	0.225	3.25	0.128	50.80	2.000
3.30	0.130	6.20	0.244			3.96	0.156	6.20	0.244	3.66	0.144		
4.11	0.162	6.35	0.250			4.11	0.162	6.35	0.250	3.96	0.156		
4.50	0.177	6.68	0.263			4.50	0.177	6.68	0.263	4.11	0.162		
		7.19	0.283			4.78	0.188	7.19	0.283	4.62	0.182		
		7.80	0.307			4.88	0.192	7.80	0.307	4.78	0.188		
		7.95	0.313			5.26	0.207	7.95	0.313	6.35	0.250		
		9.19	0.362			5.54	0.218	9.19	0.362				
		9.53	0.375			5.72	0.225	9.53	0.375				
						6.35	0.250						
						7.92	0.312						
						9.53	0.375						

SPRINGS, CONTINUED

FORMULAS FOR COMPRESSION AND EXTENSION SPRINGS

Property	Round Wire	Square Wire
Torsional stress (τ), N/m^2(psi)	$\frac{PD}{0.393d^3}$ $\frac{Gds}{\pi nD^2}$	$\frac{PD}{0.416d^3}$ $\frac{Gds}{2.32nD^2}$
Deflection (s), m(in.)	$\frac{8PnD^3}{Gd^4}$ $\frac{\pi\tau nD^2}{Gd}$	$\frac{5.58PnD^3}{Gd^4}$ $\frac{2.32\tau nD^2}{Gd}$
Change in load ($P_2 - P_1$), N(lb)	$(L_1 - L_2)k$	$(L_1 - L_2)k$
Change in load ($P_1 - P_2$) N(lb)	$(L_2 - L_1)k$	$(L_2 - L_1)k$
Stress due to initial tension (τ_{it}), N/m^2(psi)	$\frac{\tau}{P}IT$	$\frac{\tau}{P}IT$
Rate (k), N/m (lb/in.)	P/s	P/s

COMPRESSION SPRING DIMENSIONAL CHARACTERISTICS

	Type of Ends			
Dimensional Characteristics	Open or Plain (Not Ground)	Open or Plain With Ends Ground	Squared or Closed (Not Ground)	Closed and Ground
Pitch (p)	$\frac{FL-d}{n}$	$\frac{FL}{TC}$	$\frac{FL-8d}{n}$	$\frac{FL-2d}{n}$
Solid height (SH)	$(TC+1)d$	$TC \times d$	$(TC+1)d$	$TC \times d$
Active coils (n)	TC or $\frac{FL-d}{p}$	$TC-1$ or $\frac{FL}{p}-1$	$TC-2$ or $\frac{FL-8d}{p}$	$TC-2$ or $\frac{FL-2d}{p}$
Total coils (TC)	$\frac{FL-d}{p}$	$\frac{FL}{p}$	$\frac{FL-8d}{p}+2$	$\frac{FL-2d}{p}+2$
Free length (FL)	$(p \times TC)+d$	$p \times TC$	$(p \times N)+3d$	$(p \times N)+2d$

SPRINGS, CONTINUED

SPRING RATE K AND SPRING WORK U (STRAIN ENERGY)

	Characteristic	
	General	Const.
k	$\frac{dP}{ds}$	$\frac{P}{s}$
U_1	$\int_0^{s_1} P\,ds$	$P\,\frac{s_1}{2}$

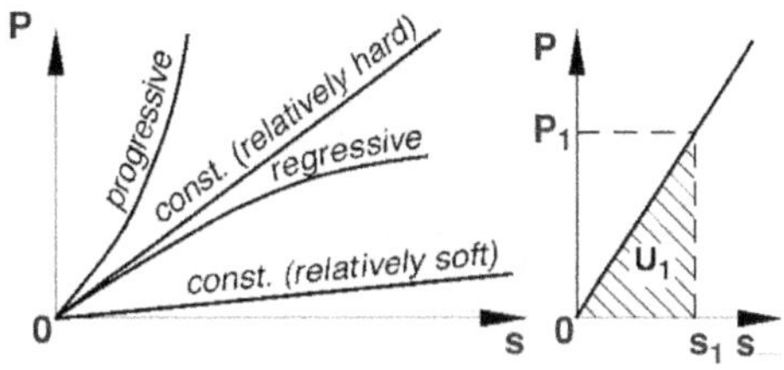

SPRINGS IN TENSION AND COMPRESSION

e.g., ring spring (Belleville spring)

	Springs in	
	Parallel	Series
	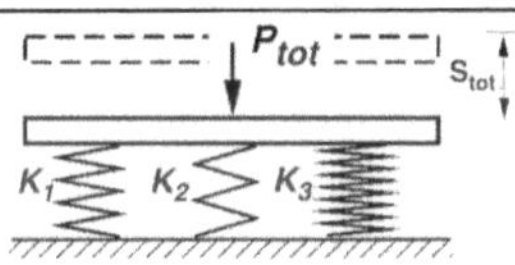	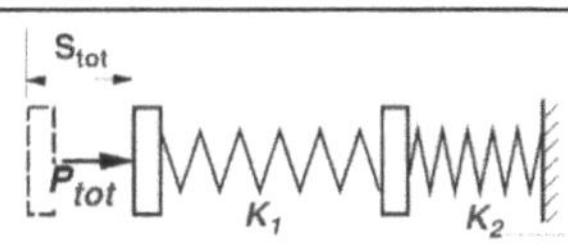
$s_{tot} =$	$s_1 = s_2 = s_3 \cdots = s_i$	$s_1 + s_2 + s_3 + \cdots + s_i$
$F_{tot} =$	$P_1 + P_2 + P_3 + \cdots + P_i$	$P_1 = P_2 = P_3 = \cdots = P_i$
k	$k_{tot} = k_1 + k_2 + k_3 + \cdots + k_i$	$\frac{1}{k_{tot}} = \frac{1}{k_1} + \frac{1}{k_2} + \cdots + \frac{1}{k_i}$

Source: The spring material appearing on pages 6-7–6-10 is from *Engineering Formulas*, 7th Ed., McGraw–Hill, New York. Copyright © 1997, Kurt Gieck, Reiner Gieck, Gieck–Verlag, Germering, Germany. Reproduced with permission of Gieck–Verlag.

SPRINGS, CONTINUED

SPRINGS IN BENDING

RECTANGULAR, TRAPEZOIDAL, TRIANGULAR SPRINGS

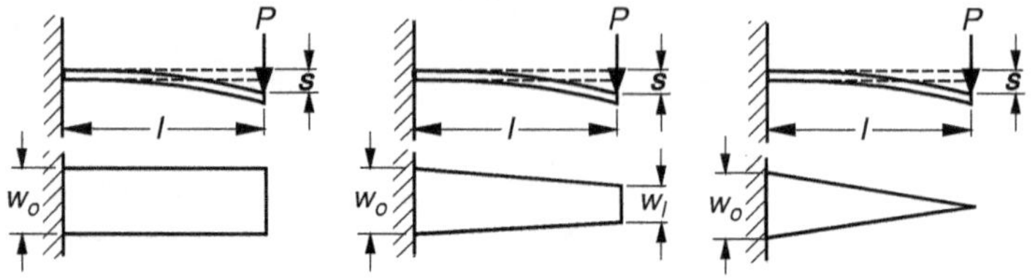

Bending stress $f_{bt} = \dfrac{6Pl}{w_0 t^2}$

Permissible load $F = \dfrac{w_0 t^2 f_{bt}}{6l}$

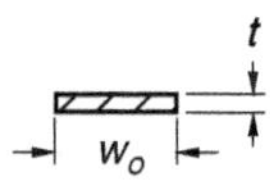

Deflection $s = 4\,\psi \dfrac{l^3}{w_0 t^3}\dfrac{P}{E}$

w_1/w_0	1[a]	0.8	0.6	0.4	0.2	0[b]
ψ	1.000	1.054	1.121	1.202	1.315	1.5

[a]Rectangular spring.
[b]Triangular spring.

f_{bt} = permissible bending stress

LAMINATED LEAF SPRINGS

Laminated leaf springs can be imagined as trapezoidal springs cut into strips and rearranged (spring in sketch can be replaced by two trapezoidal springs in parallel) of total spring width

$$w_0 = zw_1$$

where z is the number of leaves.

Then

$$P_n \approx \frac{w_0 t^2 f_{bt}}{6l}$$

If leaves 1 and 2 are the same length (as in the sketch),

$$w_1 = 2w$$

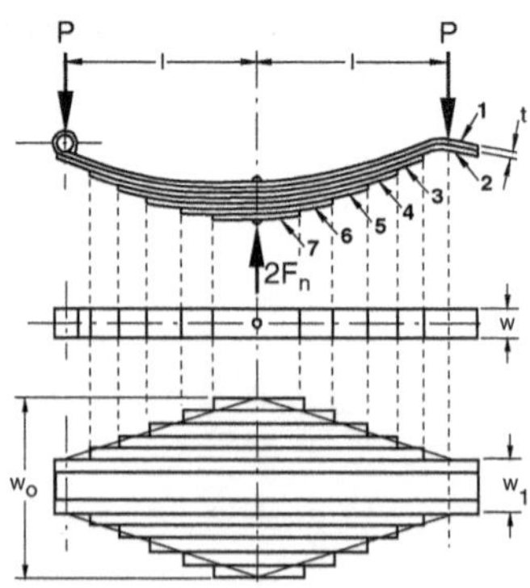

The calculation does not consider friction. In practice, friction increases the carrying capacity by between 2 and 12%.

SPRINGS, CONTINUED

SPRINGS IN BENDING, CONTINUED

DISC SPRINGS (RING SPRINGS)

Different characteristics can be obtained by combining n springs the same way and i springs the opposite way.

$$P_{tot} = nP_{single}$$

$$s_{tot} = is_{single}$$

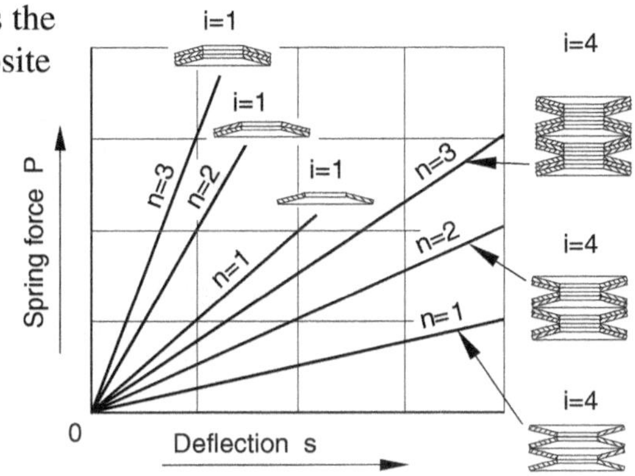

MATERIAL PROPERTIES

Hot-worked steels for springs to ASTM A 322, e.g., for leaf springs 9255; 6150—resp. to BS 970/5, e.g. 250 A 53; 735 A 50—(Modulus of elasticity: $E = 200 \times 10^9$ N/m^2 (29 × 10^6 psi)).

f_{bt}: static: 910×10^6 N/m^2 (131, 950 psi)

oscillating: $500 \pm 225 \times 10^6$ N/m^2 (72,500 ± 32,625 psi) (scale removed and tempered)

COILED TORSION SPRING

The type shown in the sketch has both ends free and must be mounted on a guide post. Positively located arms are better.

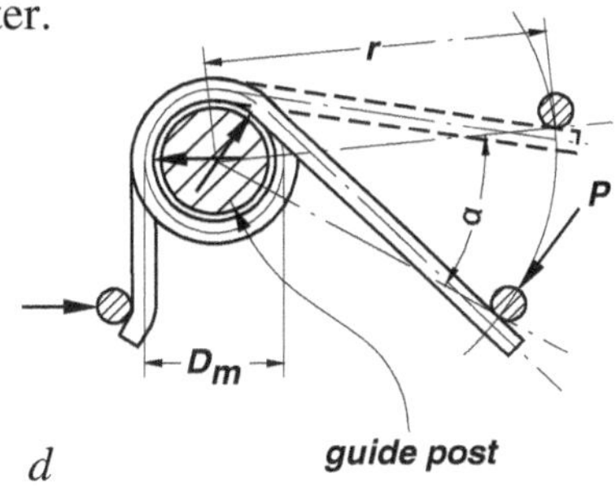

Spring force[a] $\quad P \approx \dfrac{Zf_{bt}}{r}$

Angle of deflection $\quad \alpha \approx \dfrac{Frl}{IE}$

Spring coil length (n = number of coils) $\quad l \approx D_m \pi n$

Section modulus $\quad Z = \dfrac{I}{y}$ where $y = \dfrac{d}{2}$

(Additional correction is needed for deflection of long arms.)

[a]Not allowing for the stress factor arising from the curvature of the wire.

SPRINGS, CONTINUED

SPRINGS IN TORSION

TORSION BAR SPRING

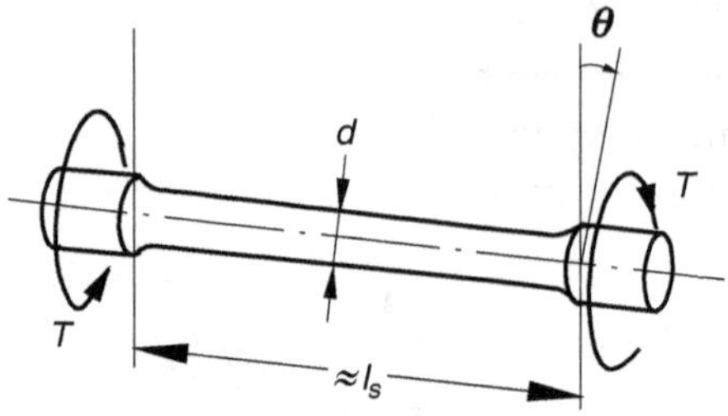

Shear Stress	Torque	Angle of Twist
$\tau = \frac{5T}{d^3}$	$T = \frac{d^3}{5}\tau_{qt}$	$\upsilon = \frac{Tls}{GI_p} \approx \frac{10Tl_s}{Gd^4}$

l_s = spring length as shown in sketch.

The stress τ_{qt} and fatigue strength τ in N/m^2 (psi) are shown in the following table:

	Static		Oscillating[a]		
τ_{qt}	not preloaded	700×10^6 (101,500)	$\tau_f = \tau_m \pm \tau_A$	$d = .02$m	$500 \pm 350 \times 10^6$ (72,500 $\pm$ 50,750)
	preloaded	1020×10^6 (147,900)		$d = .03$m	$500 \pm 240 \times 10^6$ (72,500 $\pm$ 34,800)

[a]Surface ground and shot-blasted, preloaded.
τ_m = mean stress.
τ_A = alternating stress amplitude.

MACHINE SCREWS—TAP DRILL AND CLEARANCE DRILL SIZES

Screw Size (mm)	Pitch (mm)		Drill Sizes (mm)	
	Coarse	Fine	Tap	Clear
1.6	0.35	–	1.25	1.80
2	0.40	–	1.60	2.40
2.5	0.45	–	2.00	2.90
3	0.50	–	2.50	3.40
3.5	0.60	–	2.90	3.90
4	0.70	–	3.30	4.50
5	0.80	–	4.20	5.50
6	1.00	–	5.00	6.60
8	1.25	–	6.80	9.00
8	–	1.00	7.00	9.00
10	1.50	–	8.50	12.00
10	–	1.25	8.80	11.00
12	1.75	–	10.20	14.00
12	–	1.25	10.50	14.00
14	2.00	–	12.00	16.00
14	–	1.50	12.50	16.00
15	–	1.00	14.00	17.00
16	2.00	–	14.00	18.00
16	–	1.50	14.50	18.00
17	–	1.00	15.50	19.00
18	–	1.50	16.50	20.00
20	2.50	–	17.50	22.00
20	–	1.50	18.50	22.00
22	2.50	–	19.50	25.00
22	–	1.50	20.50	25.00
24	3.00	–	21.00	27.00
24	–	2.00	22.00	27.00
27	3.00	–	24.00	30.00
27	–	2.00	25.00	30.00

MACHINE SCREWS—TAP DRILL AND CLEARANCE DRILL SIZES, CONTINUED

Screw Size (in.)		Threads per in.		Drill Sizes	
No.	O.D.	N.C.	N.F.	Tap[a]	Clear
0	0.060		80	3/64	#51
1	0.073	64		#53	#47
1	0.073		72	#53	#47
2	0.086	56		#50	#42
2	0.086		64	#50	#42
3	0.099	48		#47	#37
3	0.099		56	#46	#37
4	0.112	40		#43	#31
4	0.112		48	3/32	#31
5	0.125	40		#38	#29
5	0.125		44	#37	#29
6	0.138	32		#36	#27
6	0.138		40	#33	#27
8	0.164	32		#29	#18
8	0.164		36	#29	#18
10	0.190	24		#26	#9
10	0.190		32	#21	#9
12	0.216	24		#16	#2
12	0.216		28	#15	#2
		Fractional sizes start here			
1/4	0.250	20		#7	17/64
1/4	0.250		28	#3	17/64
5/16	0.312	18		F	21/64
5/16	0.312		24	I	21/64
3/8	0.375	16		5/16	25/64
3/8	0.375		24	Q	25/64
7/16	0.437	14		U	29/64
7/16	0.437		20	25/64	29/64
1/2	0.500	13		27/64	33/64
1/2	0.500		20	29/64	33/64
9/16	0.562	12		31/64	37/64
9/16	0.562		18	33/64	37/64
5/8	0.625	11		17/32	41/64
5/8	0.625		18	37/64	41/64
3/4	0.750	10		21/32	49/64
3/4	0.750		16	11/16	49/64
7/8	0.875	9		49/64	57/64
7/8	0.875		14	13/16	57/64
1	1.000	8		7/8	1-1/64
1	1.000		14	15/16	1-1/64

[a]Tap drill sizes shown give approximately 75% depth of thread.

DECIMAL EQUIVALENTS OF DRILL SIZE

Number Sizes						Letter Sizes		
	Diameter			Diameter			Diameter	
No.	in.	(mm)	No.	in.	(mm)	Letter	in.	(mm)
80	0.0135	(0.343)	40	0.0980	(2.489)	A	0.234	(5.944)
79	0.0145	(0.368)	39	0.0995	(2.527)	B	0.238	(6.045)
78	0.0160	(0.406)	38	0.1015	(2.578)	C	0.242	(6.147)
77	0.0180	(0.457)	37	0.1040	(2.642)	D	0.246	(6.248)
76	0.0200	(0.508)	36	0.1065	(2.705)	E	0.250	(6.350)
75	0.0210	(0.533)	35	0.1100	(2.794)	F	0.257	(6.528)
74	0.0225	(0.572)	34	0.1110	(2.819)	G	0.261	(6.629)
73	0.0240	(0.610)	33	0.1130	(2.870)	H	0.266	(6.756)
72	0.0250	(0.635)	32	0.1160	(2.946)	I	0.272	(6.909)
71	0.0260	(0.660)	31	0.1200	(3.048)	J	0.277	(7.036)
70	0.0280	(0.711)	30	0.1285	(3.264)	K	0.281	(7.137)
69	0.0292	(0.742)	29	0.1360	(3.454)	L	0.290	(7.366)
68	0.0310	(0.787)	28	0.1405	(3.569)	M	0.295	(7.493)
67	0.0320	(0.813)	27	0.1440	(3.658)	N	0.302	(7.671)
66	0.0330	(0.838)	26	0.1470	(3.734)	O	0.316	(8.026)
65	0.0350	(0.889)	25	0.1495	(3.797)	P	0.323	(8.204)
64	0.0360	(0.914)	24	0.1520	(3.861)	Q	0.332	(8.433)
63	0.0370	(0.940)	23	0.1540	(3.912)	R	0.339	(8.611)
62	0.0380	(0.965)	22	0.1570	(3.988)	S	0.348	(8.839)
61	0.0390	(0.991)	21	0.1590	(4.039)	T	0.358	(9.093)
60	0.0400	(1.016)	20	0.1610	(4.089)	U	0.368	(9.347)
59	0.0410	(1.041)	19	0.1660	(4.216)	V	0.377	(9.576)
58	0.0420	(1.067)	18	0.1695	(4.305)	W	0.386	(9.804)
57	0.0430	(1.092)	17	0.1730	(4.394)	X	0.397	(10.084)
56	0.0465	(1.181)	16	0.1770	(4.496)	Y	0.404	(10.262)
55	0.0520	(1.321)	15	0.1800	(4.572)	Z	0.413	(10.490)
54	0.0550	(1.397)	14	0.1820	(4.623)			
53	0.0595	(1.511)	13	0.1850	(4.699)			
52	0.0635	(1.613)	12	0.1890	(4.801)			
51	0.0670	(1.702)	11	0.1910	(4.851)			
50	0.0700	(1.778)	10	0.1935	(4.915)			
49	0.0730	(1.854)	9	0.1960	(4.978)			
48	0.0760	(1.930)	8	0.1990	(5.055)			
47	0.0785	(1.994)	7	0.2010	(5.105)			
46	0.0810	(2.057)	6	0.2040	(5.182)			
45	0.0820	(2.083)	5	0.2055	(5.220)			
44	0.0860	(2.184)	4	0.2090	(5.309)			
43	0.0890	(2.261)	3	0.2130	(5.410)			
42	0.0935	(2.375)	2	0.2210	(5.613)			
41	0.0960	(2.438)	1	0.2280	(5.791)			

STANDARD GAGES

This table shows the standard gages and the names of major commodities for which each is used. To determine the gage used for any commodity, note the number in parentheses opposite the commodity named and find the gage column below bearing the same number in parentheses. Standard gage dimensions are in inches and equivalent millimeters are given as reference.

		(1)		(2)		(3)		(4)		(5)		(6)	
Commodity	Ga. No.	Birmingham or Stubs		American or Brown & Sharp		U.S. Standard		Washburn & Moen		Music Wire (std.)		Mfgrs. std. ga. For Sheet Metal	
		in.	(mm)	in.	(mm)	in.	(mm)	in.	(mm)	in.	(mm)	in.	(mm)
Aluminum (2) except tubing (1)	0000000	—	—	—	—	0.5000	(12.70)	0.4900	(12.45)	—	—	—	—
Bands (1)	000000	—	—	0.5800		0.4688	(11.91)	0.4615	(11.72)	0.004	(0.10)	—	—
Brass tubing (3/8″ O.D. and larger) (1)	00000	0.500	(12.70)	0.5165	(13.12)	0.4375	(11.11)	0.4305	(10.93)	0.005	(0.13)	—	—
Brass tubing (smaller than 3/8″ O.D.) (2)	0000	0.454	(11.53)	0.4600	(11.68)	0.4063	(10.32)	0.3938	(10.00)	0.006	(0.15)	—	—
Brass sheets (2)	000	0.425	(10.80)	0.4096	(10.40)	0.3750	(9.53)	0.3625	(9.21)	0.007	(0.18)	—	—
Brass strips (2)	00	0.380	(9.65)	0.3648	(9.27)	0.3438	(8.73)	0.3310	(8.41)	0.008	(0.20)	—	—
Brass wire (2)	0	0.340	(8.64)	0.3249	(8.25)	0.3125	(7.94)	0.3065	(7.79)	0.009	(0.23)	—	—
Copper sheets (2)	1	0.300	(7.62)	0.2893	(7.35)	0.2813	(7.14)	0.2830	(7.19)	0.010	(0.25)	—	—
Copper wire (2)	2	0.284	(7.21)	0.2576	(6.54)	0.2656	(6.75)	0.2625	(6.67)	0.011	(0.28)	—	—
Flat wire (1)	3	0.259	(6.58)	0.2294	(5.83)	0.2500	(6.35)	0.2437	(6.19)	0.012	(0.30)	0.2391	(6.07)
Hoops (1)	4	0.238	(6.05)	0.2043	(5.19)	0.2344	(5.95)	0.2253	(5.72)	0.013	(0.33)	0.2242	(5.69)
Iron wire (4)	5	0.220	(5.59)	0.1819	(4.62)	0.2188	(5.56)	0.2070	(5.26)	0.014	(0.36)	0.2092	(5.31)
Monel metal sheets (3)	6	0.203	(5.16)	0.1620	(4.11)	0.2031	(5.16)	0.1920	(4.88)	0.016	(0.41)	0.1943	(4.94)
Music wire (5)	7	0.180	(4.57)	0.1442	(3.66)	0.1875	(4.76)	0.1770	(4.50)	0.018	(0.46)	0.1793	(4.55)
Nickel sheets (3)	8	0.165	(4.19)	0.1284	(3.26)	0.1719	(4.37)	0.1620	(4.11)	0.020	(0.51)	0.1644	(4.18)
Nickel silver sheets (2)	9	0.148	(3.76)	0.1144	(2.91)	0.1563	(3.97)	0.1483	(3.77)	0.022	(0.56)	0.1495	(3.80)
Nickel silver wire (2)	10	0.134	(3.40)	0.1018	(2.59)	0.1406	(3.57)	0.1350	(3.43)	0.024	(0.61)	0.1345	(3.42)
Phosphor bronze strip (2)	11	0.120	(3.05)	0.0907	(2.30)	0.1250	(3.18)	0.1205	(3.06)	0.026	(0.66)	0.1196	(3.04)
Spring steel (1)	12	0.109	(2.77)	0.0808	(2.05)	0.1094	(2.78)	0.1055	(2.68)	0.029	(0.74)	0.1046	(2.66)
Stainless steel (3)	13	0.095	(2.41)	0.0719	(1.83)	0.0938	(2.38)	0.0915	(2.32)	0.031	(0.79)	0.0897	(2.28)

(continued)

STANDARD GAGES, CONTINUED

Commodity	Ga. No.	(1) Birmingham or Stubs in.	(1) (mm)	(2) American or Brown & Sharp in.	(2) (mm)	(3) U.S. Standard in.	(3) (mm)	(4) Washburn & Moen in.	(4) (mm)	(5) Music Wire (std.) in.	(5) (mm)	(6) Mfgrs. std. ga. For Sheet Metal in.	(6) (mm)
Steel plates (6)	14	0.083	(2.11)	0.0640	(1.63)	0.0781	(1.98)	0.0800	(2.03)	0.033	(0.84)	0.0747	(1.90)
Steel sheets (6)	15	0.072	(1.83)	0.0570	(1.45)	0.0703	(1.79)	0.0720	(1.83)	0.035	(0.89)	0.0673	(1.71)
Steel tubing, seamless and welded (1)	16	0.065	(1.65)	0.0508	(1.29)	0.0625	(1.59)	0.0625	(1.59)	0.037	(0.94)	0.0598	(1.52)
Steel wire (4) exception:	17	0.058	(1.47)	0.0452	(1.15)	0.0563	(1.43)	0.0540	(1.37)	0.039	(0.99)	0.0538	(1.37)
Music wire (5)	18	0.049	(1.24)	0.0403	(1.02)	0.0500	(1.27)	0.0475	(1.21)	0.041	(1.04)	0.0478	(1.21)
Armature binding wire (2)	19	0.042	(1.07)	0.0358	(0.91)	0.0438	(1.11)	0.0410	(1.04)	0.043	(1.09)	0.0418	(1.06)
Flat wire (1)	20	0.035	(0.89)	0.0319	(0.81)	0.0375	(0.95)	0.0348	(0.88)	0.045	(1.14)	0.0359	(0.91)
Strip steel (1)	21	0.032	(0.81)	0.0284	(0.72)	0.0344	(0.87)	0.0317	(0.81)	0.047	(1.19)	0.0329	(0.84)
	22	0.028	(0.71)	0.0253	(0.64)	0.0313	(0.79)	0.0286	(0.73)	0.049	(1.24)	0.0299	(0.76)
	23	0.025	(0.64)	0.0225	(0.57)	0.0281	(0.71)	0.0258	(0.66)	0.051	(1.30)	0.0269	(0.68)
	24	0.022	(0.56)	0.0201	(0.51)	0.0250	(0.64)	0.0230	(0.58)	0.055	(1.40)	0.0239	(0.61)
	25	0.020	(0.51)	0.0179	(0.45)	0.0219	(0.56)	0.0204	(0.52)	0.059	(1.50)	0.0209	(0.53)
	26	0.018	(0.46)	0.0159	(0.40)	0.0188	(0.48)	0.0181	(0.46)	0.063	(1.60)	0.0179	(0.45)
	27	0.016	(0.41)	0.0141	(0.36)	0.0172	(0.44)	0.0173	(0.44)	0.067	(1.70)	0.0164	(0.42)
	28	0.014	(0.36)	0.0126	(0.32)	0.0156	(0.40)	0.0162	(0.41)	0.071	(1.80)	0.0149	(0.38)
	29	0.013	(0.33)	0.0112	(0.28)	0.0141	(0.36)	0.0150	(0.38)	0.075	(1.91)	0.0135	(0.34)
	30	0.012	(0.30)	0.0100	(0.25)	0.0125	(0.32)	0.0140	(0.36)	0.080	(2.03)	0.0120	(0.30)
	31	0.010	(0.25)	0.0089	(0.23)	0.0109	(0.28)	0.0132	(0.34)	0.085	(2.16)	0.0105	(0.27)
	32	0.009	(0.23)	0.0079	(0.20)	0.0102	(0.26)	0.0128	(0.33)	0.090	(2.29)	0.0097	(0.25)
	33	0.008	(0.20)	0.0070	(0.18)	0.0094	(0.24)	0.0118	(0.30)	0.095	(2.41)	0.0090	(0.23)
	34	0.007	(0.18)	0.0063	(0.16)	0.0086	(0.22)	0.0104	(0.26)	0.100	(2.54)	0.0082	(0.21)
	35	0.005	(0.13)	0.0056	(0.14)	0.0078	(0.20)	0.0095	(0.24)	0.106	(2.69)	0.0075	(0.19)
	36	0.004	(0.10)	0.0050	(0.13)	0.0070	(0.18)	0.0090	(0.23)	0.112	(2.84)	0.0067	(0.17)

(continued)

STANDARD GAGES, CONTINUED

Commodity	Ga. No	(1) Birmingham or Stubs		(2) American or Brown & Sharp		(3) U.S. Standard		(4) Washburn & Moen		(5) Music Wire (std)		(6) Mfgrs. std. ga. For Sheet Metal	
		in.	(mm)	in.	(mm)	in.	(mm)	in.	(mm)	in.	(mm)	in.	(mm)
	37	—	—	0.0044	(0.11)	0.0066	(0.17)	0.0085	(0.22)	0.118	(3.00)	0.0064	(0.16)
	38	—	—	0.0039	(0.10)	0.0063	(0.16)	0.0080	(0.20)	0.124	(3.15)	0.0060	(0.15)
	39	—	—	0.0035	(0.09)	—	—	0.0075	(0.19)	0.130	(3.30)	—	—
	40	—	—	0.0031	(0.08)	—	—	0.0070	(0.18)	0.138	(3.51)	—	—
	41	—	—	0.0028	(0.07)	—	—	0.0066	(0.17)	0.146	(3.71)	—	—
	42	—	—	0.0025	(0.06)	—	—	0.0062	(0.16)	0.154	(3.91)	—	—
	43	—	—	0.0022	(0.06)	—	—	0.0060	(0.15)	0.162	(4.11)	—	—
	44	—	—	0.0020	(0.05)	—	—	0.0058	(0.15)	0.170	(4.32)	—	—
	45	—	—	0.0018	(0.04)	—	—	0.0055	(0.14)	0.180	(4.57)	—	—
	46	—	—	0.0016	(0.04)	—	—	0.0052	(0.13)	—	—	—	—
	47	—	—	0.0014	(0.04)	—	—	0.0050	(0.13)	—	—	—	—
	48	—	—	0.0012	(0.03)	—	—	0.0048	(0.12)	—	—	—	—
	49	—	—	0.0011	(0.03)	—	—	0.0046	(0.12)	—	—	—	—
	50	—	—	0.0010	(0.03)	—	—	0.0044	(0.11)	—	—	—	—

Coefficients of Static and Sliding Friction

	Static		Sliding	
Materials	Dry	Greasy	Dry	Greasy
Hard steel on hard steel	0.78	0.11	0.42	0.029
	—	—	—	0.12
Mild steel on mild steel	0.74	—	0.57	0.09
Hard steel on graphite	0.21	0.09		
Hard steel on babbitt (ASTM No. 1)	0.70	0.23	0.33	0.16
Hard steel on babbitt (ASTM No. 8)	0.42	0.17	0.35	0.14
Hard steel on babbitt (ASTM No. 10)	—	0.25	—	0.13
Mild steel on cadmium silver	—	—	—	0.097
Mild steel on phosphor bronze	—	—	0.34	0.173
Mild steel on copper lead	—	—	—	0.145
Mild steel on cast iron	—	0.183	0.23	0.133
Mild steel on lead	0.95	0.5	0.95	0.3
Nickel on mild steel	—	—	0.64	0.178
Aluminum on mild steel	0.61	—	0.47	
Magnesium on mild steel	—	—	0.42	
Magnesium on magnesium	0.6	0.08		
Teflon on Teflon	0.04	—	—	0.04
Teflon on steel	0.04	—	—	0.04
Tungsten carbide on tungsten carbide	0.2	0.12		
Tungsten carbide on steel	0.5	0.08		
Tungsten carbide on copper	0.35			
Tungsten carbide on iron	0.8			
Bonded carbide on copper	0.35			
Bonded carbide on iron	0.8			
Cadmium on mild steel	—	—	0.46	
Copper on mild steel	0.53	—	0.36	0.18
Nickel on nickel	1.10	—	0.53	0.12
Brass on mild steel	0.51	—	0.44	
Brass on cast iron	—	—	0.30	
Zinc on cast iron	0.85	—	0.21	
Magnesium on cast iron	—	—	0.25	
Copper on cast iron	1.05	—	0.29	
Tin on cast iron	—	—	0.32	
Lead on cast iron	—	—	0.43	
Aluminum on aluminum	1.05	—	1.4	
Glass on glass	0.94	0.01	0.40	0.09
Carbon on glass	—	—	0.18	
Garnet on mild steel	—	—	0.39	
Glass on nickel	0.78	—	0.56	
Copper on glass	0.68	—	0.53	
Cast iron on cast iron	1.10	—	0.15	0.070
Bronze on cast iorn	—	—	0.22	0.077

Note: Static friction between like materials (e.g., aluminum on aluminum, nickel on nickel, corrosion resistant steel on corrosion resistant steel) is high, and they often gall or sieze when used together dry.

Source: *Marks Standard Handbook for Mechanical Engineers*, 6th Ed., pages 3–40, by T. Baumeister and E. Avallone. Copyright © 1958, McGraw–Hill, New York. Reproduced with permission of The McGraw–Hill Companies.

GEARS

BASIC FORMULAS FOR INVOLUTE GEARS

To Obtain	Symbol	Spur Gears	Helical Gears
Pitch diameter	D	$D = \frac{N}{P}$	$D = \frac{N}{P_n \cos \psi}$
Circular pitch	p	$p = \frac{\pi}{P} = \frac{\pi D}{N}$	$P_n = \frac{\pi}{P_n}$
			$p_t = \frac{\pi D}{N} = \frac{p_n}{\cos \psi}$
			$p_x = \frac{1}{N} = \frac{p_n}{\sin \psi}$
Diametral pitch	P	$P = \frac{\pi}{p} = \frac{N}{D}$	$P_n = \frac{N}{D \cos \psi}$
Number of teeth	N	$N = PD = \frac{\pi D}{p}$	$N = P_n \cos \psi \, D$
Outside diameter	D_o	$D_o = D + 2a = \frac{N+2}{P}$	$D_o = D + 2a$
Root diameter	D_r	$D_r = D_o - 2h_t$	$D_r = D_o - 2h_t$
Base diameter	D_b	$D_b = D \cos \phi$	$D_b = D \cos \phi_t$
Base pitch	p_b	$P_b = p \cos \phi$	$P_b = p_t \cos \phi_t$
Circular tooth thickness at D	t	$t = 0.5p = \frac{\pi D}{2N}$	$t_t = 0.5p_t$
			$t_n = 0.5p_n$

Source: *Marks Standard Handbook for Mechanical Engineers*, 6th Ed., pages 3–40, by T. Baumeister and E. Avallone. Copyright © 1958, McGraw–Hill, New York. Reproduced with permission of The McGraw–Hill Companies.

GEARS, CONTINUED

BEVEL GEARS

a = standard center distance
b = tooth width
h_{aO} = addendum of cutting tool
h_{aP} = addendum of reference profile
h_{fP} = dedendum of reference profile
k = change of addendum factor
p_e = normal pitch ($p_e = p \cos \alpha$, $p_{et} = p_t \cos a_t$)
z = number of teeth
z_{nx} = equivalent number of teeth
(C_{perm}), C = (permissible) load coefficient
F_t = peripheral force on pitch cylinder (plane section)
K_I = operating factor (external shock)
K_V = dynamic factor (internal shock)
$K_{F\alpha}$ = end load distribution factor }
$K_{F\beta}$ = face load distribution factor } for root stress
K_{FX} = size factor }
$K_{H\alpha}$ = end load distribution factor }
$K_{H\beta}$ = face load distribution } for flank stress
R_e = total pitch cone length (bevel gears)
R_m = mean pitch cone length (bevel gears)
T = torque
Y_f, (Y_S) = form factor, (stress concentration factor)
Y_β = skew factor
Y_ε = load proportion factor
Z_H = flank form factor
Z_ε = engagement factor
Z_R = roughness factor
Z_V = velocity factor
α_P = reference profile angle (DIN 867: $\alpha P = 20°$)
α_W = operating angle

β	skew angle for helical gears	pitch cylinder
β_b		base cylinder

ρ = sliding friction angle ($\tan \rho = \mu$)
ρ_{a0} = tip edge radius of tool
$\sigma_{F\,\lim}$ = fatigue strength
$\sigma_{H\,\lim}$ = Hertz pressure (contact pressure)

Source: The gear material appearing on pages 6-19–6-24 is from *Engineering Formulas*, 6th Ed., McGraw–Hill, New York.

GEARS, CONTINUED

BEVEL GEARS, CONTINUED

BEVEL GEARS, GEOMETRY

Cone angle δ:

$$\tan\delta_1 = \frac{\sin\Sigma}{\cos\Sigma + u}$$

$$\left(\Sigma = 90° \Rightarrow \tan\delta_1 = \frac{1}{u}\right)$$

$$\tan\delta_2 = \frac{\sin\Sigma}{\cos\Sigma + 1/u}$$

$$(\Sigma = 90° \Rightarrow \tan\delta_2 = u)$$

angle between shafts $\Big\}$ $\Sigma = \delta_1 + \delta_2$

external pitch cone distance $\Big\}$ $R_e = \dfrac{d_e}{2\sin\delta}$

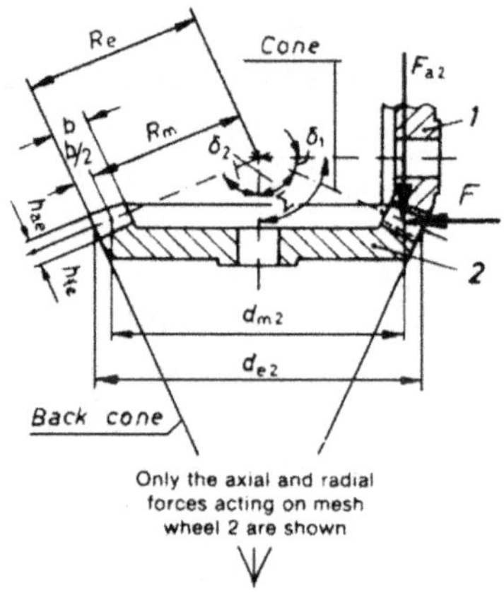

Development of the back cone to examine the meshing conditions gives the virtual cylindrical gear (suffix "v" = virtual) with the values

straight	bevel gears	$z_v = \dfrac{z}{\cos\delta}$	$u_v = \dfrac{z_{v2}}{z_{v1}}$
spiral		$z_v \approx \dfrac{z}{\cos\delta \times \cos^3\beta}$	

BEVEL GEARS, DESIGN

The design is referred to the midpoint of the width b (suffix "m") with the values

$R_m = R_e - \dfrac{b}{2}$	$m_m = \dfrac{d_m}{z}$
$d_m = 2R_m \sin\delta$	$F_{tm} = \dfrac{2T}{d_m}$

GEARS, CONTINUED

BEVEL GEARS, CONTINUED

AXIAL AND RADIAL FORCES IN MESH

axial force $F_a = F_{tm} \tan \alpha_n \times \sin \delta$
radial force $F_r = F_{tm} \tan \alpha_n \times \cos \delta$

LOAD CAPACITY OF TOOTH ROOT (APPROXIMATE CALCULATION)

Safety factor S_F against fatigue failure of tooth root:

$$SF = \frac{\sigma_{F\lim}}{\frac{F_{tm}}{bm_{nm}} \times Y_F \times Y_{EV} \times Y_\beta} \times \frac{Y_S \times K_{FX}}{K_I \times K_V \times K_{F\alpha} \times K_{F\beta}} \geq S_{F\min}$$

Giving the approximate formula

$$m_{nm} \geq \frac{F_{tm}}{b} \times Y_F \times K_I \times K_V \times \underbrace{Y_{EV} \times Y_\beta \times K_{F\alpha}}_{\approx 1} \times \frac{K_{F\beta}}{\underbrace{Y_S \times K_{FX}}_{\approx 1}} \times \frac{S_{F\min}}{\delta_{F\lim}}$$

Y_F: substitute the number of teeth of the complementary spur gear z_v or, with spiral gears, $z_{vn} \approx z_v/\cos^3\beta$.

LOAD CAPACITY OF TOOTH FLANK (APPROXIMATE CALCULATION)

Safety factor S_H against pitting of tooth surface.

$$S_H = \frac{\sigma_{H\lim}}{\sqrt{\frac{u+1}{u} \times \frac{F_{tm}}{bd_1} \times Z_H \times Z_M \times Z_{EV}}} \times \frac{Z_V \times K_{HX} \times Z_R \times K_L}{\sqrt{K_I \times K_V \times K_{H\alpha} \times K_{H\beta}}} \geq S_{H\min}$$

For metals the material factor Z_M is simplified to

$$Z_M = \sqrt{0.35E} \quad \text{with} \quad E = \frac{2E_1E_2}{E_1 + E_2}$$

where E, E_1, and E_2 represent the modulus of elasticity.

Giving the approximate formula

$$d_{vm1} \geq \sqrt{\frac{2T_1}{b} \times \frac{u_v + 1}{u_v} 0.35E \times Z_{HV} \times \underbrace{Z_{EV} \times \sqrt{K_{H\alpha}}}_{\approx 1}}$$

$$\times \frac{\sqrt{K_I \times K_V} \times \sqrt{K_{H\beta}}}{Z_V \times K_{HX} \times Z_R \times K_L} \times \frac{S_{H\min}}{\sigma_{H\lim}}$$

GEARS, CONTINUED

WORM GEARING

WORM GEARING, GEOMETRY

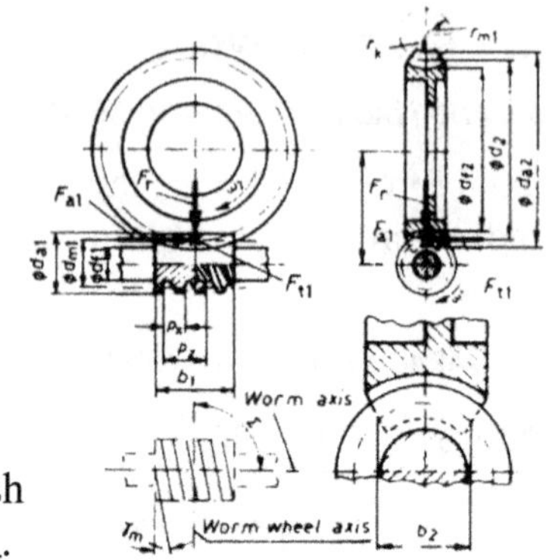

(Cylindrical worm gearing, normal module in axial section, BS 2519, angle between shafts $\Sigma = 90°$.)

DRIVE WORM

All the forces acting on the teeth in mesh are shown by the three arrows F_a, F_t, and F_r.

In the example,

$$z_1 = 2, \text{ right-hand helix}$$

	Worm, Suffix 1	Worm Wheel, Suffix 2
Module	$m_x = m = m_t$	
Pitch	$p_x = m\pi = p_2 = d_2\pi/z_2$	
Mean diameter	$d_{m1} = 2r_{m1}$	
(free to choose, for normal values see DIN 3976)		
Form factor	$q = d_{m1}/m$	
Center helix angle	$\tan\gamma_m = \dfrac{mz_1}{d_{m1}} = \dfrac{z_1}{q}$	
Pitch diameter		$d_2 = mz_2$
Addendum	$h_{a1} = m$	$h_{a2} = m(1 + x)$[a]
Dedendum	$h_{f1} = m(1 + c_1^*)$	$h_{f2} = m(1 - x + c_2^*)$
Tip clearance factor	$c_2^* = (0.167 \ldots \underline{0.2} \ldots 0.3) = c_2^*$	
Outside diameter	$d_{a1} = d_{m1} + 2h_{a1}$	$d_{a2} = d_2 + 2h_{a2}$
Tip groove radius		$r_k = a - d_{a2}/2$
Tooth width	$b_1 \geq \sqrt{d_{a2}^2 - d_2^2}$	$b_2 \approx 0.9\, d_{m1} - 2m$
Root diameter	$d_{f1} = d_{m1} - 2h_{f1}$	$d_{f2} = d_2 - h_{f2}$
Center distance	$a = (d_{m1} + d_2)/2 + xm$	

[a]Profile offset factor x for check of a preset center distance. Otherwise, $x = 0$.

GEARS, CONTINUED

WORM GEARING, CONTINUED

WORM GEARING, DESIGN (WORM DRIVING)

	Worm	Worm Wheel
Peripheral force	$F_{t1} = \frac{2T_1}{d_{m1}} K_I \times K_V$	$F_{t2} = F_{a1}$
Axial force	$F_{a1} = F_{t1} \times \frac{1}{\tan(\gamma + \rho)}$	$F_{a2} = F_{t1}$
Radial force	$F_{r1} = F_{t1} \times \frac{\cos\rho \times \tan a_n}{\sin(\gamma + \rho)}$	$= F_r = F_{r2}$
Rubbing speed	$v_g = \frac{d_{m1}}{2} \times \frac{\omega_1}{\cos\gamma_m}$	

EFFICIENCY

Worm Driving	Worm Wheel Driving
$\eta = \tan\gamma_m / \tan(\gamma_m + \rho)$	$\eta' = \tan(\gamma_m - \rho) / \tan\gamma_m$
	$(\gamma_m < \rho) \Rightarrow$ self-locking

COEFFICIENT OF FRICTION (TYPICAL VALUES) $\mu = \tan\rho$

	$v_g \approx 1$ m/s	$v_g \approx 10$ m/s
Worm teeth hardened and ground	0.04	0.02
Worm teeth tempered and machine cut	0.08	0.05

GEARS, CONTINUED

WORM GEARING, CONTINUED

CALCULATION OF MODULUS *m*

Load capacity of teeth root and flanks and temperate rise are combined in the approximate formula

$$F_{t2} = C\, b_2\, p_2 \quad \text{where } b_2 \approx 0.8\, d_{m1} \text{ and } p_2 = m\pi$$

$$m \approx \sqrt[3]{\frac{0.8T_2}{C_{\text{perm}\, q\, z_2}}} \quad \Bigg| \quad \begin{array}{l} F_{t2} = 2T_2/d_2 = 2T_2/(m\, z_2) \\ q \approx 10 \text{ for } i = 10, 20, 40 \\ q \approx 17 \text{ for } i = 80, \text{self-locking} \end{array}$$

Assumed values for normal, naturally cooled worm gears (worm of hardened and ground steel, worm wheel of bronze)

v_g	m/s	1	2	5	10	15	20
$C_{\text{perm}}(\times 10^6)$	N/m^2	8	8	5	3.5	2.4	2.2

When cooling is adequate this value can be used for all speeds

$$C_{\text{perm}} \geq 8 \times 10^6 \text{ N/m}^2$$

EPICYCLIC GEARING

Presented here are a velocity diagram and angular velocities (referred to fixed space).

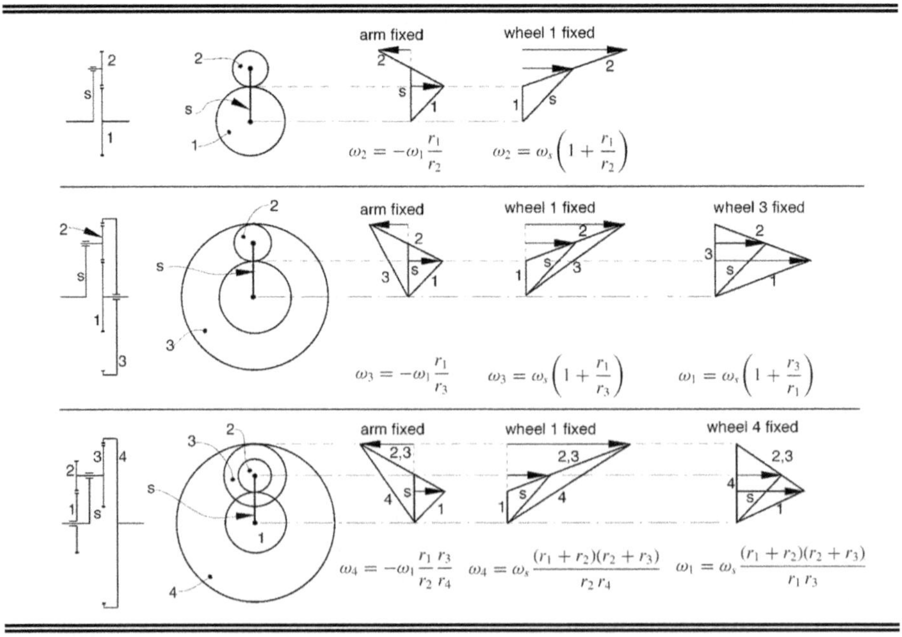

FORMULAS FOR BRAKES, CLUTCHES, AND COUPLINGS

The kinetic energy E of a rotating body is

$$E = \frac{I_g \omega^2}{2} = \frac{Wk^2}{g} \times \frac{N^2}{182.5} J(\text{ft-lb})$$

where

I_g = mass moment of inertia of the body, kg/m^2 (lb-ft-s^2)
k = radius of gyration of the body, m (ft)
W = weight of the body, N (lb)
N = rotational speed, rpm
g = gravitational acceleration, m/s^2 (ft/s^2)

If the angular velocity, ω rad/s, of the body changes by ΔN rpm in t seconds, the angular acceleration is

$$\alpha = \frac{2\pi \Delta N}{60t} = \frac{\Delta \omega}{t}$$

The torque T necessary to impart this angular acceleration to the body is

$$T = I_g \alpha = \frac{Wk^2}{g} \times \frac{2\pi \Delta N}{60t} \text{N-m(ft-lb)}$$

When a torque source drives a load inertia through a gear train, the equivalent inertia of the load must be used when calculating the torque required to accelerate the load.

$$I_{\text{equivalent}} = \left(\frac{N_2}{N_1}\right)^2 I_{\text{load}}$$

where

$I_{\text{equivalent}}$ = equivalent moment of inertia
I_{load} = moment of inertia of the load
N_1 = rotational speed of the torque source
N_2 = rotational speed of the load

For example, in the following figure, the equivalent inertia seen by the clutch is

$$I_{\text{clutch}} = I_1 + I_2\left(\frac{N_2}{N_1}\right)^2 + I_3\left(\frac{N_3}{N_1}\right)^2$$

FORMULAS FOR BRAKES, CLUTCHES, AND COUPLINGS, CONTINUED

where

I_1 = moment of inertia of clutch and attached shaft 1 and gear
I_2 = moment of inertia of shaft 2 and attached gears
I_3 = moment of inertia of load and attached shaft and gear
N_1 = rotational speed of clutch and attached shaft and gear
N_2 = rotational speed of shaft 2 and attached gears
N_3 = rotational speed of load and attached shaft 3 and gear

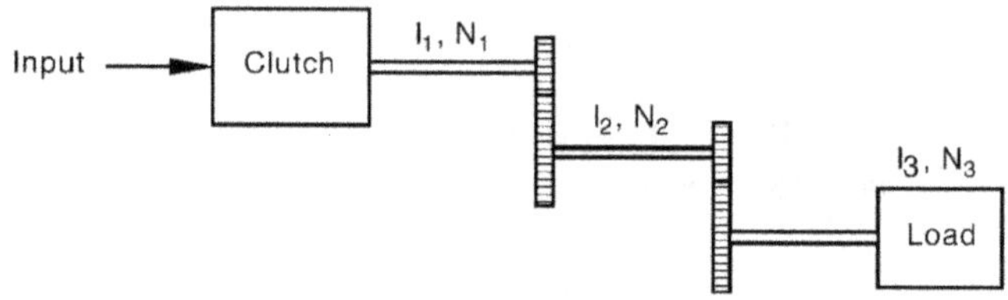

Work is the product of the magnitude of a force and the distance moved in the direction of the force. Power is the time rate at which the work is performed with respect to the figure below:

$$\text{Power} = \frac{P2r\pi N}{60} = \frac{TN}{9.55} W$$

The English unit of power is

$$1 \text{ hp} = 550 \text{ ft lb/s} = 33{,}000 \text{ ft lb/min} = 745.7\ W$$

Hence,

$$\text{hp} = \frac{P2r\pi N}{33{,}000} = \frac{PrN}{5252} = \frac{TN}{5252} \cdots$$

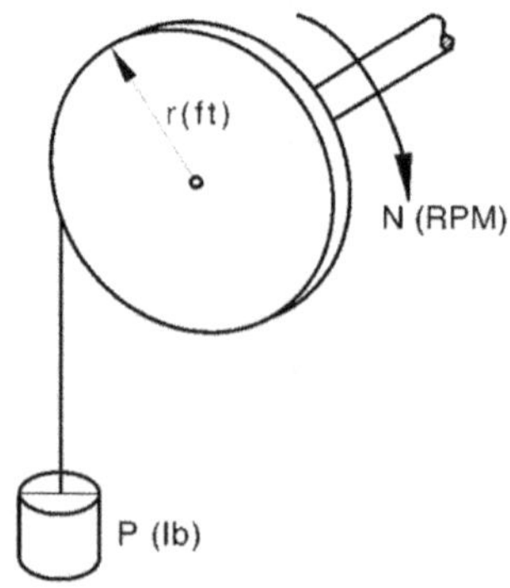

PUMPS

PUMP RELATIONSHIPS

b.p. = brake horsepower, kw (hp)
D = impeller diameter, mm (in.)
f.p. = fluid horsepower, kw (hp)
g = gravitational acceleration, m/s^2 (ft/s^2)
H_p = fluid static head, m (ft)
H_{sv} = suction head above vapor pressure, m (ft)
H_t = fluid total head, m (ft)
H_v = fluid velocity head, m (ft)
H_{vp} = fluid vapor pressure head, m (ft)
N = rotational speed, rpm
N_s = pump specific speed, L/S
Q = volume flow rate, (gpm)
S = suction specific speed
V = fluid velocity, m/s (ft/s)
δ = fluid specific gravity
η = overall efficiency, %
Φ = overall head rise coefficient at point of maximum efficiency

Pump specific speed:

$$N_s = N\frac{\sqrt{Q}}{H^{3/4}}$$

Suction specific speed:

$$S = \frac{N\sqrt{Q}}{H_{sv}^{3/4}}$$

Fluid velocity head:

$$H_v = \frac{V^2}{2g}$$

Overall efficiency:

$$\eta = \frac{\text{f.p.}}{\text{b.p.}}$$

Fluid total head:

$$H_t = H_p + H_v$$

Fluid horsepower:

$$\text{f.p.} = -\frac{QH\delta}{3960}$$

Suction head above vapor pressure:

$$H_{sv} = H_p + H_v - H_{vp}$$

Impeller diameter:

$$D = \frac{1840\Phi\sqrt{H}}{N}$$

PUMPS, CONTINUED

APPROXIMATE RELATIVE IMPELLER SHAPES AND EFFICIENCIES AS RELATED TO SPECIFIC SPEED

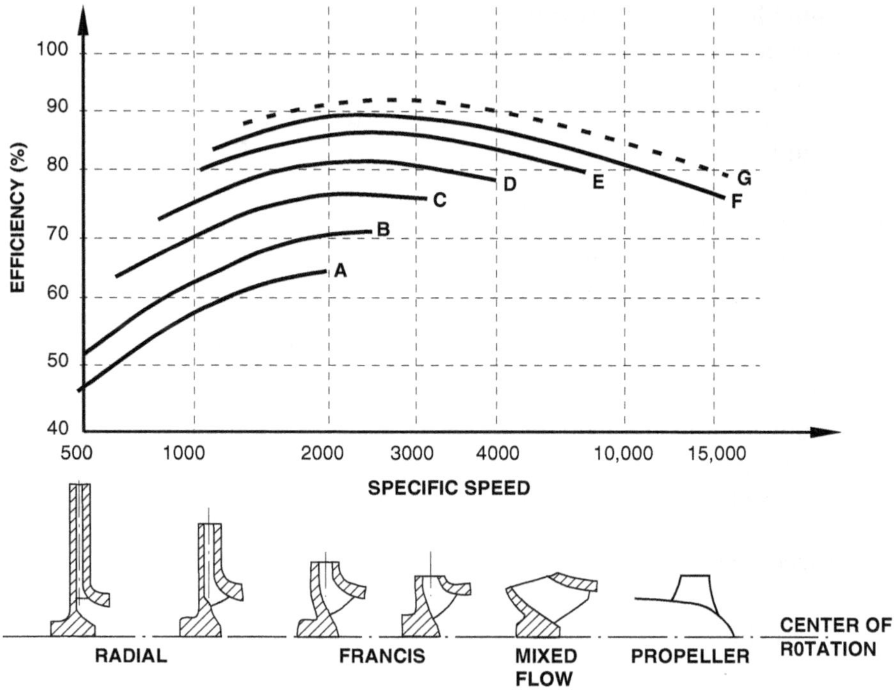

Curve	Flow Rate	
	gpm	L/S
A	Below 100	Below 6.3
B	100–200	6.3–13
C	200–500	13–32
D	500–1000	32–63
E	1000–3000	63–190
F	3000–10,000	190–630
G	Over 10,000	Over 630

CONTROL SYSTEM LOADS

Control System		Type	Ult. Design Load[a] (1.5 Factor of Safety Included), N (lb)
Elevator	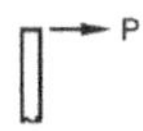	Stick	1668 (375)
	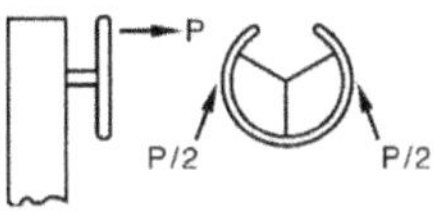	Wheel	2002 (450)
Aileron		Stick	667 (150)
		Wheel	1066 (240)
Rudder & brake		Pedal	2002 (450)
Flap, tab, stabilizer, spoiler, landing gear, arresting hook,		Crank, wheel, or lever operated by push or pull	$\left(\frac{1+R}{3}\right)^{b}(222)(1.5)^{c,d}$ N $\left[\left(\frac{1+R}{3}\right)^{b}(50)(1.5)^{c,d}\text{ lbs}\right]$
wing-fold controls		Small wheel or knob	23 N-m (200 in.-lb)[c,e] 1066 (150)[f]

[a]For dual systems, design of 75% of two-pilot control loads from control bus to control surface connection.

[b]R = Radius of wheel or length of lever.

[c]Applied at circumference of wheel, or grip of crank, or lever, and allowed to be active at any angle within 20 deg of plane of control.

[d]But not less than 334 N (75 lb) or more than 1000 N (225 lb).

[e]If operated only by twist.

[f]If operated by push or pull.

Source: *Federal Aviation Regulation* (FAR) Part 25 and MIL-A-8865B.

DYNAMIC-STOP LOADS

Loads due to stopping a moving mass can be estimated from the following:

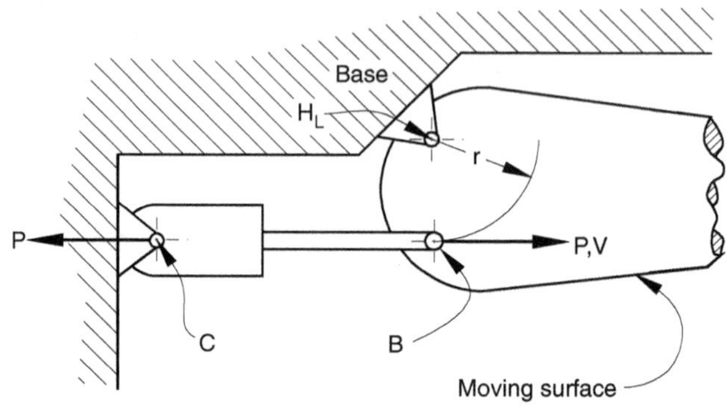

where $P = V(k \cdot m)^{1/2}$ and where

B = actuator to surface pivot
C = actuator to base pivot
H_L = surface hinge line
P = dynamic-stop load
k = linear spring rate, N/m (lb/in.), includes actuator and moving surface
m = equivalent mass at $B = I_p/386r^2$ (for linear movement, m = weight/386)
V = linear velocity of B in direction of stop reaction, m/s (in./s)
I_p = polar moment of inertia of mass about H_L, N-m^2 (lb-in.^{2})

FRICTION OF PULLEYS AND ROLLERS

A study of the following formula shows friction moment (M_F) will be a minimum when the diameter of the bearing is as small as practical relative to the diameter of the pulley.

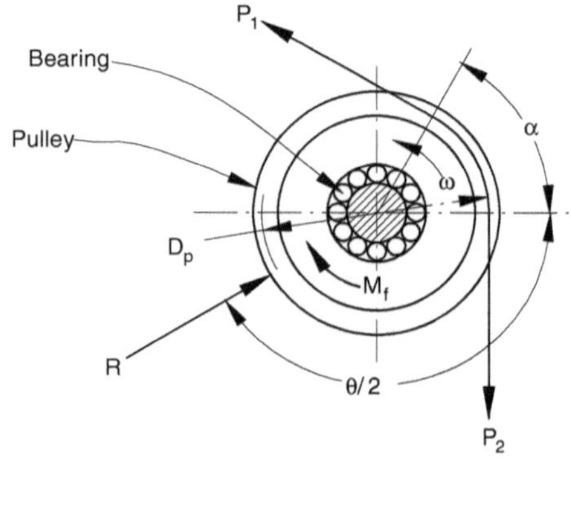

FRICTION OF PULLEYS AND ROLLERS, CONTINUED

$$M_F = \frac{R\mu_{br}d_{br}}{2}$$

$$= \frac{(P_1 - P_2)D_p}{2}$$

$$R = (P_1 + P_2)\sin\frac{\theta}{2}$$

$\alpha =$ wrap angle

$d_{br} =$ bearing diameter

$\theta = 2\pi - \alpha$

$\mu_{br} =$ bearing coefficient of friction

$\omega =$ relative motion

If the bearing is frozen, friction of the cable on the pulley governs. Then,

$$P_1 = P_2(e^{\mu\alpha}) = P_2(10^{0.4343\mu\alpha})$$

where

$\mu =$ coefficient of friction of cable to pulley
$\alpha =$ angle of wrap, rad

COEFFICIENTS OF FRICTION

The following values are recommended for Teflon-lined lubricated metal, plain spherical bearings, and sliding surfaces.

Material in Contact with Polished or Chromed Steel	μ
Teflon-lined	0.10
Alum. bronze, beryllium copper	0.15
Steel	0.30
Dry nonlubricated bushings or bearings	≦0.60
Ball and roller bearings in the unjammed or noncorroded condition	0.01

CABLES

Aircraft cable is an efficient means for transmitting control loads over long distances. To ensure that a control loop will carry load in both legs and to reduce cable load deformation, preloading or rigging load is applied.

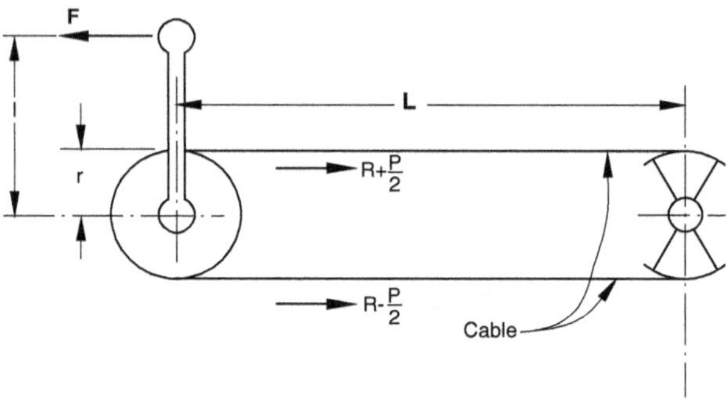

where

d = cable extension or contraction between supports due to load application

$$= \frac{PL}{2A_c E_c} \qquad R - \frac{P}{2} > 0$$

$P = F(\ell/r)$, active cable load
R = rigging load
A_c = cable cross-sectional area
E_c = cable modulus of elasticity

As shown, the rigging load must be greater than one-half of the active load to reduce cable deflection and ensure that the cable will not become slack and cause a different load distribution.

AIRFRAME DEFORMATION LOADS

Usually the cable run cannot be located at the airframe structure neutral axis, and so rigging loads will be affected by structure deformations.

An estimate of the change in rigging load ΔR due to structure deformation is

$$\Delta R \approx \frac{f}{E_s} A_c E_c$$

where

f = average airframe material working stress along the line of cable supports (if compression, P will be a negative load)
E_s = modulus of elasticity of structure

Rigging loads are also affected by temperature changes when steel cables are used with aluminum airframes.

CABLES, CONTINUED

CABLE-SHEAVE FRICTION WITH GROOVE RADIUS = (d/2)

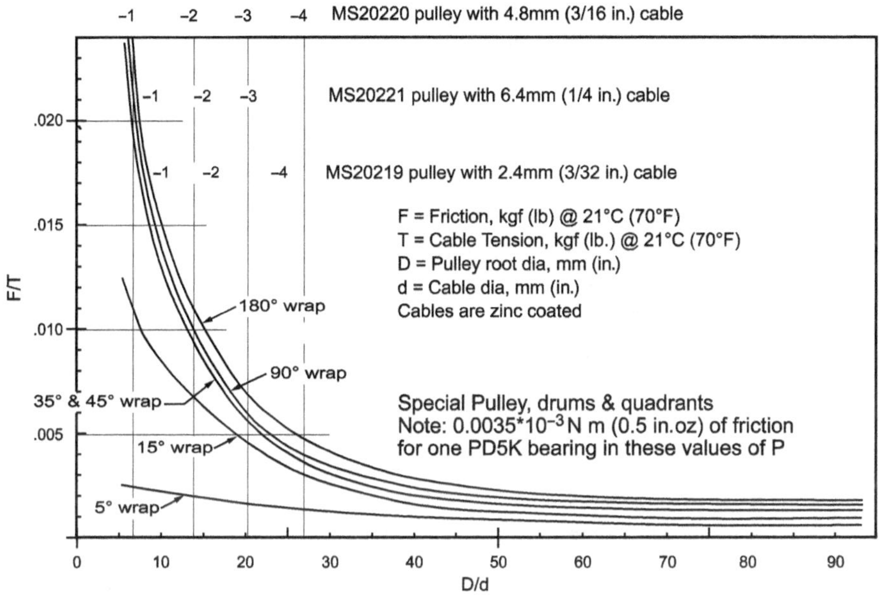

CABLE-SHEAVE FRICTION WITH GROOVE RADIUS = (d/2) + 0.02

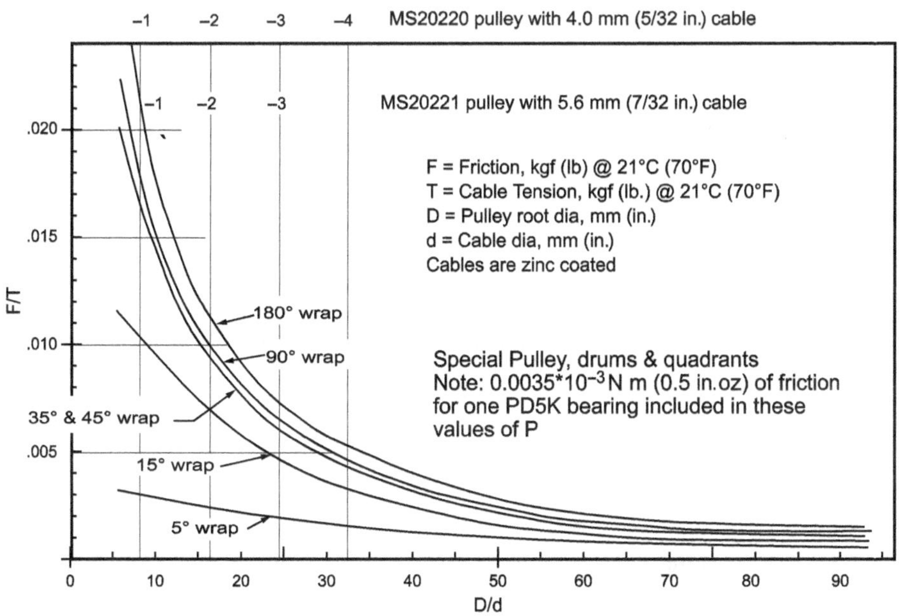

BEARINGS AND BUSHINGS

The radial limit load P_s for ball bearings may be obtained from the following expression:

$$P_s = knD^2 \text{ N(lb)}$$

where

k = a design factor (see table)
n = number of balls
D = ball diameter, m (in.)

Bearing Type	Design Factor, k $\text{N/m}^2 \times 10^3$	(lb/in.^2)
Deep groove	68,950	(10,000)
Single row, self-aligning	33,100	(4800)
Double row, self-aligning	26,200	(3800)
Rod end	22,060	(3200)

The basic dynamic capacity is the constant radial load at which 10% of the bearings tested fail within 2000 cycles. (For rating purposes a cycle is defined as a 90-deg rotation from a fixed point and return.) The dynamic load capacity P_d for N cycles is given by:

$$P_d = \frac{D_0}{L}$$

where D_0 = basic dynamic capacity and

$$L = \left(\frac{N}{2000}\right)^{1/3.6} \quad \text{life factor}$$

VARIABLE DYNAMIC LOADS

If the dynamic loads on a bearing vary greatly, the following equation may be used to estimate the representative design load:

$$P_e = \sqrt[3.6]{\frac{\Sigma N(P)^{3.6}}{\Sigma N}}$$

where

P_e = equivalent dynamic design load to give the same life as the variable loads
N = number of revolutions for a particular value of P
P = load acting for a particular value of N

BEARINGS AND BUSHINGS, CONTINUED

COMBINED LOADS

For combined radial and thrust loading, either static or dynamic, the equivalent radial load P_{e_c} is

$$P_{e_c} = R + YT$$

where

R = applied radial load
Y = (radial load rating/thrust load rating)
T = applied thrust load

CONTACT STRESSES

Contact stress due to spheres or cylinders on various surfaces can be calculated from the following expressions where

E = modulus of elasticity
μ = Poisson's ratio
R = reaction
D = diameter of cylinder or sphere
D_b = diameter of base
K, k = constant dependent on material and shape of the race to be found by references not in this handbook

COMPRESSIVE STRESS ON CONTACT POINT

$$f_{c_{(cyl)}} = k\sqrt{\frac{pE}{D}}$$

where p = load/in.

$$f_{c_{(sph)}} = K\sqrt[3]{P\left(\frac{E}{D}\right)^2}$$

where P = total load.

MAXIMUM TENSILE STRESS IN REGION OF CONTACT POINT

$$\tau_{max} = -f_c\left(\frac{1-2\mu}{3}\right)$$

MAXIMUM SHEAR STRESS IN REGION OF CONTACT POINT

$$f_{s_{max}} = 0.33 f_c$$

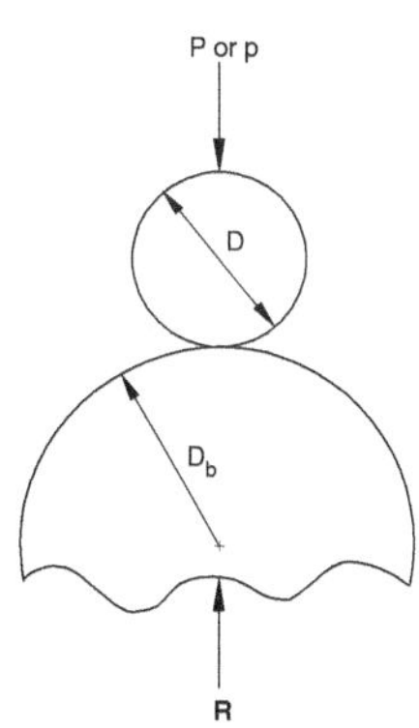

(Spheres or end view of cylinders)—convex case shown.

BEARINGS AND BUSHINGS, CONTINUED

GENERAL BEARING CHARACTERISTICS

Operating Parameter	Rolling Element	Sliding Surface
Load		
static	Low (ball) to medium (needle)	High
oscillating	Low	High
rotating	High	Low
dynamic	Poor	Good
vibration & impact	Poor	Good
Life	—[a]	—[b]
Speed	High	Low
Friction	Low	High
Noise	—[c]	Low
Damping	Poor	Medium
Envelope restrictions		
radial	Large	Small
axial	Small	Small
Lubricant type	Oil or grease	Grease, solid dry film, or none
Cost	Medium	Low
Type of failure	—[d]	—[e]
Power requirement	Low	High

[a]Limited by properties of bearing metal, lubrication, and seals.
[b]Limited by resistance to wear, galling, fretting, seizing, and lubrication.
[c]Depends on bearing quality and mounting.
[d]Quite rapid, with operations severely impaired.
[e]Gradual wear, bearing operable unless seizure occurs.

BEARINGS AND BUSHINGS, CONTINUED

FRICTION CALCULATION FOR AIRFRAME BEARINGS

The total running friction torque in an airframe bearing may be estimated by means of the following equation:

$$F_t \approx ST + \frac{PD}{B}$$

where

F_t = total running friction torque, N-mm (in.-oz)
S = speed factor from Bearing Friction Speed Factor chart
T = torque, N-mm (in.-oz), from Airframe Bearings Friction Torque chart
P = applied radial load, N (lb)
D = *OD* of bearing, mm (in.)
B = empiricle factor based on test results
= 440 (62.5)

BEARING FRICTION SPEED FACTOR

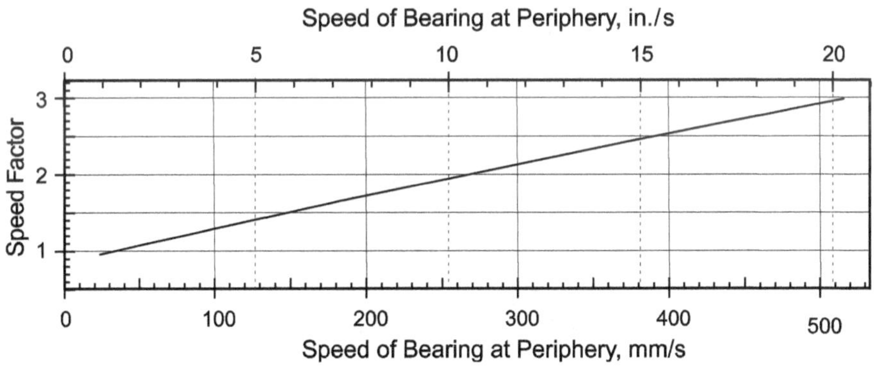

BEARINGS AND BUSHINGS, CONTINUED

AIRFRAME BEARINGS FRICTION TORQUE

For maximum expected torque, multiply values from curve by 1.50.

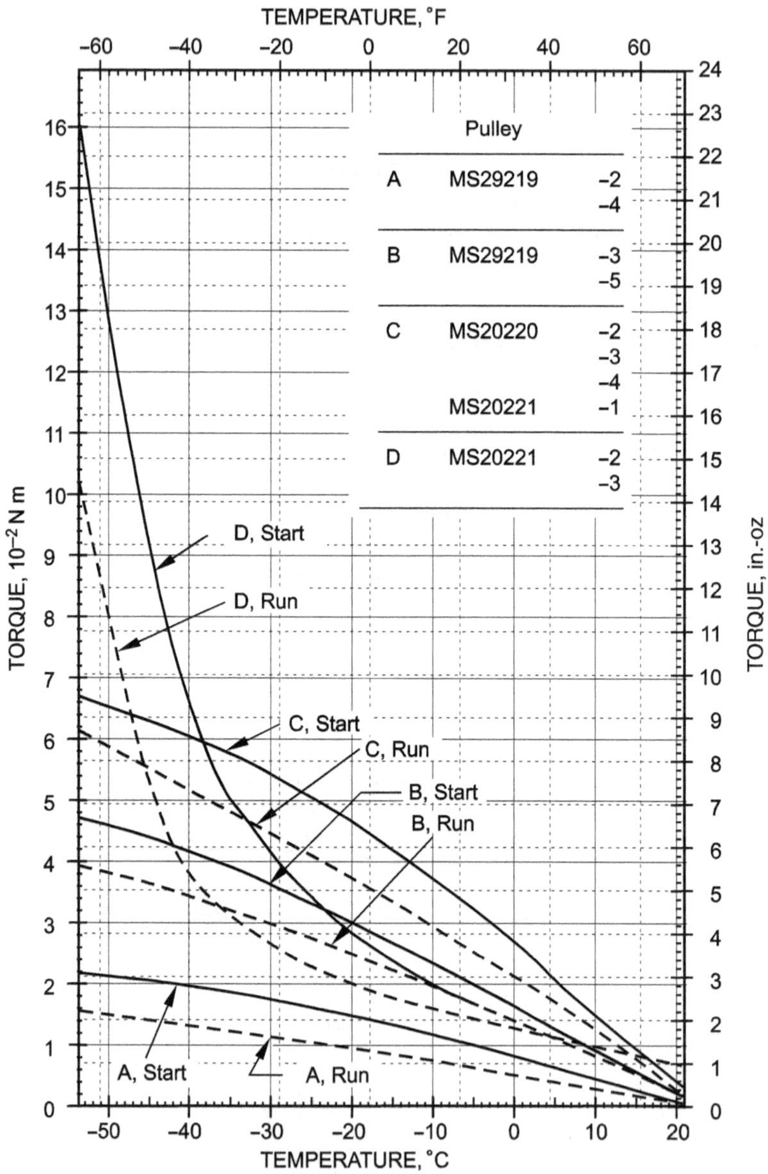

BEARINGS AND BUSHINGS, CONTINUED

MACHINERY BEARINGS FRICTION TORQUE

The friction of rolling element bearings will vary with the type of lubrication, temperature, bearing material, and bearing type. The following formula and table can be used to estimate friction torque.

$$F_t \approx \mu r_b P, \quad r_b = \text{bore radius}$$

Values of μ are given in the following table, which shows friction coefficients applicable to rolling element machinery bearings.

Bearing Type	Friction Coefficient,[a] μ
Self-aligning ball	0.0010
Cylindrical roller	0.0011
Thrust ball	0.0013
Single-row ball	0.0015
Tapered roller	0.0018
Full-complement needle	0.0025

[a]Friction coefficient referred to bearing bore radius.

RODS AND LINKS

Rods and links used as tension members are designed for the most severe combination of loads. These include rigging, structure deformation, temperature, and any beam effect loads applied in addition to operation loads. Any rod or link that it is possible to grasp or step on at any time during construction, maintenance, or inspection should be checked independently for a 667 N (150 lb) ultimate lateral load applied at any point along its length. Transverse vibration and fatigue checks should also be made on all control rods.

Rods and links used as compression members should be avoided if possible. When they are necessary, a beam column analysis should be made. Buckling due to column or transverse load conditions is the most critical mode of failure. The stiffness of a link support will also affect the stability of the link under compression. The equation presented below enables an instability load to be determined based on existing spring rates at each of the link ends. In a system consisting of several links, it is necessary to check for the instability of the whole system.

RODS AND LINKS, CONTINUED

LINK STABILITY

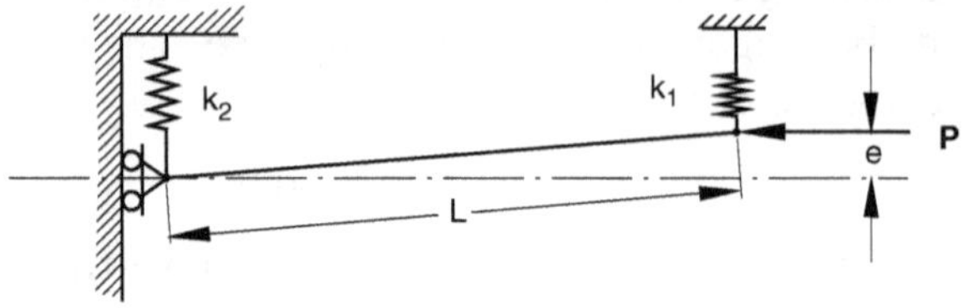

Critical buckling load P_{cr} for the link is

$$P_{cr} = \frac{k_1 k_2 L}{k_1 + k_2}\left[1 - \left(\frac{e}{L}\right)^{2/3}\right]^{3/2}$$

where

k_1, k_2 = spring rates in lb/in.
e = initial eccentricity
L = link length

TEMPERATURE LOADS

Change in temperature that will cause buckling in rods or links is

$$\Delta T_{cr} = (C\pi^2 E_t I - P_a L^2) / E_s \alpha A L^2$$

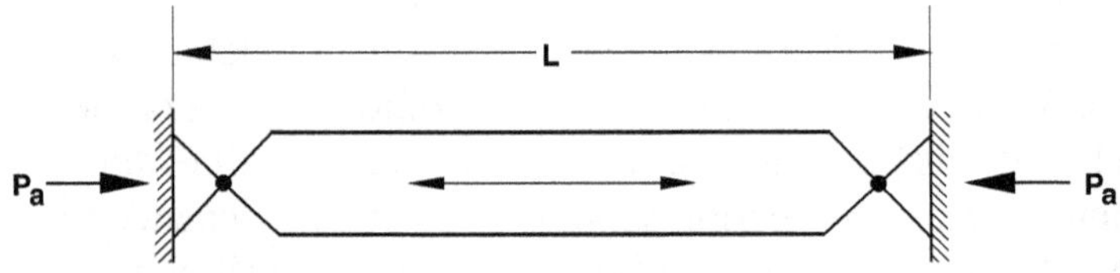

where

α = coefficient of thermal expansion
C = end restraint constant
E_t = tangent modulus
E_s = secant modulus
P_a = applied mechanical load

RODS AND LINKS, CONTINUED

TEMPERATURE LOADS, CONTINUED

Change in tension in a rod, link, or cable due to a change in temperature is

$$\Delta P_t = \Delta T(\alpha_1 - \alpha_2) \Big/ \left(\frac{1}{A_1 E_1} + \frac{1}{A_2 E_2} \right)$$

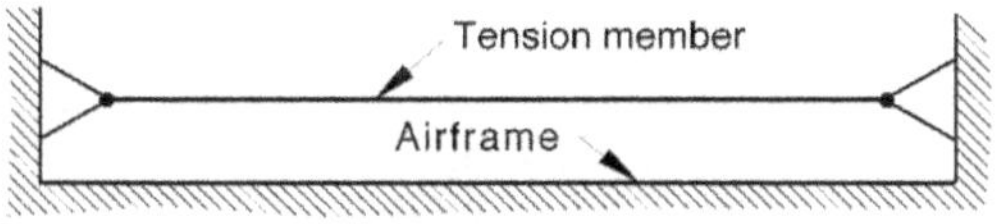

where

ΔT = change in temperature
α_1, α_1 = coefficient of linear expansion of airframe and member, respectively
A_1E_1, A_2E_2 = cross-sectional area times modulus of elasticity of airframe and member, respectively

A_1 (effective) for the airframe is usually quite large compared to A_2; therefore, in many cases, $1/A_1E_1$ can be neglected without serious error. In this case, $\Delta P_t \approx \Delta T(\alpha_1 - \alpha_2)A_2E_2$.

SHAFTS

AXLES AND SHAFTS (APPROXIMATE CALCULATION)

AXLES

Axle Type	Required Section Modulus for Bending	Solid Axle of Circular Cross Section ($Z \approx d^3/10$)	Permissible Bending Stress[a]
fixed[b]	$Z = \dfrac{M}{f_{bt}}$	$d = \sqrt[3]{\dfrac{10M}{f_{bt}}}$	$f_{bt} = \dfrac{f_{btU}}{(3...5)}$ [c]
rotating			$f_{bt} = \dfrac{f_{btA}}{(3...5)}$ [c]

[a]f_{bt} allows for stress concentration-, roughness-, size-, safety-factor, and combined stresses—see Gieck reference.
[b]Formulas are restricted to load classes I and II.
[c]3…5 means a number between 3 and 5.

Source: The shaft material appearing on pages 6-41 and 6-42 is from *Engineering Formulas,* 7th Ed., McGraw–Hill, New York. Copyright © 1997, Kurt Gieck, Reiner Gieck, Gieck–Verlag, Germering, Germany. Reproduced with permission of Gieck–Verlag.

SHAFTS, CONTINUED

AXLES AND SHAFTS (APPROXIMATE CALCULATION), CONTINUED

SHAFTS

Stress	Diameter for Solid Shaft	Permissible Torsional Stress[a]
pure torsion	$d = \sqrt[3]{\frac{5T}{\tau_{qt}}}$	$\tau_{qt} = \frac{\tau_t U}{(3...5)}$ [b]
torsion and bending		$\tau_{qt} = \frac{\tau_t U}{(10...15)}$ [c]

[a] τ_{qt} allows for stress concentration-, roughness-, size-, safety-factor and combined stresses—see Gieck reference.
[b] 3…5 means a number between 3 and 5.
[c] 10…15 means a number between 10 and 15.

BEARING STRESS

on shaft extension } $f_{bm} = \frac{F}{db} \leq f_b$

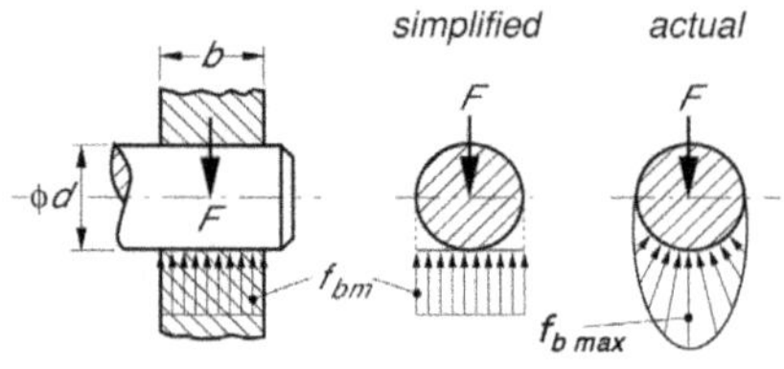

SHEAR DUE TO LATERAL LOAD

Calculation unnecessary when

$l > d/4$	for all shafts	with	circular	cross sections
$l > 0.325h$	for fixed axles		rectangular	

where

l = moment arm of force F
M, T = bending moment, torque, respectively
$f_{bm}, (f_b)$ = mean, (permissible) bearing stress

PINS

For hollow pins, keep D_0/t at 7.0 or below to avoid local pin deformation.

Check pins for shear and bending as shown for double shear symmetrical joints.

Shear stress:

$$\tau_{au} = \frac{P}{1.571(D_0^2 - D_1^2)}$$

Bending stress:

$$f_{bu} = \frac{M_{max}\ D_0}{\frac{\pi}{32}k_b(D_0^4 - D_1^4)} = \frac{M_{max}\ D_0}{0.0982k_b(D_0^4 - D_1^4)}$$

where

k_b = plastic bending coefficient (1.0 to 1.7) (usually 1.56 for pins made from material with more than 5% elongation)

$$M_{max} = \frac{P}{2}\left(\frac{t_1}{2} + \frac{t_2}{4} + g\right)$$

Pin clamp-up stress will be additive to bending and tension stress.

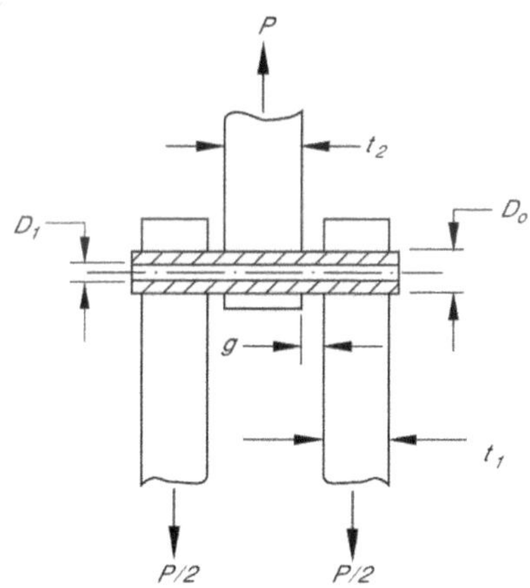

ACTUATORS

In general, hydraulic or pneumatic actuators are strength checked as pressure vessels—with inherent discontinuities and stress risers—and as stepped columns. Jackscrew actuators are also checked as columns. In jackscrews, combined compression and torque stresses at the thread root along with the reduced section make the effective length longer than the physical support length.

The system loaded by actuators must withstand maximum actuator output load. Design pressures for hydraulic and pneumatic actuators are defined in SAE AS5440 and MIL-P-5518, respectively. Those pressures are defined as a percentage of nominal system pressure as shown in the following table:

ACTUATORS, CONTINUED

	Load Factor, %	
Pressure	Hydraulic	Pneumatic
Operating	100	100
Proof	150[a]	200
Burst	300	400

[a]Factors apply to nominal pressure except the 150 applies to relief valve setting (usually 120% of nominal).

Cylindrical bodies are subject to hoop and axial tension due to internal pressure. Minimum thickness of the cylinder wall is

$$t_c = \frac{pD}{2f_{tU}}$$

where

p = burst pressure
D = inside diameter of the cylinder
f_{tU} = ultimate tensile strength

t_c should be such that $d/t_c < 30$ to preclude excessive deflection and piston seal leakage.

The actuator piston rod is subject to an external pressure. The required wall thickness of a hollow piston rod is

$$t_p = D\sqrt[3]{\frac{p(1-\mu)^2}{2E}}$$

where

p = burst pressure
D = piston rod outside diameter
μ = Poisson's ratio (0.3 for metals)

Stress concentrations at cylinder ends, ports, and section changes should be carefully considered and stress kept at a low enough level to meet fatigue requirements.

ACTUATORS, CONTINUED

MARGIN OF SAFETY

To determine whether an actuator will fail under the combined action of the applied compressive load, end moment, and moment due to eccentricities, the following two margins of safety must be obtained. To preclude failure, the resulting margins of safety must be greater than or equal to zero.

Piston rod with bending and column action:

$$MS = \frac{1}{R_c + R_b} - 1$$

where

$$R_c = \frac{P}{P_{cr}} \qquad R_b = \frac{f_b}{f_{bU}}$$

f_b = bending stress including beam column effect
f_{bU} = bending modulus of rupture
P = applied comprehensive load
P_{cr} = column critical load

Cylinder with combined bending and longitudinal and hoop tension:

$$\mathrm{MS} = \frac{1}{\sqrt{R_b^2 + R_{gt}^2 + R_b R_{ht}}} - 1$$

where

$$R_b = \frac{f_b}{f_{by}} \quad R_{ht} = \frac{f_{htU}}{f_{ty}}$$

f_{ty} = tensile yield stress
f_{htU} = applied ultimate hoop tension stress
f_{by} = bending modulus of yield
f_b = applied bending stress including beam column effect

CYLINDRICAL FITS

Tolerances and allowances for cylindrical fits are given in the tables on the following pages, based on a more comprehensive treatment contained in ASME B4.2 for metric units and ASME B4.1 for English units. The hole diameter should be selected in accordance with standard reamer sizes or the shaft size should be selected using standard dowel pin and shafting material sizes and called out on drawings as decimal conversions. Define tolerances using the data in the tables.

Always specify appropriate surface finish on drawings and provide appropriate lubrication for all rotating applications as well as venting for all interference fit applications. Ensure associated loads, especially from interference fits, do not over-stress the mating parts.

DESCRIPTION OF FITS

RUNNING AND SLIDING FITS

Running and sliding fits are intended to provide for rotational applications, with suitable lubrication allowance, throughout the range of sizes.

Sliding fits are intended for accurate location. Parts made to this fit move and turn easily but are not intended to run freely, and in the larger sizes may seize with small temperature changes.

Medium or free running fits are intended for higher running speeds, or heavy journal pressures, or both.

Loose running fits are intended for use where wide commercial tolerances may be necessary, together with an allowance, on the external member.

LOCATIONAL FITS

Locational fits are intended to determine only the location of mating parts.

Locational transition fits are a compromise between clearance and interference locational fits, for application where accuracy of location is important, but either a small amount of clearance or interference is permissible.

FORCE OR SHRINK FITS

Force or shrink fits constitute a special type of interference fit, normally characterized by maintenance of constant bore pressures throughout the range of sizes.

Medium drive fits are suitable for ordinary steel parts or for shrink fits on light sections. They are about the tightest fits that can be used with high-grade cast-iron external members.

Force fits are suitable for parts that can be highly stressed or for shrink fits where the heavy pressing forces required are impractical.

CYLINDRICAL FITS, CONTINUED

Nominal Size Range (mm)	Clearance Fits									Transition Fits						Interference Fits					
	Loose Running			Free Running			Sliding			Locational Transition (H7/k6)			Locational Transition (H7/n6)			Medium Drive			Force		
		Standard Tolerance			Standard Tolerance			Standard Tolerance			Standard Tolerance			Standard Tolerance			Standard Tolerance			Standard Tolerance	
	Limits of Clearance	Hole	Shaft	Limits of Clearance	Hole	Shaft	Limits of Clearance	Hole	Shaft	Fit	Hole	Shaft	Fit	Hole	Shaft	Limits of Interference	Hole	Shaft	Limits of Interference	Hole	Shaft
	0.060	0.000	–0.060	0.020	0.000	–0.020	0.002	0.000	–0.002	–0.006	0.000	0.006	–0.010	0.000	0.010	–0.020	0.000	0.020	–0.024	0.000	0.024
0–3.0	0.180	0.060	–0.120	0.070	0.025	–0.045	0.018	0.010	–0.008	0.010	0.010	0.000	0.006	0.010	0.004	–0.004	0.010	0.014	–0.008	0.010	0.018
	0.070	0.000	–0.070	0.030	0.000	–0.030	0.004	0.000	–0.004	–0.009	0.000	0.009	–0.016	0.000	0.016	–0.027	0.000	0.027	–0.031	0.000	0.031
3.0–6.0	0.220	0.075	–0.145	0.090	0.030	–0.060	0.024	0.012	–0.012	0.011	0.012	0.001	0.004	0.012	0.008	–0.007	0.012	0.019	–0.011	0.012	0.023
	0.095	0.000	–0.095	0.040	0.000	–0.040	0.005	0.000	–0.005	–0.010	0.000	0.010	–0.019	0.000	0.019	–0.032	0.000	0.032	–0.037	0.000	0.037
6.0–10.0	0.260	0.090	–0.170	0.112	0.036	–0.076	0.029	0.015	–0.014	0.014	0.015	0.001	0.005	0.015	0.010	–0.008	0.015	0.023	–0.013	0.015	0.028
	0.095	0.000	–0.095	0.050	0.000	–0.050	0.006	0.000	–0.006	–0.012	0.000	0.012	–0.023	0.000	0.023	–0.039	0.000	0.039	–0.044	0.000	0.044
10.0–16.0	0.315	0.110	–0.205	0.136	0.043	–0.093	0.035	0.018	–0.017	0.017	0.018	0.001	0.006	0.018	0.012	–0.010	0.018	0.028	–0.015	0.018	0.033
	0.110	0.000	–0.110	0.065	0.000	–0.065	0.007	0.000	–0.007	–0.015	0.000	0.015	–0.028	0.000	0.028	–0.048	0.000	0.048	–0.054	0.000	0.054
16.0–20.0	0.370	0.130	–0.240	0.169	0.052	–0.117	0.041	0.021	–0.020	0.019	0.021	0.002	0.006	0.021	0.015	–0.014	0.021	0.035	–0.020	0.021	0.041
	0.110	0.000	–0.110	0.065	0.000	–0.065	0.007	0.000	–0.007	–0.015	0.000	0.015	–0.028	0.000	0.028	–0.048	0.000	0.048	–0.061	0.000	0.061
20.0–30.0	0.370	0.130	–0.240	0.169	0.052	–0.117	0.041	0.021	–0.020	0.019	0.021	0.002	0.006	0.021	0.015	–0.014	0.021	0.035	–0.027	0.021	0.048
	0.120	0.000	–0.120	0.080	0.000	–0.080	0.009	0.000	–0.009	–0.018	0.000	0.018	–0.033	0.000	0.033	–0.059	0.000	0.059	–0.076	0.000	0.076
30.0–40.0	0.440	0.160	–0.280	0.204	0.062	–0.142	0.050	0.025	–0.025	0.023	0.025	0.002	0.008	0.025	0.017	–0.018	0.025	0.043	–0.035	0.025	0.060
	0.130	0.000	–0.130	0.080	0.000	–0.080	0.009	0.000	–0.009	–0.018	0.000	0.018	–0.033	0.000	0.033	–0.059	0.000	0.059	–0.086	0.000	0.086
40.0–50.0	0.450	0.160	–0.290	0.204	0.062	–0.142	0.050	0.025	–0.025	0.023	0.025	0.002	0.008	0.025	0.017	–0.018	0.025	0.043	–0.045	0.025	0.070

Note: Nominal sizes are in millimeters. Standard tolerances for hole and shaft are applied algebraically to the nominal size to obtain the limits of size for the parts. "Fit" represents the maximum interference (negative values) and the maximum clearance (positive values).

CYLINDRICAL FITS, CONTINUED

	Sliding Fit			Running Fit						Location Transition Fit						Force or Shrink Fit					
	RC2			RC5			RC8			LT2			LT6			FN2			FN4		
		Standard Tolerance			Standard Tolerance			Standard Tolerance			Standard Tolerance			Standard Tolerance			Standard Tolerance			Standard Tolerance	
Nominal Size Range (inches)	Limits of Clearance	Hole	Shaft	Limits of Clearance	Hole	Shaft	Limits of Clearance	Hole	Shaft	Fit	Hole	Shaft	Fit	Hole	Shaft	Limits of Interference	Hole	Shaft	Limits of Interference	Hole	Shaft
	0.1	0	−0.1	0.6	0	−0.6	2.5	0	−2.5	−0.2	0	0.2	−0.65	0	0.65	0.2	0	0.85	0.3	0	0.95
0–0.12	0.55	0.25	−0.3	1.6	0.6	−1	5.1	1.6	−3.5	0.8	0.6	−0.2	0.15	0.4	0.25	0.85	0.4	0.6	0.95	0.4	0.7
	0.15	0	−0.15	0.8	0	−0.8	2.8	0	−2.8	−0.25	0	0.25	−0.8	0	0.8	0.2	0	1	0.4	0	1.2
0.12–0.24	0.65	0.3	−0.35	2	0.7	−1.3	5.8	1.8	−4	0.95	0.7	−0.25	0.2	0.5	0.3	1	0.5	0.7	1.2	0.5	0.9
	0.2	0	−0.2	1	0	−1	3	0	−3	−0.3	0	0.3	−1	0	1	0.4	0	1.4	0.6	0	1.6
0.24–0.40	0.85	0.4	−0.45	2.5	0.9	−1.6	6.6	2.2	−4.4	1.2	0.9	−0.3	0.2	0.6	0.4	1.4	0.6	1	1.6	0.6	1.2
	0.25	0	−0.25	1.2	0	−1.2	3.5	0	−3.5	−0.35	0	0.35	−1.2	0	1.2	0.5	0	1.6	0.7	0	1.8
0.40–0.56	0.95	0.4	−0.55	2.9	1	−1.9	7.9	2.8	−5.1	1.35	1	−0.35	0.2	0.7	0.5	1.6	0.7	1.2	1.8	0.7	1.4
	0.25	0	−0.25	1.2	0	−1.2	3.5	0	−3.5	−0.35	0	0.35	−1.2	0	1.2	0.5	0	1.6	0.7	0	1.8
0.56–0.71	0.95	0.4	−0.55	2.9	1	−1.9	7.9	2.8	−5.1	1.35	1	−0.35	0.2	0.7	0.5	1.6	0.7	1.2	1.8	0.7	1.4
	0.3	0	−0.3	1.6	0	−1.6	4.5	0	−4.5	−0.4	0	0.4	−1.4	0	1.4	0.6	0	1.9	0.8	0	2.1
0.71–0.95	1.2	0.5	−0.7	3.6	1.2	−2.4	10	3.5	−6.5	1.6	1.2	−0.4	0.2	0.8	0.6	1.9	0.8	1.4	2.1	0.8	1.6
	0.3	0	−0.3	1.6	0	−1.6	4.5	0	−4.5	−0.4	0	0.4	−1.4	0	1.4	0.6	0	1.9	1	0	2.3
0.95–1.19	1.2	0.5	−0.7	3.6	1.2	−2.4	10	3.5	−6.5	1.6	1.2	−0.4	0.2	0.8	0.6	1.9	0.8	1.4	2.3	0.8	1.8
	0.4	0	−0.4	2	0	−2	5	0	−5	−0.5	0	0.5	−1.7	0	1.7	0.8	0	2.4	1.5	0	3.1
1.19–1.58	1.4	0.6	−0.8	4.6	1.6	−3	11.5	4	−7.5	2.1	1.6	−0.5	0.3	1	0.7	2.4	1	1.8	3.1	1	2.5
	0.4	0	−0.4	2	0	−2	5	0	−5	−0.5	0	0.5	−1.7	0	1.7	0.8	0	2.4	1.8	0	3.4
1.58–1.97	1.4	0.6	−0.8	4.6	1.6	−3	11.5	4	−7.5	2.1	1.6	−0.5	0.3	1	0.7	2.4	1	1.8	3.4	1	2.8

Note: Nominal sizes are in inches. Limits and tolerances are in thousands of an inch. Standard tolerances for hole and shaft are applied algebraically to the nominal size to obtain the limits of size for the parts. "Fit" represents the maximum interference (negative values) and the maximum clearance (positive values).

KEY JOINTS

TAPERED KEYS (COTTER PINS)

Safety margin for loaf F: In view of the additional load produced when the key is tightened, a 25% margin should be added in the following.

CONTACT PRESSURE BETWEEN

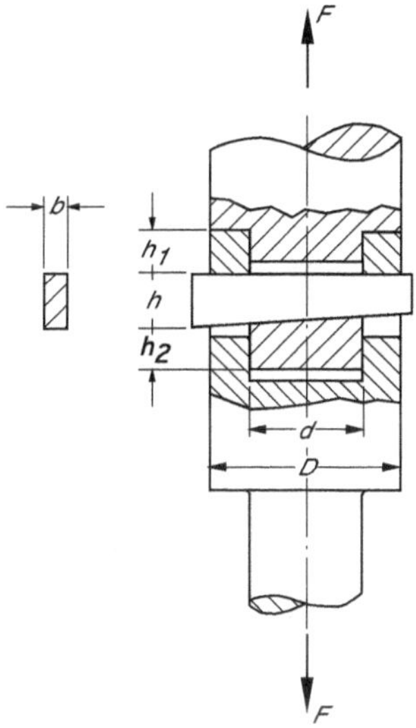

rod and key $f = \frac{1.25F}{bd} \leq f_{bU}$

socket and key $f = \frac{1.25F}{b(D-d)} \leq f_{bU}$

REQUIRED KEY WIDTH

$$h = 0.87\sqrt{\frac{0.625F(D+d)}{bf_{bU}}} \quad h_1 = h_2 = k \quad 0.5h < k < 0.7h$$

SHEAR STRESS

$$\tau = \frac{1.5F}{bh} \leq \tau_{qU}$$

Source: The key joint and kinematics material appearing on pages 6-49–6-51 is from *Engineering Formulas*, 7th Ed., McGraw–Hill, New York. Copyright © 1997, Kurt Gieck, Reiner Gieck, Gieck–Verlag, Germering, Germany. Reproduced with permission of Gieck–Verlag.

Key Joints, continued

Load on Screw

Required root diameter d_k when load is applied

a) during screwing operation

$$d_k = \sqrt{\frac{4F}{\pi \times 0.75 f_U}}$$

b) after

$$d_k = \sqrt{\frac{4F}{\pi \times f_U}}$$

where F = load to be moved and f_{bU}, f_U, and τ_{qU} are allowable stresses.

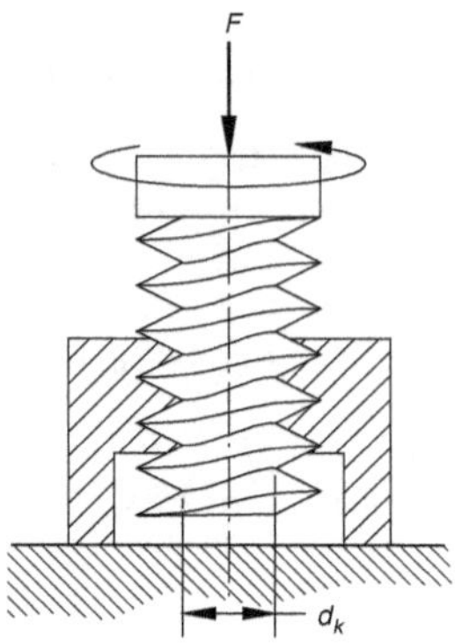

Kinematics

Simple Connecting-Rod Mechanism

$$s = r(1 - \cos\varphi) + \frac{\lambda}{2} r \sin^2\varphi$$

$$v = \omega r \sin\varphi(1 + \lambda\cos\varphi)$$

$$a = \omega^2 r(\cos\varphi + \lambda\cos 2\varphi)$$

$$\lambda = \frac{r}{l} = \frac{1}{4} \text{ to } \frac{1}{6} \quad (\lambda \text{ is called the crank ratio})$$

$$\varphi = \omega t = 2\pi n t$$

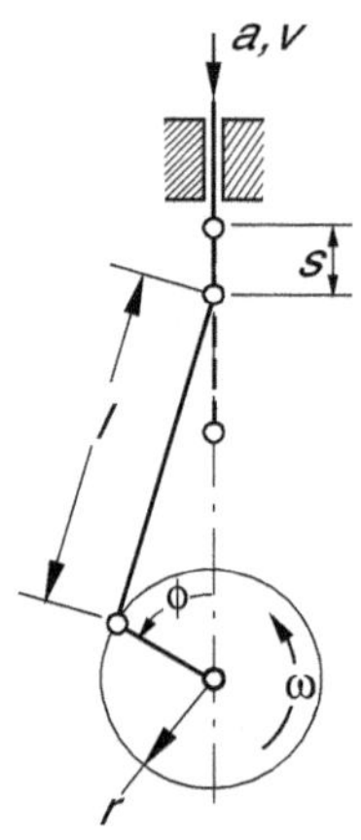

Scotch-Yoke Mechanism

$$s = r\sin(\omega t)$$

$$v = \omega r\cos(\omega t)$$

$$a = -\omega^2 r\sin(\omega t)$$

$$\omega = 2\pi n$$

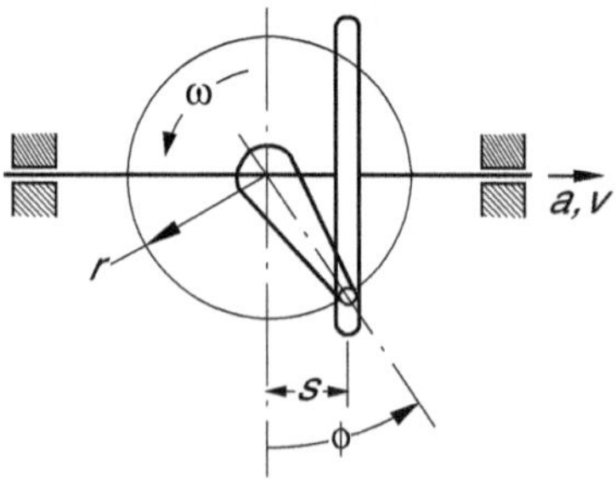

KINEMATICS, CONTINUED

CARDAN JOINT

For constant input speed, the output speed will be as follows.

Variable	Constant (Due to Auxiliary Shaft *H*)	

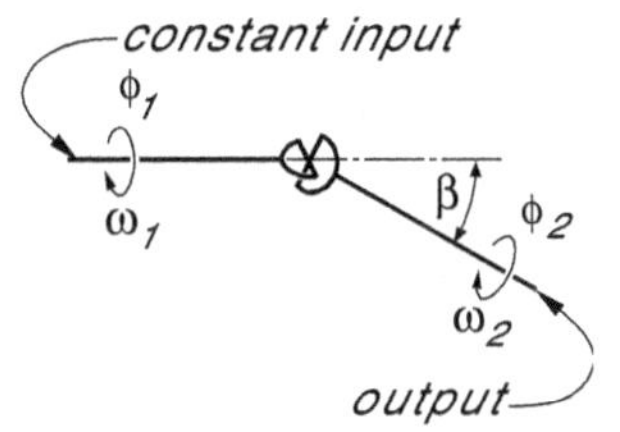

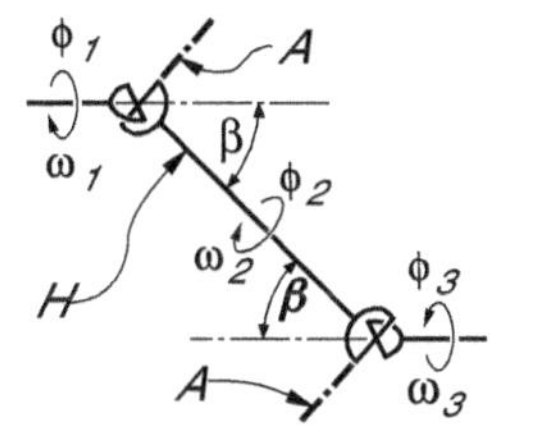

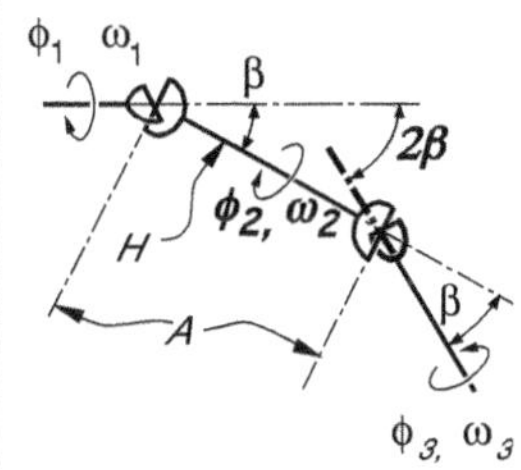

For all shafts located in one plane, the following relations apply:

$\tan \varphi_2 = \tan \varphi_1 \times \cos \beta$	$\tan \varphi_3 = \tan \varphi_1$	$\tan \varphi_3 = \tan \varphi_1$
$\omega_2 = \omega_1 \dfrac{\cos \beta}{1 - \sin^2 \beta \times \sin^2 \varphi_1}$	$\omega_3 = \omega_1$	$\omega_3 = \omega_1$

$$\alpha_2 = \omega_1^2 \frac{\sin^2 \beta \times \cos \beta \times \sin 2\varphi_1}{(1 - \sin^2 \beta \times \sin^2 \varphi_1)^2}$$

Both axes A of the auxiliary shaft joints must be parallel.

Note: The more the angle of inclination β increases, the more the maximum acceleration α and the accelerating moment M_α become; therefore, in practice, $\beta \leq 45$ deg.

SECTION 7

GEOMETRIC DIMENSIONING AND TOLERANCING (ASME Y14.5 – 2009)

ASME Y14.5 – 2009

The ASME Y14.5, revised in 1994, is an accepted geometric dimensioning standard followed throughout the industry. It is based on a philosophy of establishing datums and measuring features following the same procedures one would use to inspect the physical part.

DEFINITIONS

Datum—a theoretically exact point, axis, line, plane, or combination thereof derived from the theoretical datum feature simulator.

Datum Feature—a feature that is identified with either a datum feature symbol or a datum target symbol.

Dimension, Basic—a theoretically exact dimension.

Feature—a physical portion of a part such as a surface, pin, hole, or slot or its representation on drawings, models, or digital data files.

Least Material Condition (LMC)—the condition in which a feature of size contains the least amount of material within the stated limits of size (e.g., maximum hole diameter, minimum shaft diameter).

Maximum Material Condition (MMC)—the condition in which a feature of size contains the maximum amount of material within the stated limits of size (e.g., minimum hole diameter, maximum shaft diameter).

Regardless of Feature Size (RFS)—indicates a geometric tolerance applies at any increment of size of the actual mating envelope of the feature of size.

Tolerance, Geometric—the general term applied to the category of tolerances used to control size, form, profile, orientation, location, and runout.

True Position—the theoretically exact location of a feature of size, as established by basic dimensions.

Virtual Condition—a constant boundary generated by the collective effects of a considered feature of the size's specified MMC or LMC and the geometric tolerance for that material condition.

GEOMETRIC SYMBOLS AND MODIFIERS

GEOMETRIC CHARACTERISTIC SYMBOLS

APPLICATION	TYPE OF TOLERANCE	CHARACTERISTIC	SYMBOL
INDIVIDUAL FEATURES	FORM	STRAIGHTNESS	—
		FLATNESS	⏥
		CIRCULARITY	○
		CYLINDRICITY	⌭
INDIVIDUAL OR RELATED FEATURES	PROFILE	PROFILE OF A LINE	⌒
		PROFILE OF A SURFACE	⌓
RELATED FEATURES	ORIENTATION	ANGULARITY	∠
		PERPENDICULARITY	⊥
		PARALLELISM	//
	LOCATION	POSITION **	⌖
		CONCENTRICITY	◎
		SYMMETRY	⌯
	RUNOUT	CIRCULAR RUNOUT	↗
		TOTAL RUNOUT	⌰

BASIC DIMENSION SYMBOL

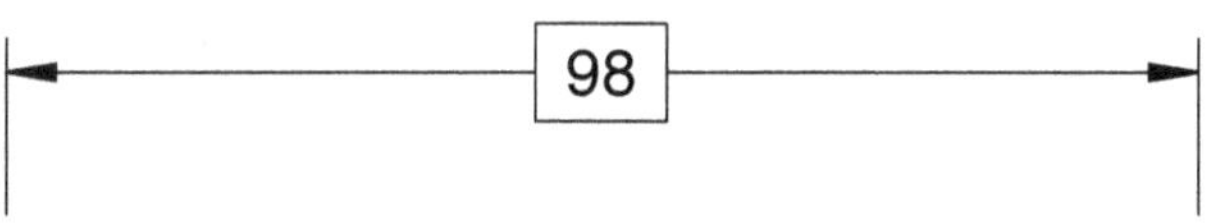

DATUM FEATURE SYMBOL

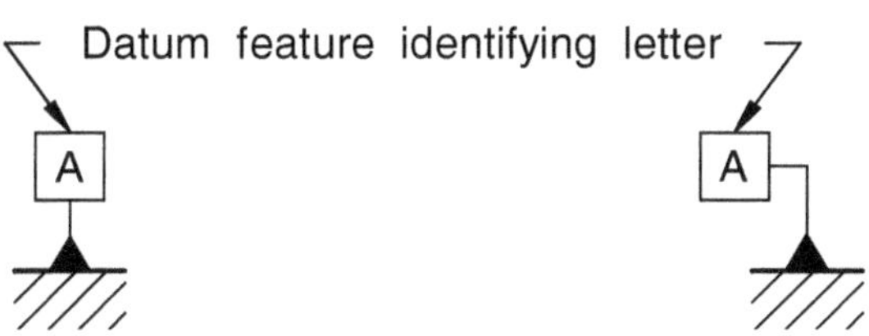

GEOMETRIC SYMBOLS AND MODIFIERS, CONTINUED

DATUM TARGET SYMBOL

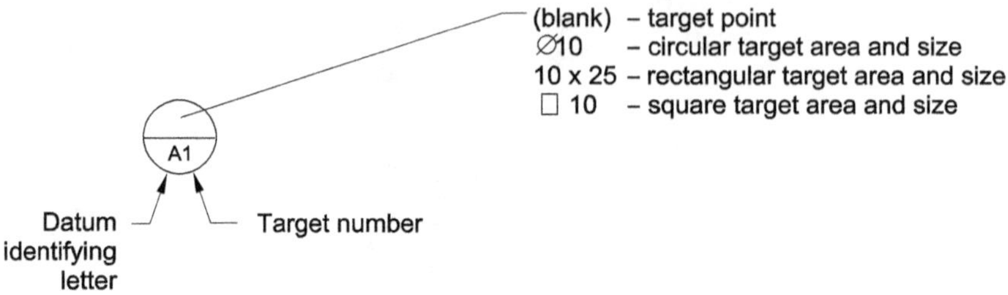

FEATURE CONTROL FRAME

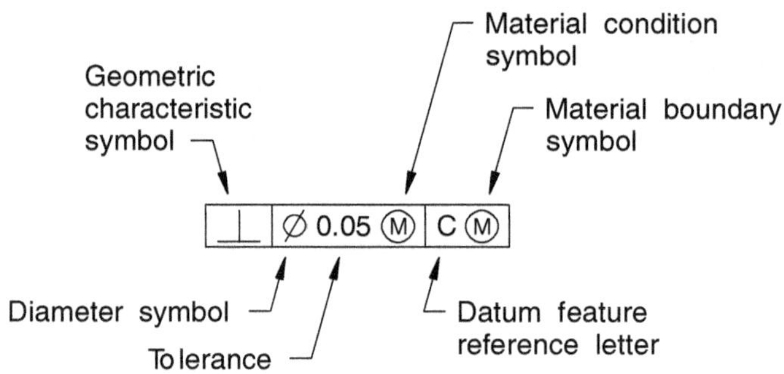

Figures reprinted from *ASME Y14.5 – 2009*, by permission of the American Society of Mechanical Engineers.

GEOMETRIC SYMBOLS AND MODIFIERS, CONTINUED

MODIFYING SYMBOLS

TERM	SYMBOL	TERM	SYMBOL
AT MAXIMUM MATERIAL CONDITION (When applied to a tolerance value) AT MAXIMUM MATERIAL BOUNDARY (When applied to a datum reference)	Ⓜ	DIAMETER	Ø
		SPHERICAL DIAMETER	SØ
AT LEAST MATERIAL CONDITION (When applied to a tolerance value) AT LEAST MATERIAL BOUNDARY (When applied to a datum reference)	Ⓛ	RADIUS	R
		SPHERICAL RADIUS	SR
TRANSLATION	▷	CONTROLLED RADIUS	CR
PROJECTED TOLERANCE ZONE	Ⓟ	SQUARE	□
FREE STATE	Ⓕ	REFERENCE	( )
TANGENT PLANE	Ⓣ	ARC LENGTH	⌒
UNEQUALLY DISPOSED PROFILE	Ⓤ	DIMENSION ORIGIN	⌽→
INDEPENDENCY	Ⓘ	BETWEEN	◄►
STATISTICAL TOLERANCE	⟨ST⟩	ALL AROUND	↗○—
CONTINUOUS FEATURE	⟨CF⟩	ALL OVER	↗◎—

MATERIAL CONDITION MODIFIERS

Ⓜ *MAXIMUM MATERIAL CONDITION (MMC)*

Applying MMC to a geometric tolerance has the effect of increasing the amount of allowable tolerance equal to the amount of departure from MMC. Unless otherwise specified, the maximum allowable tolerance occurs at the feature's LMC.

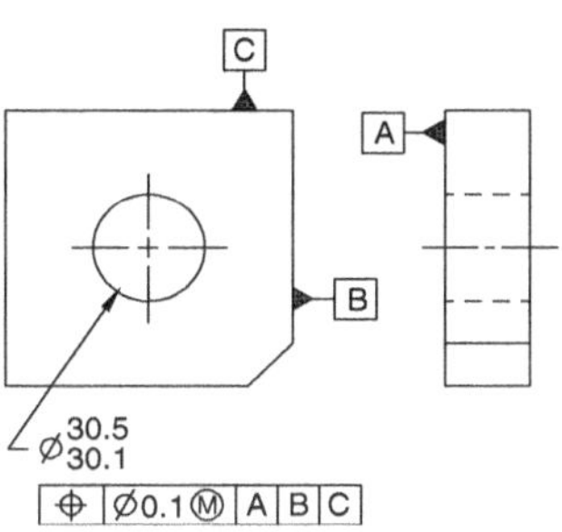

	Ø Hole	Ø Tol	VC
MMC	30.1	0.1	
	30.2	0.2	
	30.3	0.3	30.0
	30.4	0.4	
LMC	30.5	0.5	

Figures reprinted from *ASME Y14.5 – 2009*, by permission of the American Society of Mechanical Engineers.

MATERIAL CONDITION MODIFIERS, CONTINUED

Ⓛ LEAST MATERIAL CONDITION (LMC)

Applying LMC to a geometric tolerance has the effect of increasing the amount of allowable tolerance equal to the amount of departure from the feature's LMC. Unless otherwise specified, the maximum allowable tolerance occurs at the feature's MMC. When a geometric tolerance is applied at LMC, perfect form at MMC is not required.

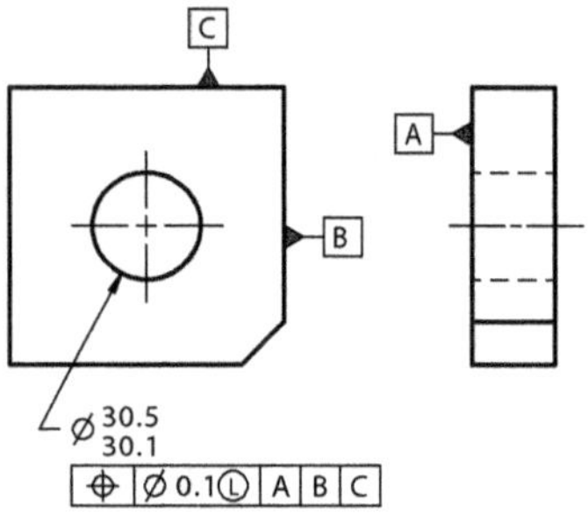

	Ø Hole	Ø Tol	VC
MMC	30.1	0.5	30.6
	30.2	0.4	
	30.3	0.3	
	30.4	0.2	
LMC	30.5	0.1	

REGARDLESS OF FEATURE SIZE (RFS)

In the absence of a material condition modifier, RFS is applied to the geometric tolerance. The geometric tolerance remains constant and is independent of the actual size of the feature.

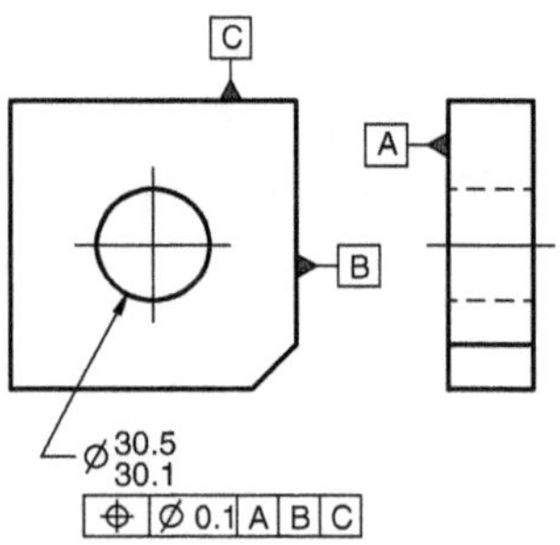

	Ø Hole	Ø Tol	VC
MMC	30.1	0.1	30.6
	30.2	0.1	
	30.3	0.1	
	30.4	0.1	
LMC	30.5	0.1	

STANDARD RULES

LIMITS OF SIZE

The limits of size of a feature prescribe the extent within which variations of geometric form, as well as size, are allowed. Unless otherwise specified, perfect form at MMC of a feature is required (Rule #1).

TOLERANCE RULE

For all applicable geometric tolerances, RFS applies in the absence of a material condition modifier (Rule #2).

PITCH DIAMETER RULE

Unless otherwise specified, each tolerance of orientation or position and datum reference specified for a screw thread applies to the axis of the thread pitch diameter.

RIGHT ANGLE FEATURE

Centerlines and surfaces of a feature that intersect at right angles are implied to be 90-deg angles and are not specified on the drawing.

TOLERANCES OF FORM

STRAIGHTNESS

Straightness is a condition where an element of a surface, or derived median line, is a straight line. A straightness tolerance specifies a tolerance zone within which the considered element of a surface or derived median line must lie. A straightness tolerance is applied in the view where the elements to be controlled are represented by a straight line.

Straightness of a Surface

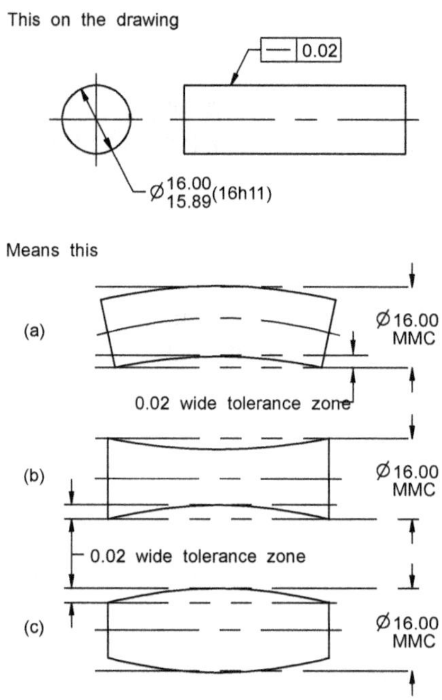

Straightness of an Axis

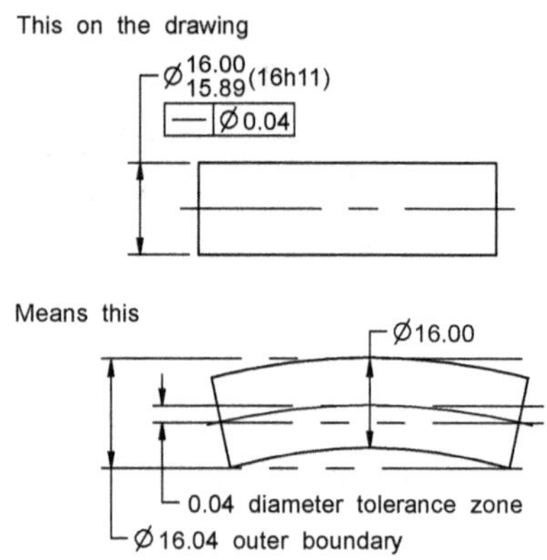

TOLERANCES OF FORM, CONTINUED

FLATNESS

Flatness is the condition of a surface or derived median plane having all elements in one plane. A flatness tolerance specifies a tolerance zone defined by two parallel planes within which the surface or derived median plane must lie. A flatness tolerance is applied in a view where the surface to be controlled is represented as a line.

Flatness of a Surface

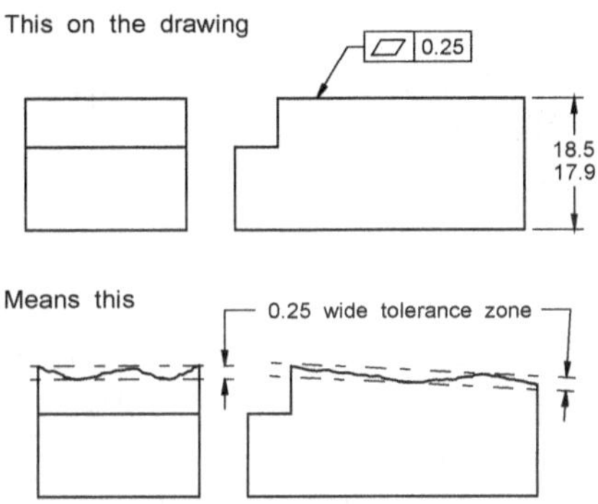

Flatness of a Median Plane

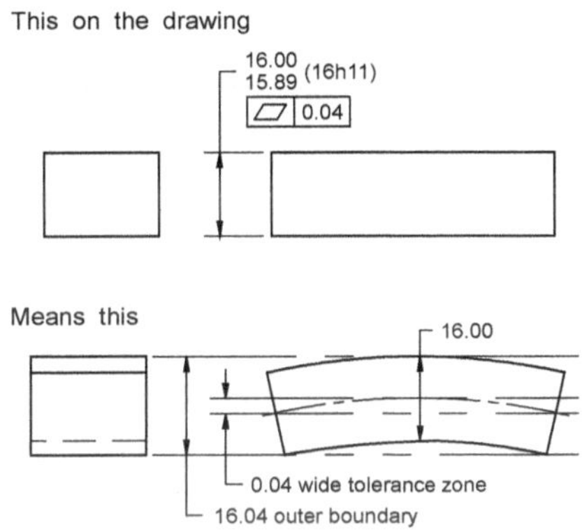

CIRCULARITY (ROUNDNESS)

Circularity is a condition of a surface where:

a) For a feature other than a sphere, all points of the surface intersected by any plane perpendicular to an axis or spine (curved line) are equidistant from that axis or spine.

b) For a sphere, all points of the surface intersected by any plane passing through a common center are equidistant from that center.

A circularity tolerance specifies a tolerance zone bounded by two concentric circles within which each circular element of the surface must lie.

TOLERANCES OF FORM, CONTINUED

CIRCULARITY (ROUNDNESS), CONTINUED

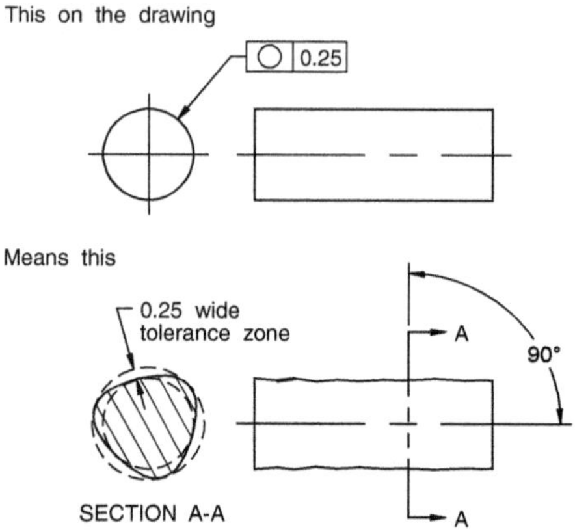

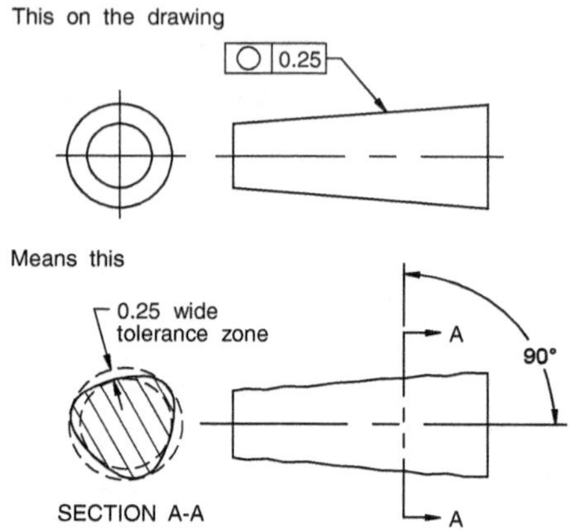

CYLINDRICITY

Cylindricity is a condition of a surface of revolution in which all points of the surface are equidistant from a common axis. A cylindricity tolerance specifies a tolerance zone bounded by two concentric cylinders within which the surface must lie. In the case of cylindricity, unlike that of circularity, the tolerance applies simultaneously to both circular and longitudinal elements of the surface (the entire surface).

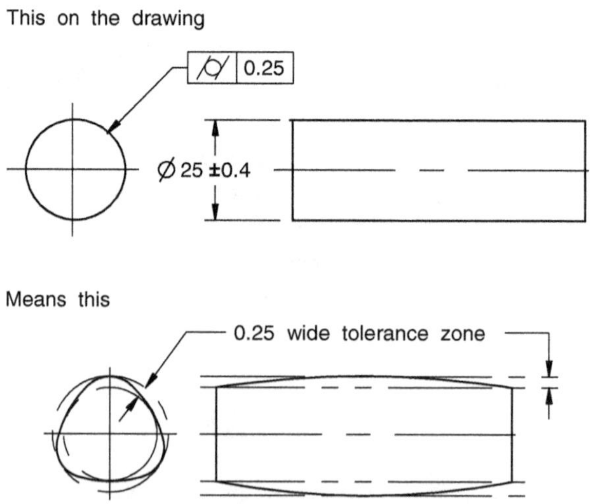

TOLERANCES OF ORIENTATION

An orientation tolerance controls parallel, perpendicular, and all other angular relationships. Note that an orientation tolerance, when applied to a plane surface, controls flatness to the extent of the orientation tolerance. When the flatness control in the orientation tolerance is not sufficient, a separate flatness tolerance should be considered.

ANGULARITY

Angularity is the condition of a surface, feature's center plane, or feature's axis at any specified angle from a datum plane or datum axis.

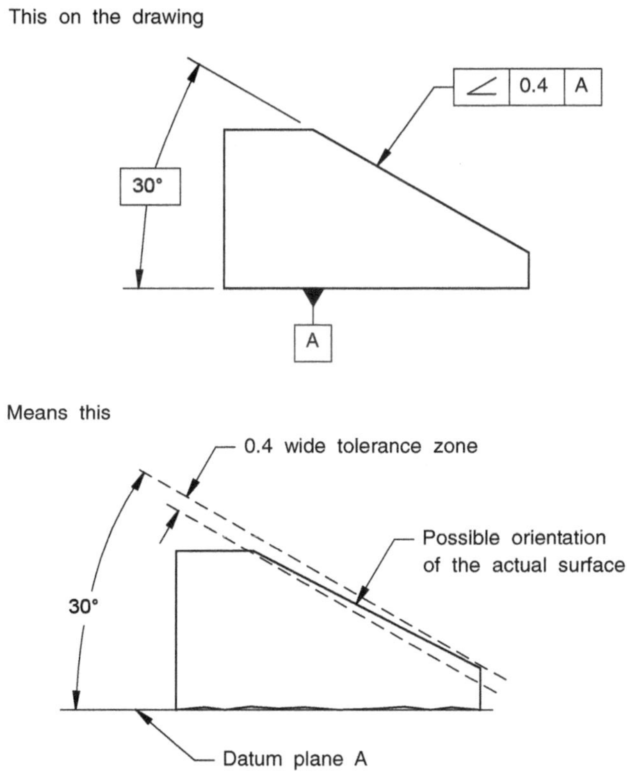

TOLERANCES OF ORIENTATION, CONTINUED

PARALLELISM

Parallelism is the condition of a surface or feature's center plane, equidistant at all points from a datum plane; or a feature's axis, equidistant along its length from one or more datum planes or datum axis.

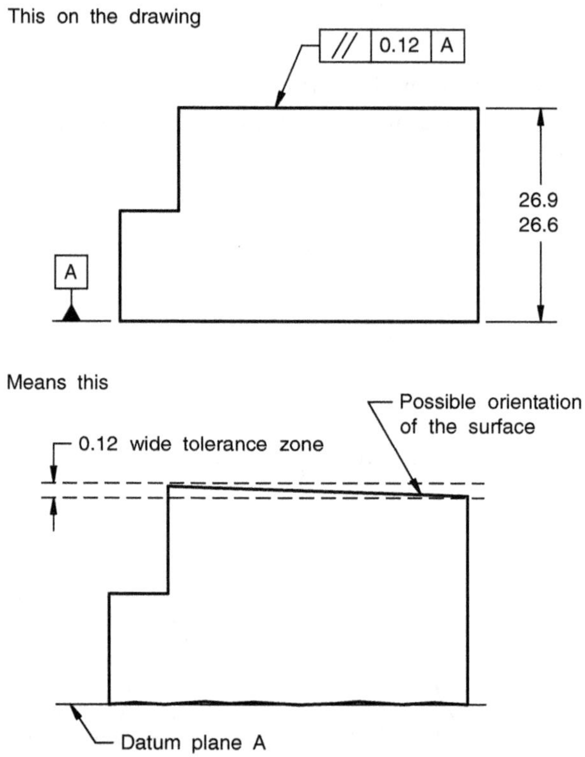

TOLERANCES OF ORIENTATION, CONTINUED

PERPENDICULARITY

Perpendicularity is the condition of a surface, feature's center plane, or feature's axis at a right angle to a datum plane or datum axis.

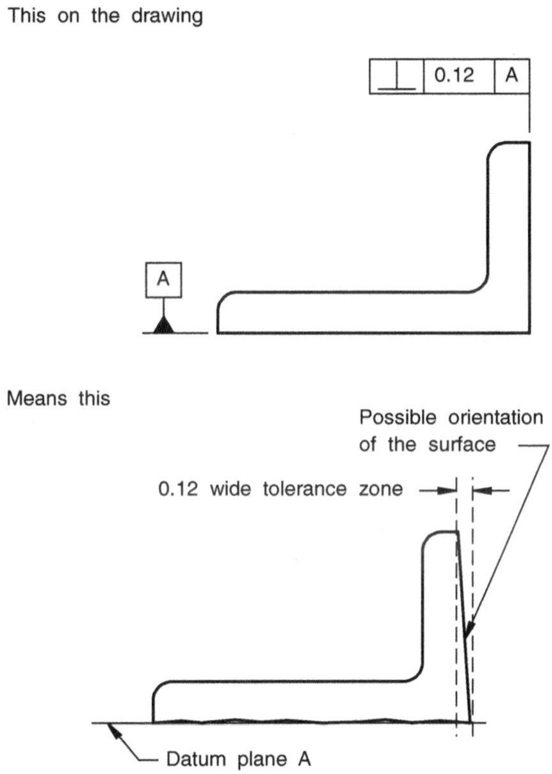

TOLERANCES OF LOCATION

Tolerances of location include position, concentricity, and symmetry and are used to control the following relationships:

a) center distance between features of size such as holes, slots, bosses, and tabs
b) location of features of size as a group, from datum features, such as plane and cylindrical surfaces
c) coaxiality of features of size
d) concentricity or symmetry of features of size—center distances of correspondingly located feature elements equally disposed about a datum axis or plane

POSITION

Position is the location of one or more features of size relative to one another or to one or more datums. A positional tolerance defines either of the following:

a) a zone within which the center, axis, or center plane of a feature of size is permitted to vary from a true (theoretically exact) position
b) (where specified on an MMC or LMC basis) a boundary, defined as the virtual condition, located at the true (theoretically exact) position, that may not be violated by the surface or surfaces of the considered feature of size

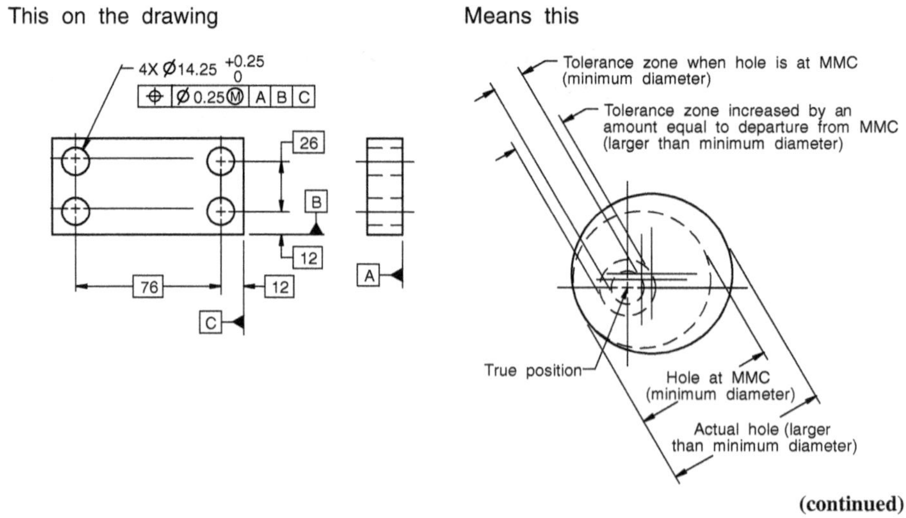

(continued)

Figures reprinted from *ASME Y14.5 – 2009*, by permission of the American Society of Mechanical Engineers.

TOLERANCES OF LOCATION, CONTINUED

POSITION, CONTINUED

Projected tolerance zones are used to prevent interference of fixed fasteners, studs, or press-fit pins with mated parts. The amount of projected tolerance should be sufficient to clear the through-hole of the mating part or the height of the stud or pin.

This on the drawing

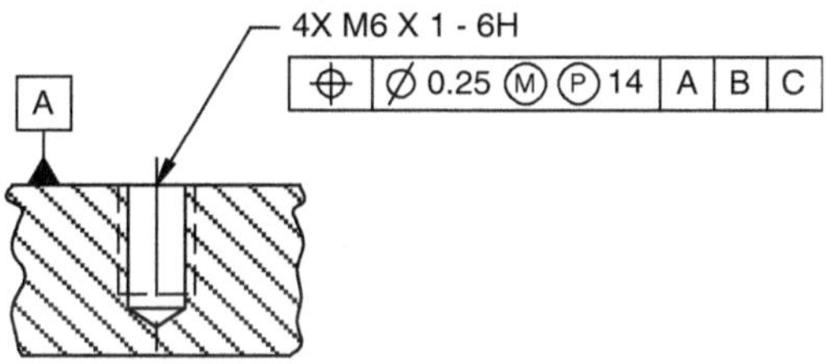

Means this

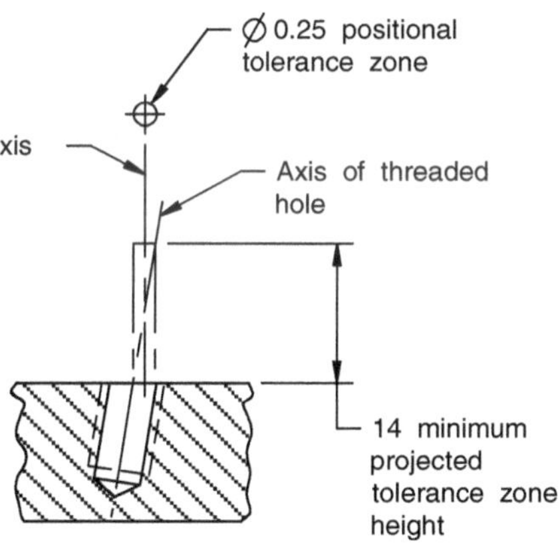

(continued)

TOLERANCES OF LOCATION, CONTINUED

POSITION, CONTINUED

Composite positional tolerancing provides an application of positional tolerancing for the location of a feature of size patterns as well as the interrelation of features of size within these patterns.

This on the drawing

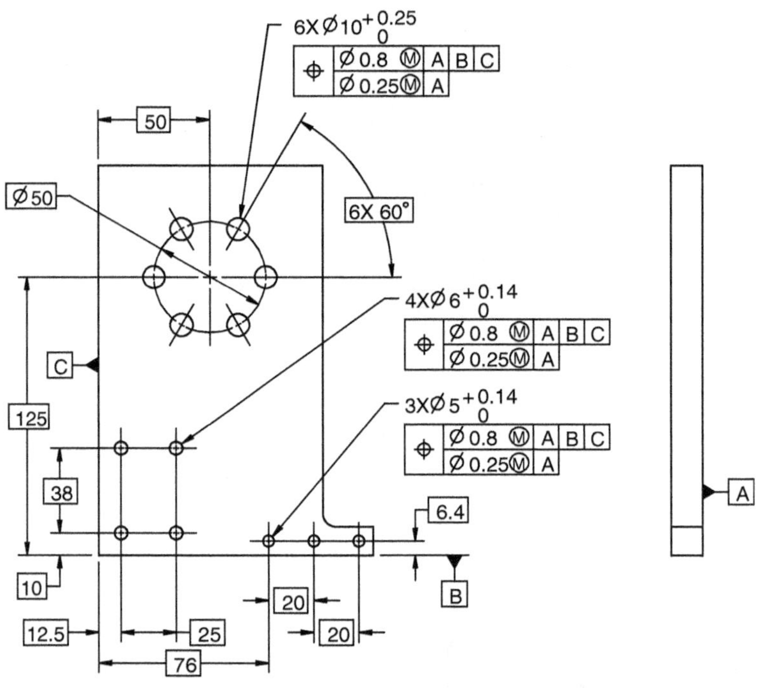

Means this for the feature pattern

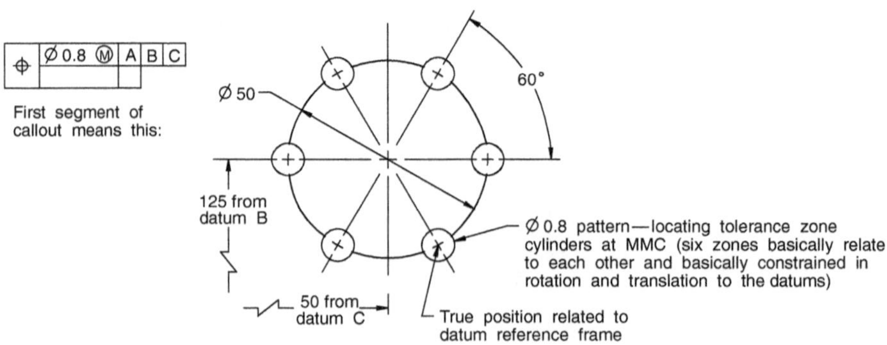

(continued)

Figures reprinted from *ASME Y14.5 – 2009*, by permission of the American Society of Mechanical Engineers.

TOLERANCES OF LOCATION, CONTINUED

POSITION, CONTINUED

And means this for the features within the pattern

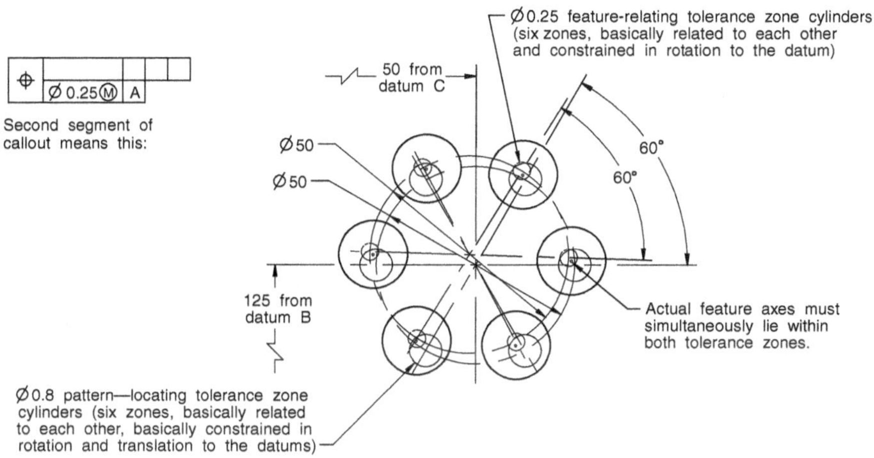

CONCENTRICITY

The condition where the median points of all diametrically opposed elements of a surface of revolution are congruent with a datum axis.

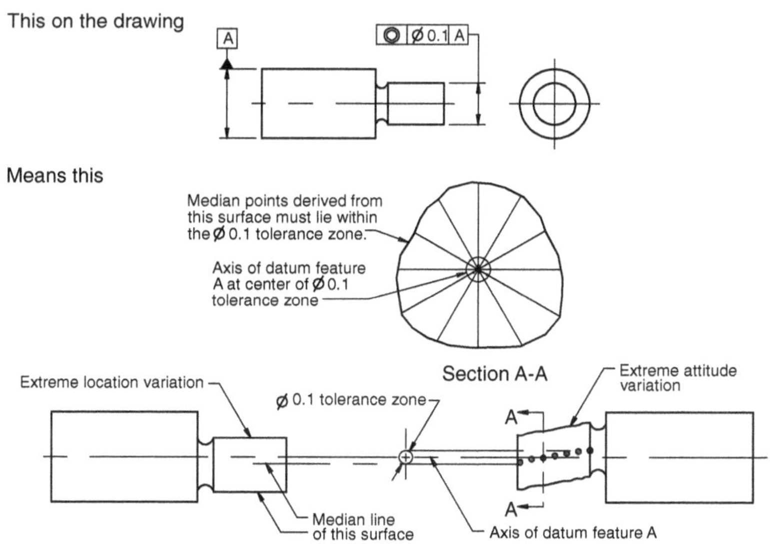

TOLERANCES OF LOCATION, CONTINUED

SYMMETRY

The condition where the median points of all opposed elements of two or more feature surfaces are congruent with a datum axis or center plane.

This on the drawing

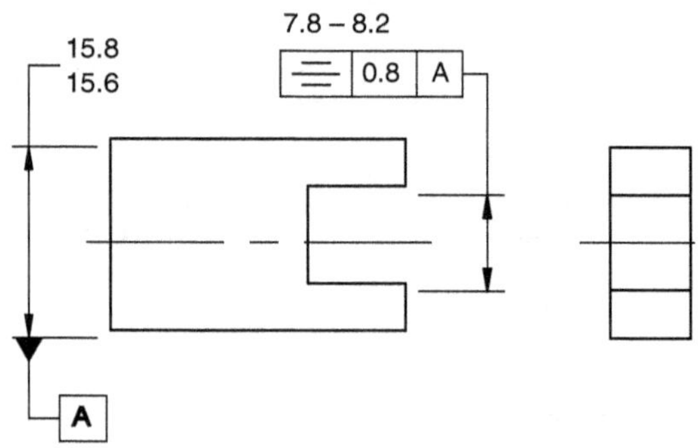

Means this

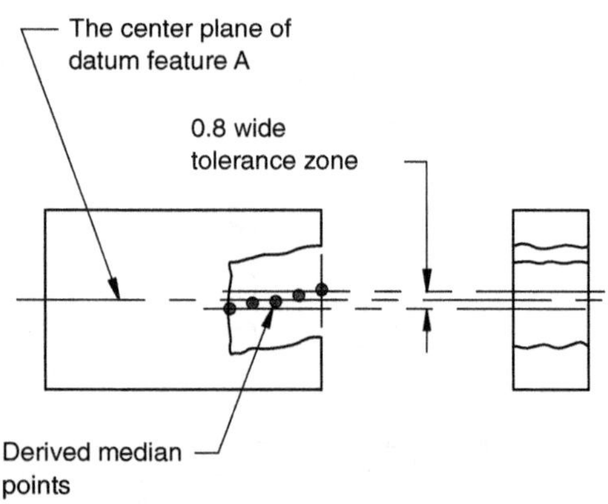

TOLERANCES OF PROFILE

A profile is an outline of a surface, a shape made up of one or more features, or a two-dimensional element of one or more features. Profile tolerances are used to define a tolerance zone to control form or combinations of size, form, orientation, and location of a feature(s) relative to a true profile.

PROFILE OF A LINE

Each line element tolerance zone established by the profile of a line tolerance requirement is two-dimensional (an area) and the tolerance zone is normal to the true profile of the feature at each line element.

This on the drawing

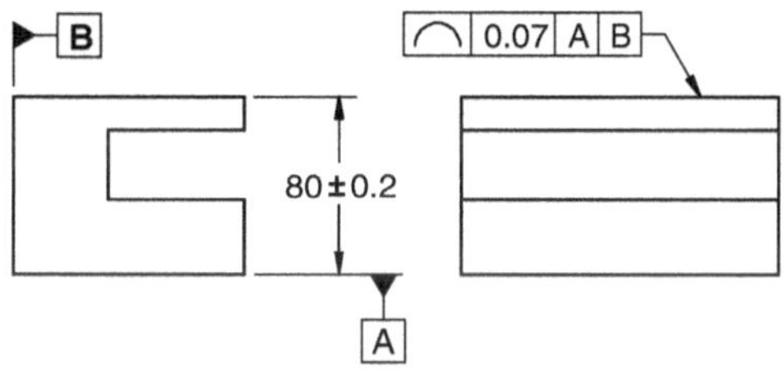

Means this

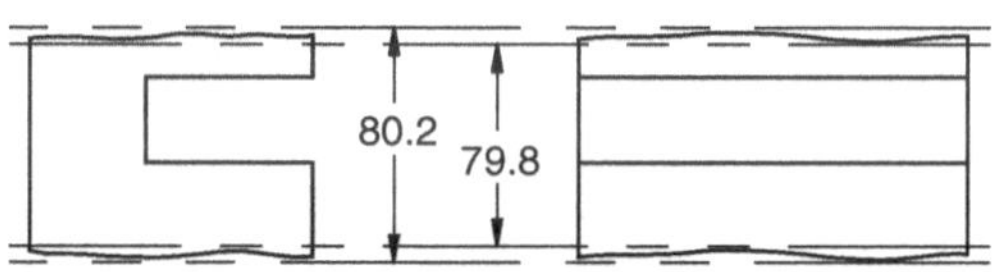

80±0.2 size tolerance zone
The surfaces must be within
the specified limits of size.

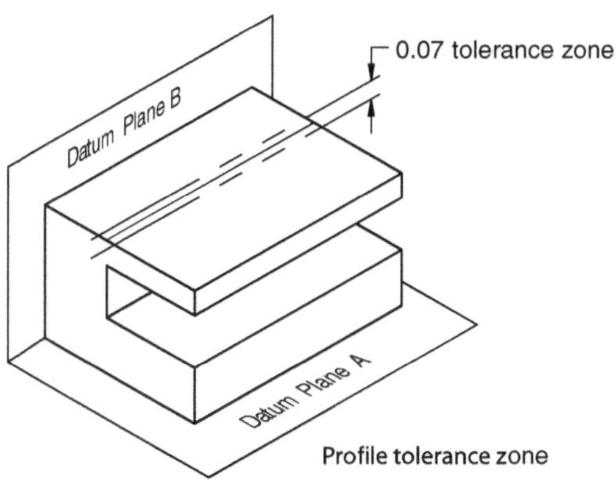

Each logitudinal element of the surface must
be between two parallel lines 0.07 apart relative
to datum planes A and B.

TOLERANCES OF PROFILE, CONTINUED

PROFILE OF A SURFACE

The tolerance zone established by the profile of a surface tolerance is three-dimensional (a volume), extending along the length and width (or circumference) of the considered feature or features. The profile tolerance can be applied as a bilateral or unilateral tolerance. In the absence of any modifiers, the profile tolerance is implied as bilateral and divided equally on both sides of the true profile. In the case of an unequally disposed profile tolerance, the modifier is placed in the feature control frame followed by the amount of profile tolerance in the direction that has the effect of adding material to the true profile.

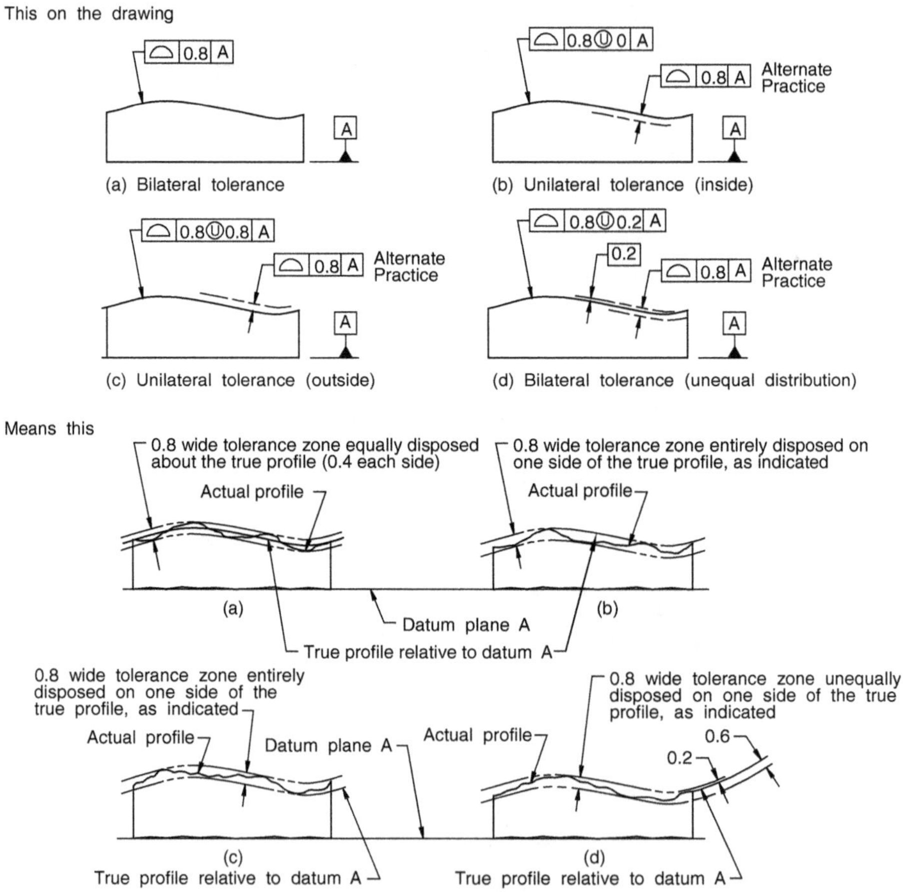

TOLERANCES OF RUNOUT

Runout is a tolerance used to control the functional relationship of one or more features to a datum axis.

CIRCULAR RUNOUT

Circular runout provides control of circular elements of a surface. The tolerance is applied independently at each circular measuring position as the part is rotated the full angular extent of the surface about the simulated datum axis.

This on the drawing

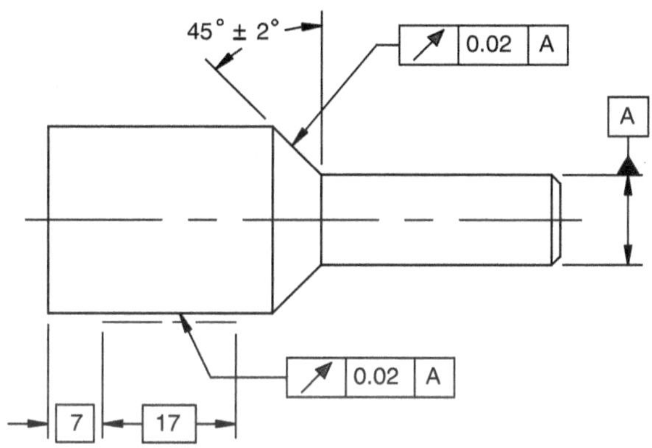

Means this

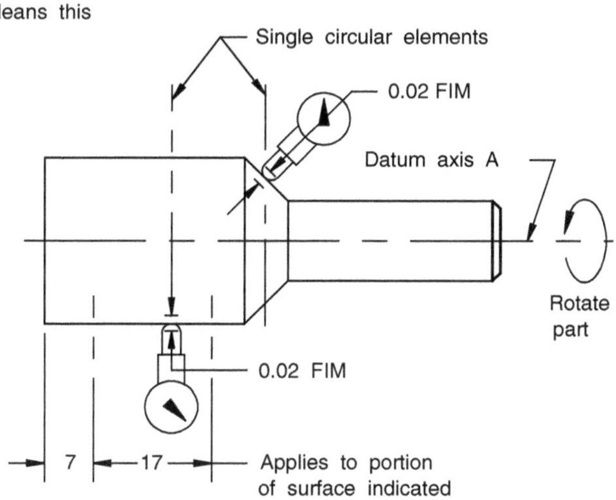

Figures reprinted from *ASME Y14.5 – 2009*, by permission of the American Society of Mechanical Engineers.

TOLERANCES OF RUNOUT, CONTINUED

TOTAL RUNOUT

Total runout provides control of all surface elements. The tolerance is applied simultaneously to all circular and profile measuring positions as the part is rotated 360 deg about the datum axis.

This on the drawing

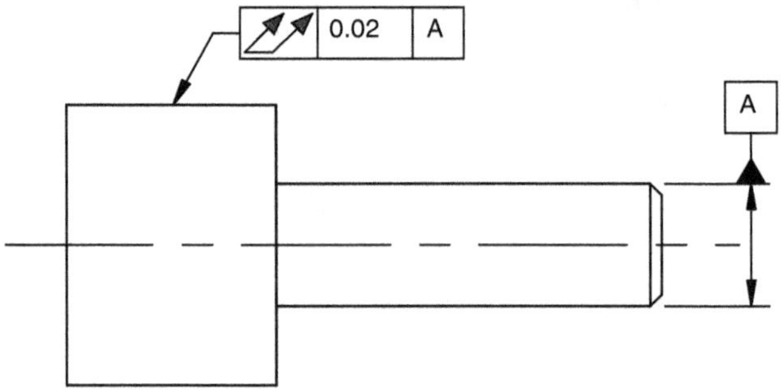

Means this

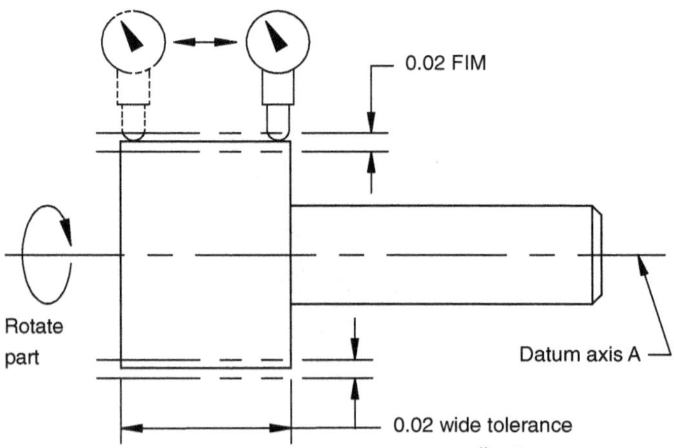

Figures reprinted from *ASME Y14.5 – 2009*, by permission of the American Society of Mechanical Engineers.

CALCULATION OF POSITIONAL TOLERANCES

CONVERTING FROM CARTESIAN COORDINATES TO TRUE POSITIONAL TOLERANCES

$$2\sqrt{X^2 + Y^2} = \oplus$$

where
X, Y = coordinate tolerance (±)
$\oplus$ = diametral true position tolerance

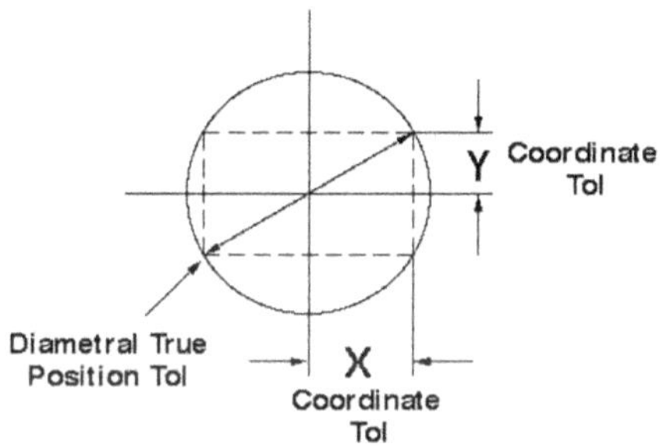

This conversion may be used only if the designer takes into account the tolerance across corners of a square zone. If not, a conversion of this type could result in a round tolerance zone much larger than is really allowed.

FLOATING FASTENER TOLERANCES

Where two or more parts are assembled with fasteners, such as bolts and nuts, and all parts have clearance holes for the bolts, it is termed the "floating fastener case." Where the fasteners are of the same diameter, and it is desired to use the same clearance hole diameters and the same positional tolerances for the parts to be assembled, the following formula applies:

$$T = H - F$$

where:

T = positional tolerance diameter
H = minimum diameter of clearance hole (MMC limit)
F = maximum diameter of fastener (MMC limit)

CALCULATION OF POSITIONAL TOLERANCES, CONTINUED

FIXED FASTENER TOLERANCES

Where one of the parts to be assembled has restrained fasteners, such as screws in tapped holes or studs, it is termed the "fixed fastener case." Where the fasteners are of the same diameter and it is desired to use the same positional tolerance in each of the parts to be assembled, the following formula applies:

$$T = \frac{H - F}{2}$$

SECTION 8

ELECTRICAL/ELECTRONIC/ ELECTROMAGNETIC DESIGN

ELECTRICAL SYMBOL DEFINITION

These symbols are generally accepted as standard to represent electrical quantities.

Symbol	Definition
C	capacitance, farads
E	electromotive force, volts
E_{eff} or E_{rms}	effective or rms voltage
E_{max}	peak voltage
f	frequency, hertz
fr	resonant frequency, hertz
G	conductance, siemens
I	current, amperes
i	instantaneous current, amperes
I_{eff} or I_{rms}	effective or rms current
I_{max}	peak current
L	inductance, henries
λ	wavelength
M	mutual inductance, henries
O	instantaneous voltage
P	power, watts
PF	power factor
Q	figure of merit (quality)
R	resistance, ohms
θ	phase angle, degrees
T	time, seconds
X	reactance, ohms
X_c	capacitive reactance, ohms
X_L	inductive reactance, ohms
VA	apparent power, volt-amperes
Z	impedance, ohms

Ohm's Law

For AC Circuits

Known Values	Formulas for Determining Unknown Values of I	Z	E	P
$I\&Z$			IZ	$I^2 Z \cos\theta$
$I\&E$		$\frac{E}{I}$		$IE\cos\theta$
$I\&P$		$\frac{P}{I^2\cos\theta}$	$\frac{P}{I\cos\theta}$	
$Z\&E$	$\frac{E}{Z}$			$\frac{E^2\cos\theta}{Z}$
$Z\&P$	$\sqrt{\frac{P}{Z\cos\theta}}$		$\sqrt{\frac{PZ}{\cos\theta}}$	
$E\&P$	$\frac{P}{E\cos\theta}$	$\frac{E^2\cos\theta}{P}$		

For DC Circuits

Known Values	Formulas for Determining Unknown Values of I	R	E	P
$I\&R$			IR	I^2R
$I\&E$		$\frac{E}{I}$		EI
$I\&P$		$\frac{P}{I^2}$	$\frac{P}{I}$	
$R\&E$	$\frac{E}{R}$			$\frac{E^2}{R}$
$R\&P$	$\sqrt{\frac{P}{R}}$		$\sqrt{PR}$	
$E\&P$	$\frac{P}{E}$	$\frac{E^2}{P}$		

WIRE CHART

CURRENT CAPACITY AND LENGTH VS VOLTAGE DROP

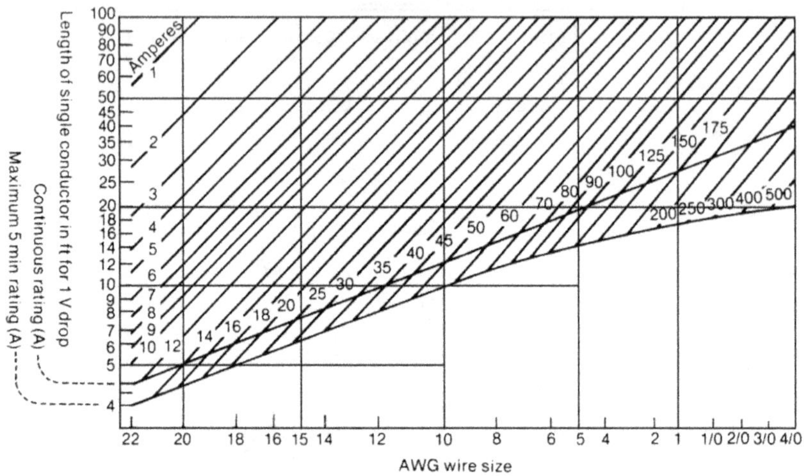

Source: *Specialties Handbook, 1964.* Copyright © 1964, Avionics Specialties, Inc., Charlottesville, VA. Reproduced with permission of Avionics Specialties.

RESISTOR COLOR CODES

COLOR CODE FOR SMALL RESISTORS—MILITARY AND EIA

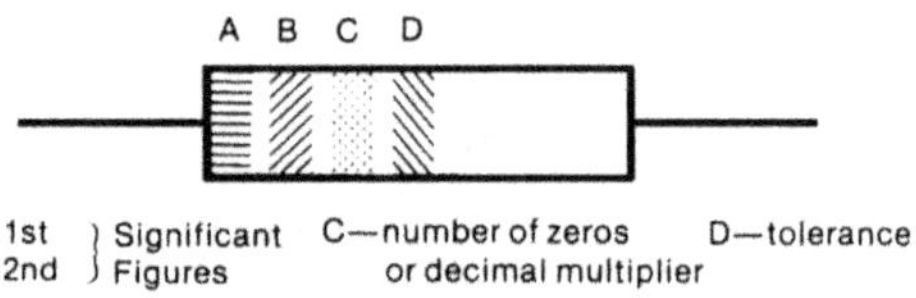

Color	Significant Figure	Multiplying Value
Black	0	1
Brown	1	10
Red	2	100
Orange	3	1000
Yellow	4	10,000
Green	5	100,000
Blue	6	1,000,000
Violet	7	10,000,000
Gray	8	100,000,000
White	9	1,000,000,000
Gold	±5% tolerance	0.1
Silver	±10% tolerance	0.01
No color	±20% tolerance	——

Resistor Color Codes, continued

Resistors Available

Decimal Multiples of						
5%				10%		20%
1.0	1.8	3.3	5.6	1.0	3.3	1.0
1.1	2.0	3.6	6.2	1.2	3.9	1.5
1.2	2.2	3.9	6.3	1.5	4.7	2.2
1.3	2.4	4.3	7.5	1.8	5.6	3.3
1.5	2.7	4.7	8.2	2.2	6.8	4.7
1.6	3.0	5.1	9.1	2.7	8.2	6.8

Properties of Insulating Materials

Insulating Material	Dielectric Constant (60 Hz)	Dielectric Strength, KV mm^{-1}	Resistivity, ohm cm
Air normal pressure	1	0.78–0.90	
Amber	2.7–2.9	90	Very high
Asphalts	2.7–3.1	1.0–1.2	——
Casein—moulded	6.4	16–28	Poor
Cellulose—acetate	6–8	10–40	4.5×10^{10}
Ceresin wax	2.5–2.6	——	——
Fibre	2.5–5	5.9–7.1	5×10^{9}
Glass—electrical	4–5	79	8×10^{14}
Hallowax	3.4–3.8	——	$10^{13} \times 10^{14}$
Magnesium silicate	5.9–6.4	7.9–9.5	$>10^{14}$
Methacrylic resin	2.8	——	——
Mica	2.5–8	40–160	2×10^{17}
Micalex 364	6–8	14	——
Nylon	3.6	12	10^{13}
Paper	2–2.6	49	——
Paraffin oil	2.2	15	——
Paraffin wax	2.25	8–12	10^{16}
Phenol—black moulded	5.5	16–20	——
Phenol—paper base	5.5	26–30	$10^{10} \times 10^{13}$
Phenol—yellow	5.3	20	——
Polyethylene	2.25	39	10^{17}
Polystyrene	2.5	20–28	10^{17}
Porcelain—dry process	6.2–7.5	1.5–3.9	5×10^{8}
Porcelain—wet process	6.5–7	5.9	——
Polyvinyl chloride	2.9–3.2	16	10^{14}
Quartz—fused	3.5–4.2	8	10^{14}, 10^{18}

(continued)

Properties of Insulating Materials, continued

Insulating Material	Dielectric Constant (60 Hz)	Dielectric Strength, KV mm^{-1}	Resistivity, ohm cm
Rubber—hard	2–3.5	18	10^{12}, 10^{15}
Shellac	2.5–4	35	10^{16}
Steatite—commercial	4.9–6.5	——	——
Steatite—low-loss	4.4	6–12	10^{14}, 10^{15}
Titanium dioxide	90–170	4–8	10^{13}, 10^{14}
Varnished cloth	2–2.5	18–22	——
Vinyl resins	4	16–20	10^{14}

Connectors

Frequently Used Connectors

Type	Governing Spec	Coupling	Max Freq.,[a] GHz	Voltage Rating[b]	Relative Cost	Overall Size
SMC	MIL-C-39012	Thread	10	500	Low	Micro
SMB	MIL-C-39012	Snap-on	4	500	Low	Micro
SMA	MIL-C-39012	Thread	12.4	500	Med	Submin
TPS	MIL-C-55235	Bayonet	10	500	Med	Submin
TNC	MIL-C-39012	Thread	11	500	Med	Min
BNC	MIL-C-39012	Bayonet	4	500	Low	Min
N	MIL-C-39012	Thread	11	1000	Med-Low	Medium
SC	MIL-C-39012	Thread	11	1500	Med	Medium
C	MIL-C-39012	Bayonet	11	1500	Med	Medium
QDS	MIL-C-18867	Snap-on	11	1500	Med	Medium
HN	MIL-C-3643	Thread	2.5	5000	Med	Medium
LT(LC)	MIL-C-26637	Thread	4	5000	High	Large
QL	MIL-C-39012	Thread	5	5000	High	Large

[a]Maximum recommended operating frequency.
[b]Volts rms at sea level, tested at 60 Hz and 5 MHz (derate by a factor of 4 at 70,000-ft altitude).

High-Precision Connectors

Size	Governing Spec	Coupling	Max Freq.,[a] GHz
2.75 mm	NIST	Thread	40.0
3.5 mm	NIST	Thread	26.5
7 mm	IEEE 287	Thread	18
14 mm	IEEE 287	Thread	8.5

[a]Maximum recommended operating frequency.

INTERFACE CONNECTOR PIN ASSIGNMENTS PER RS-232-C

Pin Number	Circuit	Description
1	AA	Protective ground
2	BA	Transmitted data
3	BB	Received data
4	CA	Request to send
5	CB	Clear to send
6	CC	Data set ready
7	AB	Signal ground (common return)
8	CF	Received line signal detector
9	——	(Reserved for data set testing)
10	——	(Reserved for data set testing)
11	——	Unassigned
12	SCF	Sec. rec'd. line sig. detector
13	SCB	Sec. clear to send
14	SBA	Secondary transmitted data
15	DB	Transmission signal element timing (DCE source)
16	SBB	Secondary received data
17	DD	Receiver signal element timing (DCE source)
18	——	Unassigned
19	SCA	Secondary request to send
20	CD	Data terminal ready
21	CG	Signal quality detector
22	CE	Ring indicator
23	CH/CI	Data signal rate selector (DTE/DCE source)
24	DA	Transmit signal element timing (DTE source)
25	——	Unassigned

RESISTOR, CAPACITOR, INDUCTANCE COMBINATIONS

PARALLEL COMBINATIONS OF RESISTORS, CAPACITORS, AND INDUCTORS

Parallel Combination	Impedance, ohms $(Z = R + jX)$	Magnitude of Impedance, ohms $\left(\lvert Z\rvert = \sqrt{R^2 + X^2}\right)$
R_1, R_2	$\frac{R_1R_2}{R_1+R_2}$	$\frac{R_1R_2}{R_1+R_2}$
C_1, C_2	$-j\frac{1}{\omega(C_1+C_2)}$	$\frac{1}{\omega(C_1+C_2)}$
L, R	$\frac{\omega^2L^2R + j\omega LR^2}{\omega^2L^2+R^2}$	$\frac{\omega LR}{\sqrt{\omega^2L^2+R^2}}$
R, C	$\frac{R - j\omega R^2C}{1+\omega^2R^2C^2}$	$\frac{R}{\sqrt{1+\omega^2R^2C^2}}$
L, C	$+j\frac{\omega L}{1-\omega^2LC}$	$\frac{\omega L}{1-\omega^2LC}$
$L_1(M)L_2$	$+j\omega\frac{L_1L_2-M^2}{L_1+L_2\mp 2M}$	$\omega\frac{L_1L_2-M^2}{L_1+L_2\mp 2M}$
L, C, R	$\frac{1}{R} - j\left(\omega C - \frac{1}{\omega L}\right) \div \left(\frac{1}{R}\right)^2 + \left(\omega C - \frac{1}{\omega L}\right)^2$	$R\Big/\sqrt{1+R^2\left(\omega C - \frac{1}{\omega L}\right)^2}$

Parallel Combination	Phase Angle, rad $\left[\phi = \tan^{-1}(X/R)\right]$	Admittance, siemens $(Y = 1/Z)$
R_1, R_2	0	$\frac{R_1+R_2}{R_1R_2}$
C_1, C_2	$-\frac{\pi}{2}$	$+j\omega\,(C_1+C_2)$
L, R	$\tan^{-1}\frac{R}{\omega L}$	$\frac{1}{R} - \frac{j}{\omega L}$
R, C	$\tan^{-1}(-\omega RC)$	$\frac{1}{R} - j\omega C$
L, C	$\pm\frac{\pi}{2}$	$j\left(\omega C - \frac{1}{\omega L}\right)$
$L_1(M)L_2$	$\pm\frac{\pi}{2}$	$-j\frac{1}{\omega}\left(\frac{L_1+L_2\mp 2M}{L_1L_2-M^2}\right)$
L, C, R	$\tan^{-1}R\left(\omega C - \frac{1}{\omega L}\right)$	$\frac{1}{R} + j\left(\omega C - \frac{1}{\omega L}\right)$

RESISTOR, CAPACITOR, INDUCTANCE COMBINATIONS, CONTINUED

SERIES COMBINATIONS OF RESISTORS, CAPACITORS, AND INDUCTORS

Series Combination	Impedance, ohms $(Z = R + jX)$	Magnitude of Impedance, ohms $\left(\lvert Z\rvert = \sqrt{R^2 + X^2}\right)$
R	R	R
L	$+j\omega L$	ωL
C	$-j(1/\omega C)$	$1/\omega C$
$R_1 + R_2$	$R_1 + R_2$	$R_1 + R_2$
$L_1(M)L_2$	$+j\omega\,(L_1 + L_2 \pm 2M)$	$\omega\,(L_1 + L_2 \pm 2M)$
$C_1 + C_2$	$-j\frac{1}{\omega}\left(\frac{C_1 + C_2}{C_1 C_2}\right)$	$\frac{1}{\omega}\left(\frac{C_1 + C_2}{C_1 C_2}\right)$
$R + L$	$R + j\omega L$	$\sqrt{R^2 + \omega^2 L^2}$
$R + C$	$R - j\frac{1}{\omega C}$	$\sqrt{\frac{\omega^2 C^2 R^2 + 1}{\omega^2 C^2}}$
$L + C$	$+j\left(\omega L - \frac{1}{\omega C}\right)$	$\left(\omega L - \frac{1}{\omega C}\right)$
$R + L + C$	$R + j\left(\omega L - \frac{1}{\omega C}\right)$	$\sqrt{R^2 + \left(\omega L - \frac{1}{\omega C}\right)^2}$

Series Combination	Phase Angle, rad $\left[\phi = \tan^{-1}(X/R)\right]$	Admittance, siemens $(Y = 1/Z)$
R	0	$1/R$
L	$+\pi/2$	$-j(1/\omega L)$
C	$-\pi/2$	$j\omega C$
$R_1 + R_2$	0	$1/(R_1 + R_2)$
$L_1(M)L_2$	$+\pi/2$	$-j/\omega(L_1 + L_2 \pm 2M)$
$C_1 + C_2$	$-\frac{\pi}{2}$	$j\omega\left(\frac{C_1 C_2}{C_1 + C_2}\right)$
$R + L$	$\tan^{-1}\frac{\omega L}{R}$	$\frac{R - j\omega L}{R^2 - \omega^2 L^2}$
$R + C$	$-\tan^{-1}\frac{1}{\omega RC}$	$\frac{\omega^2 C^2 R + j\omega C}{\omega^2 C^2 R^2 + 1}$
$L + C$	$\pm\frac{\pi}{2}$	$-\frac{j\omega C}{\omega^2 LC - 1}$
$R + L + C$	$\tan^{-1}\left(\frac{\omega L - 1/\omega C}{R}\right)$	$\frac{R - j(\omega L - 1/\omega C)}{R^2 + (\omega L - 1/\omega C)^2}$

DYNAMIC ELEMENTS AND NETWORKS

ELEMENT OR SYSTEM — *G(s)*

INTEGRATING CIRCUIT

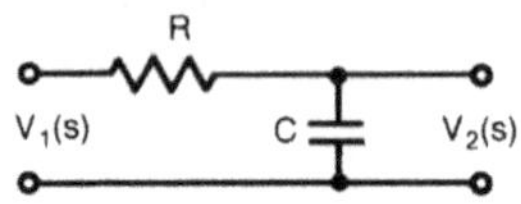

$$\frac{V_2(s)}{V_1(s)} = \frac{1}{RCs+1}$$

DIFFERENTIATING CIRCUIT

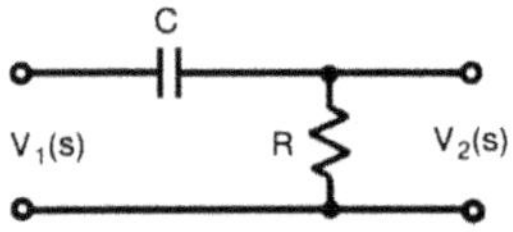

$$\frac{V_2(s)}{V_1(s)} = \frac{RCs}{RCs+1}$$

DIFFERENTIATING CIRCUIT

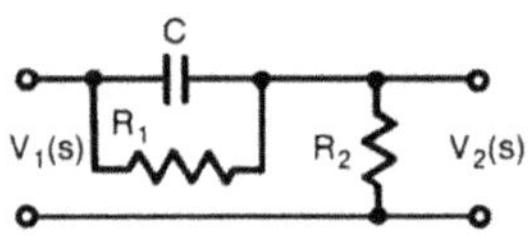

$$\frac{V_2(s)}{V_1(s)} = \frac{s+1/R_1C}{s+(R_1+R_2)/R_1R_2C}$$

LEAD-LAG FILTER CIRCUIT

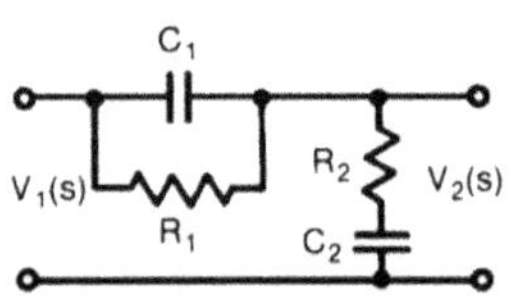

$$\frac{V_2(s)}{V_1(s)} = \frac{(1+s\tau_a)(1+s\tau_b)}{\tau_a\tau_b s^2+(\tau_a+\tau_b+\tau_{ab})s+1} = \frac{(1+s\tau_a)(1+s\tau_b)}{(1+s\tau_1)(1+s\tau_2)}$$

$$\tau_a = R_1C_1$$
$$\tau_b = R_2C_2$$
$$\tau_{ab} = R_1C_1$$
$$\tau_1\tau_2 = \tau_a\tau_b$$
$$\tau_1+\tau_2 = \tau_a+\tau_b+\tau_{ab}$$

Source: The dynamic elements and networks material appearing on pages 8-10 and 8-11 is from *Modern Control Systems*, 3rd Ed., page 46, table 2-6, figures 1–6, by R. C. Dorf. Copyright © 1980, Addison Wesley Longman, Inc., Upper Saddle River, NJ. Reprinted by permission of Pearson Education.

DYNAMIC ELEMENTS AND NETWORKS, CONTINUED

DC-MOTOR, FIELD CONTROLLED

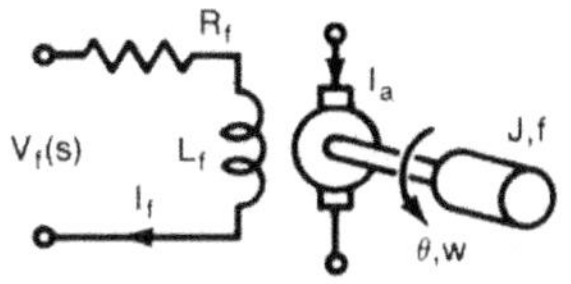

$$\frac{\theta(s)}{V_f(s)} = \frac{K_m}{s(Js+f)(L_f s+R_f)}$$

DC-MOTOR, ARMATURE CONTROLLED

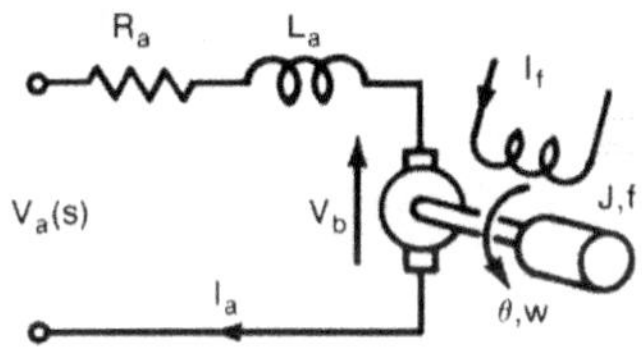

$$\frac{\theta(s)}{V_a(s)} = \frac{K_m}{s[(R_a+L_a s)(Js+f)+K_b K_m]}$$

LAPLACE TRANSFORMS

The Laplace transform of a function $f(t)$ is defined by the expression

$$F(p) = \int_0^\infty f(t)e^{-pt}\,dt$$

If this integral converges for some $p = p_0$, real or complex, then it will converge for all p such that $\mathrm{Re}(p) > \mathrm{Re}(p_0)$.

The inverse transform may be found by

$$f(t) = (j2\pi)^{-1}\int_{c-j\infty}^{c+j\infty} F(z)e^{tz}dz \quad t > 0$$

where there are no singularities to the right of the path of integration.

Re (p) denotes real part of p
Im (p) denotes imaginary part of p

Source: The Laplace transform material appearing on pages 8-11–8-13 is from *Reference Data for Engineers: Radio, Electronics, Computer, and Communications*, 8th Ed., pages 11–36.

LAPLACE TRANSFORMS, CONTINUED

GENERAL EQUATIONS

	Function	Transform[a]
Shifting theorem	$f(t-a),\quad f(t)=0,\quad t<0$	$e^{-ap}F(p),\quad a>0$
Convolution	$\int_0^t f_1(\lambda)f_2(t-\lambda)\mathrm{d}\lambda$	$F_1(p)F_2(p)$
Linearity	$a_1f_1(t)+a_2f_2(t)$, $(a_1,a_2 \text{ const})$	$a_1F_1(p)+a_2F_2(p)$
Derivative	$df(t)/\mathrm{d}t$	$-f(0)+pF(p)$
Integral	$\int f(t)\mathrm{d}t$	$p^{-1}\left[\int f(t)\mathrm{d}t\right]_{t=0}+[f(p)/p]$, $\mathrm{Re}\,p>0$
Periodic function	$f(t)=f(t+r)$	$\int_0^r f(\lambda)e^{-p\lambda}\mathrm{d}\lambda/(1-e^{-pr}),\quad r>0$
	$f(t)=-f(t+r)$	$\int_0^r f(\lambda)e^{-\lambda}\mathrm{d}\lambda/(1+e^{-pr}),\quad r>0$
	$f(at),\quad a>0$	$F(p/a)/a$
	$e^{at}f(t)$	$F(p-a),\quad \mathrm{Re}(p)>\mathrm{Re}(a)$
	$t^n f(t)$	$(-1)^n[d^nF(p)/\mathrm{d}p^n]$
Final-value theorem	$f(\infty)$	$\lim_{p\to 0} pF(p)$
Initial-value theorem	$f(0+)$	$\lim_{p\to\infty} pF(p)$

[a]$F(p)$ denotes the Laplace transform of $f(t)$.

MISCELLANEOUS FUNCTIONS

	Function	Transform
Step	$u(t-a)=0,\quad 0\le t<a$; $=1,\quad t\ge a$	e^{-ap}/p
Impulse	$\delta(t)$	1
	$t^a,\ \mathrm{Re}(a)>-1$	$\Gamma(a+1)/p^{a+1}$
	e^{at}	$1/(p-a),\quad \mathrm{Re}(p)>\mathrm{Re}(a)$
	$t^a e^{bt},\ \mathrm{Re}(a)>-1$	$\Gamma(a+1)/(p-b)^{a+1},\quad \mathrm{Re}(p)>\mathrm{Re}(b)$
	cos *at*	$p/(p^2+a^2)$ } $\mathrm{Re}(p)>\lvert\mathrm{Im}(a)\rvert$
	sin *at*	$a/(p^2+a^2)$ } $\mathrm{Re}(p)>\lvert\mathrm{Im}(a)\rvert$

(continued)

LAPLACE TRANSFORMS, CONTINUED

MISCELLANEOUS FUNCTIONS, CONTINUED

Function	Transform
cosh *at*	$p/(p^2-a^2)$ } $\mathrm{Re}(p) > \lvert\mathrm{Re}(a)\rvert$
sinh *at*	$1/(p^2-a^2)$ } $\mathrm{Re}(p) > \lvert\mathrm{Re}(a)\rvert$
$\ln t$	$-(\gamma + \ln p)/p, \gamma$ is Euler's constant = 0.57722
$1/(t+a), \quad a>0$	$e^{ap}E_1(ap)$
e^{-at2}	$\frac{1}{2}(\pi/a)^{1/2}e^{p^2/4a}\mathrm{erfc}[p/2(a)^{1/2}]$
Bessel function $J_\nu(at)$, $\mathrm{Re}(\nu) > -1$	$r^{-1}[(r-p)/a]^\nu, \quad r=(p^2+a^2)^{1/2}, \quad \mathrm{Re}(p) > \lvert\mathrm{Re}(a)\rvert$
Bessel function $I_\nu(at)$, $\mathrm{Re}(\nu) > -1$	$R^{-1}[(R-p)/a]^\nu, \quad R=(p^2-a^2)^{1/2}, \quad \mathrm{Re}(p) > \lvert\mathrm{Re}(a)\rvert$

INVERSE TRANSFORMS

Transform	Function
1	$\delta(t)$
$1/(p+a)$	e^{-at}
$1/(p+a)^\nu, \quad \mathrm{Re}(\nu)>0$	$t^{\nu-1}e^{-at}/\Gamma(\nu)$
$1/[(p+a)(p+b)]$	$(e^{-at}-e^{-bt})/(b-a)$
$p/[(p+a)(p+b)]$	$(ae^{-at}-be^{-bt})/(a-b)$
$1/(p^2+a^2)$	$a^{-1}\sin at$
$1/(p^2-a^2)$	$a^{-1}\sinh at$
$p/(p^2+a^2)$	$\cos at$
$p/(p^2-a^2)$	$\cosh at$
$1/(p^2+a^2)^{1/2}$	$J_0(at)$
e^{-ap}/p	$u(t-a)$
$e^{-ap}/p^\nu, \mathrm{Re}(\nu)>0$	$(t-a)^{\nu-1}u(t-a)/\Gamma(\nu)$
$(1/p)e^{-a/p}$	$J_0[2(at)^{1/2}]$
$(1/p^\nu)e^{-a/p}$ } $\mathrm{Re}(\nu)>0$	$(t/a)^{(\nu-1)/2}J_{\nu-1}[2(at)^{1/2}]$
$(1/p^\nu)e^{a/p}$ } $\mathrm{Re}(\nu)>0$	$(t/a)^{(\nu-1)/2}I_{\nu-1}[2(at)^{1/2}]$
$(1/p)\ln p$	$-\gamma-\ln t, \quad \gamma=0.57722$

ELECTROMAGNETIC SYMBOL DEFINITION

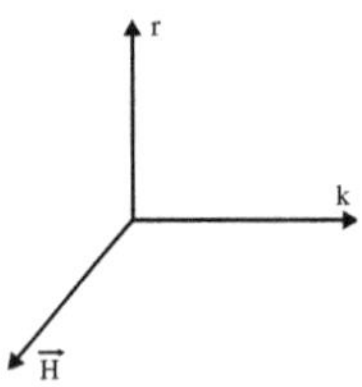

$\vec{E}$	Electric field (V m^{-1})
$\vec{H}$	Magnetic field (A m^{-1})
$f \cdot \lambda = c$	Frequency (f) and wavelength (λ) related by velocity of propagation (c)
k	Direction of propagation
$\lvert k \rvert = k = \dfrac{2\pi}{\lambda}$	Propagation constant (m^{-1})
$Z = \eta_0 = \dfrac{E}{H} = 377$ ohms	Impedance of free space
$\vec{S} = \vec{E} \times \vec{H}$	Energy flow (W m^{-2})
$W = \dfrac{1}{2}\mu \cdot H^2 + \dfrac{1}{2}\varepsilon \cdot E^2$	Energy density
μ	Complex permeability
ε	Complex permittivity

ELECTROMAGNETIC SPECTRUM (WAVE LENGTH VS FREQUENCY)

Electromagnetic Wave Spectrum Wavelength vs. Frequency

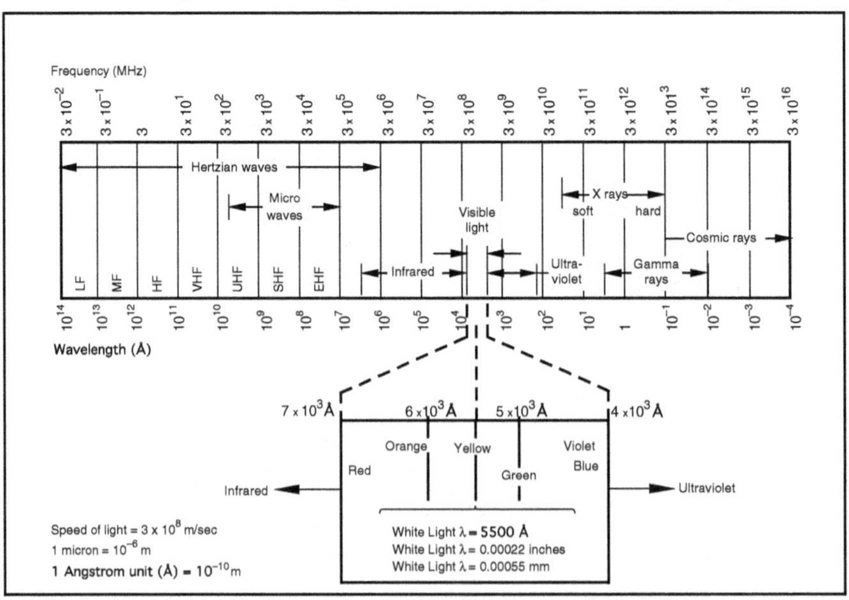

$$\text{db} = 10 \log_{10} \frac{P_1}{P_2} \qquad \text{db} = 20 \log_{10} \frac{V_1}{V_2} = 20 \log_{10} \frac{I_1}{I_2}$$

VOICE TRANSMISSION FACILITIES

The following table shows the number of voice circuits, present and future.

Transmission Method	Maximum Frequency Used	Voice Circuits per System	Maximum Voice Circuits per Route
Carrier on paired cable	260 kHz	12–24	1000–2000
Digital on paired cable	1.5 MHz	24	4800
Carrier on coaxial cable	2.7 MHz	600	1800
	8.3 MHz	1860	5580
	18.0 MHz	3600	32,900
Carrier on microwave radio	4.2 GHz	600	3000
	6.4 GHz	1800	14,000
	11.7 GHz	100	400
Satellite	30 GHz	1500	20,000
Millimeter wave guide	100 GHz	5000	250,000
Optical-guide laser	1,000,000 GHz	?	10,000,000

RADAR[a] AND ELECTRONICS COUNTERMEASURES BANDS

Standard Radar Bands		Electronics Countermeasures Bands	
Band Designation	Frequency Range	Band Designation	Frequency Range
HF	3–30 MHz	A	0–250 MHz
VHF	30–300 MHz	B	250–500 MHz
		C	500–1000 MHz
UHF	300–1000 MHz	D	1–2 GHz
		E	2–3 GHz
L	1–2 GHz	F	3–4 GHz
S	2–4 GHz	G	4–6 GHz
		H	6–8 GHz
C	4–8 GHz	I	8–10 GHz
X	8–12 GHz	J	10–20 GHz
Ku	12–18 GHz	K	20–40 GHz
		L	40–60 GHz
K	18–27 GHz	M	60–100 GHz
Ka	27–40 GHz		
Millimeter	40–300 GHz		

[a]RAdio Detection And Ranging.

Source: "The Changing Criteria for Engineering Designs," by P.S. Meyers, SAE-690303.

ANTENNA

Antenna is defined as that part of a transmitting or receiving system that is designed to radiate or to receive electromagnetic waves.

ANTENNA TYPES

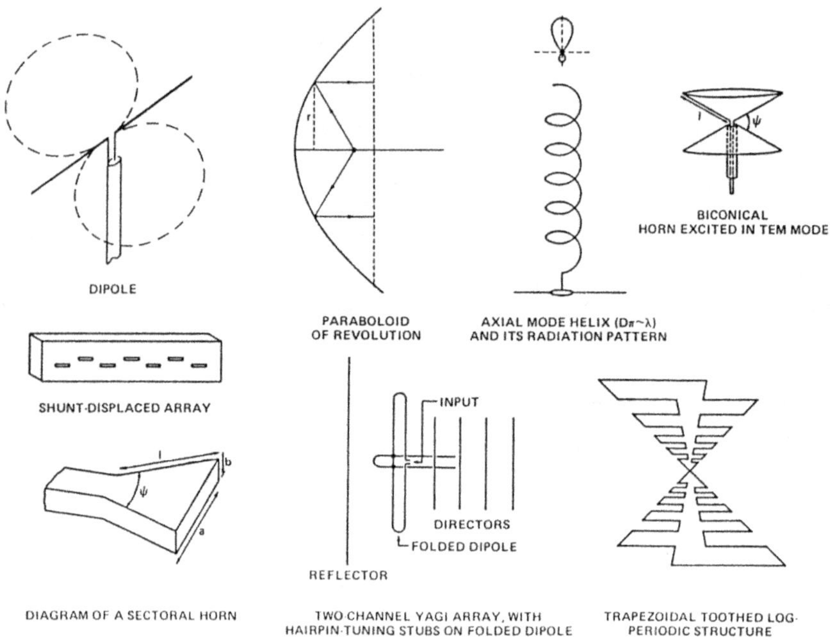

DIPOLE

PARABOLOID OF REVOLUTION

AXIAL MODE HELIX ($D\pi \sim \lambda$) AND ITS RADIATION PATTERN

BICONICAL HORN EXCITED IN TEM MODE

SHUNT-DISPLACED ARRAY

DIAGRAM OF A SECTORAL HORN

TWO-CHANNEL YAGI ARRAY, WITH HAIRPIN-TUNING STUBS ON FOLDED DIPOLE

TRAPEZOIDAL TOOTHED LOG-PERIODIC STRUCTURE

Note: Definitions for terms used in the antenna section can be found in "The IEEE Standard Definitions of Terms for Antennas" (IEEE Std. 145-1983), *IEEE Transaction on Antenna and Propagation.* Vols. AP-17, No. 3, May 1969; AP-22, No. 1, Jan. 1974; and AP-31, No. 6, Part II, Nov. 1983.

ANTENNA, CONTINUED

ANTENNA RADIATION PATTERN

An antenna radiation pattern consists of lobes, which are classified as major lobe (main beam), side lobes, and back lobes, as illustrated in the figure.

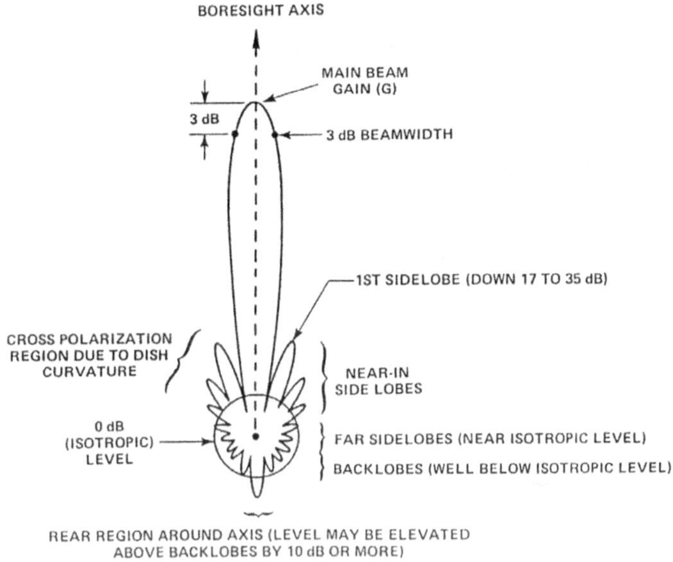

Source: *Space Vehicle Design*, page 439, by M. D. Griffin and J. R. French. Copyright © 1991, AIAA, Washington, DC. All rights reserved. Reprinted with permission of AIAA.

FIELD REGIONS

An antenna radiates electromagnetic fields in the surrounding space, which is subdivided into three regions: 1) reactive near-field, 2) radiating near-field (Fresnel), and 3) far-field (Fraunhofer), as shown in the following figure. The far-field region is defined as that region of the field of an antenna where the angular field distribution is essentially independent of the distance from a specified point in the antenna region.

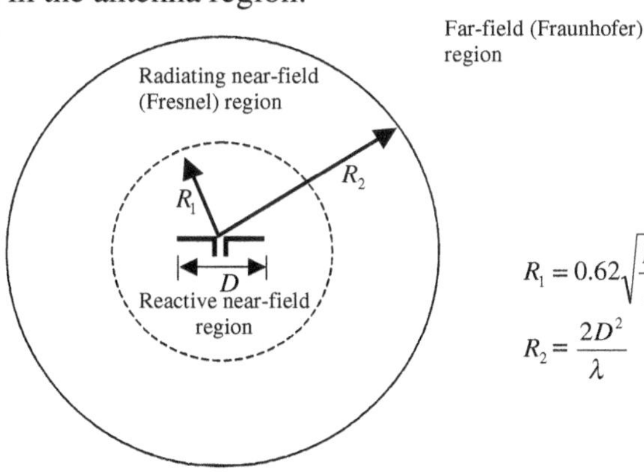

$$R_1 = 0.62\sqrt{\frac{D^3}{\lambda}}$$

$$R_2 = \frac{2D^2}{\lambda}$$

ANTENNA, CONTINUED

DIRECTIVITY

Directivity of an antenna is defined as the ratio of the radiation intensity (power radiated from an antenna per unit solid angle) in a given direction from the antenna to the radiation intensity averaged over all directions:

$$D = \frac{4\pi U}{P_{rad}}$$

For antennas with one narrow major lobe and very negligible minor lobes, the maximum directivity D_0 can be approximated by

$$D_0 = \frac{41{,}253}{\Theta_{hb1} \cdot \Theta_{hb2}} \quad \text{(dimensionless)}$$

For planar array, a better approximation is

$$D_0 = \frac{32{,}400}{\Theta_{hb1} \cdot \Theta_{hb2}} \quad \text{(dimensionless)}$$

where
Θ_{hb1}= half power beamwidth in one plane (degrees)
Θ_{hb2}= half-power beamwidth in a plane at a right angle to the other (degrees)

GAIN

The absolute gain of an antenna, in a given direction, is defined as the ratio of the radiation intensity, in a given direction, to the radiation intensity that would be obtained if the power accepted by the antenna was radiated isotropically. Note:

1) Gain does not include losses arising from impedance and polarization mismatches.
2) The radiation intensity corresponding to the isotropically radiated power is equal to the power accepted by the antenna divided by 4π.
3) If an antenna is without dissipative loss then, in any given direction, its gain is equal to its directivity.
4) If the direction is not specified, the direction of maximum radiation intensity is implied.
5) The term absolute gain is used in those instances where added emphasis is required to distinguish gain from relative gain; for example, absolute gain measurements.

EFFECTIVE AREA

The effective area (partial) of an antenna, for a given polarization and direction, is defined as follows: In a given direction, the ratio of the available power at the terminals of a receiving antenna to the power flux density of a plane wave incident on the antenna from that direction and with a specified polarization differing from the receiving polarization of the antenna.

ANTENNA, CONTINUED

PATTERNS, GAINS, AND AREAS OF TYPICAL ANTENNAS

Type	Configuration	Pattern	Power Gain Over Isotropic	Effective Area
Electric doublet		$\cos\theta$	1.5	$1.5\lambda^2/4\pi$
Magnetic doublet or loop		$\sin\theta$	1.5	$1.5\lambda^2/4\pi$
Half-wave dipole	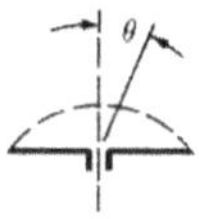	$\dfrac{\cos(\pi/2\sin\theta)}{\cos\theta}$	1.64	$1.64\lambda^2/4\pi$
Half-wave dipole and screen	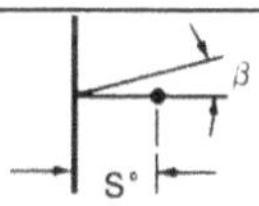	$2\sin(S°\cos\beta)$	6.5	$1.64\lambda^2/4\pi$
Turnstile array	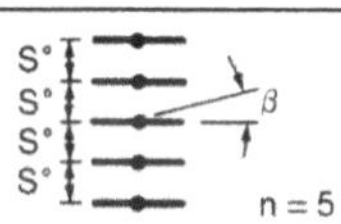	$\dfrac{\sin(nS°/2\sin\beta)}{n\sin(S°/2\sin\beta)}$	n or $2L/\lambda$	$n\lambda^2/4\pi$ or $L\lambda/2\pi$
Loop array	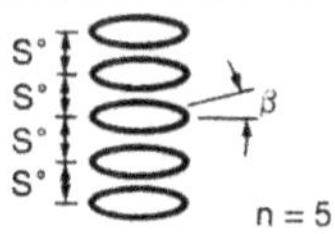	$\dfrac{\cos\beta\sin(nS°/2\sin\beta)}{n\sin(S°/2\sin\beta)}$	n or $2L/\lambda$	$n\lambda^2/4\pi$ or $L\lambda/2\pi$
Optimum horn $L \geq a^2/\lambda$	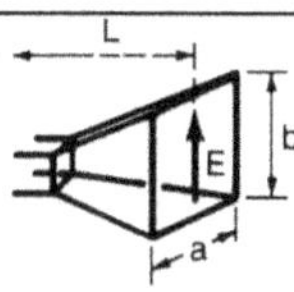	Half-power width 70 λ/a deg (H plane) 51 λ/b deg (E plane)	$10\,ab/\lambda^2$	$0.81\,ab$
Parabola		Half-power width 70 λ/d deg	$2\pi d^2/\lambda^2$	$d^2/2$

RADAR CROSS SECTION

The term "stealth" pertaining to aircraft has been associated with invisibility to radar and with radar cross section (RCS). In fact, radar is only one of several sensors that are considered in the design of low-observable aircraft platforms.

RADAR CROSS SECTION, CONTINUED

Others include infrared (IR), optical (visible), and acoustic (sound) sensors. It is also important that a low-observable target have low emissions such as low electromagnetic and thermal radiation. Stealthy targets are not completely invisible to radar. To be undetectable, it is only necessary that a target's RCS be low enough for its echo return to be below the detection threshold of the radar. RCS reduction has evolved as a countermeasure against radars and, conversely, more sensitive radars have evolved to detect lower RCS targets.

RADAR EQUATION

The radar equation describes the performance of a radar system for a given set of operational, environmental, and target parameters. In the most general case, the radar transmitter and receiver can be at different locations when viewed from the target, as shown in the following figure. This is referred to as bistatic radar.

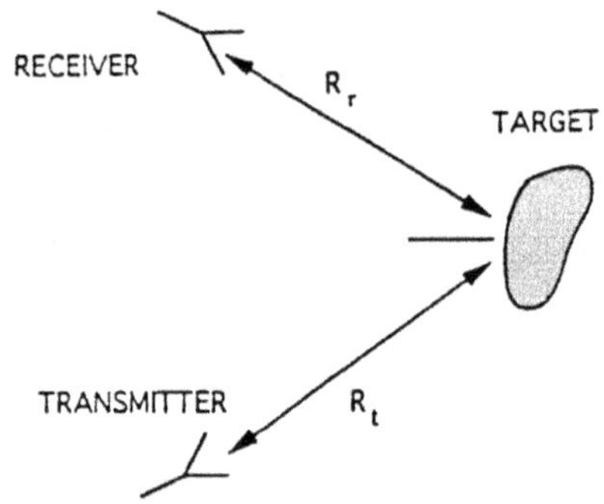

BISTATIC RADAR

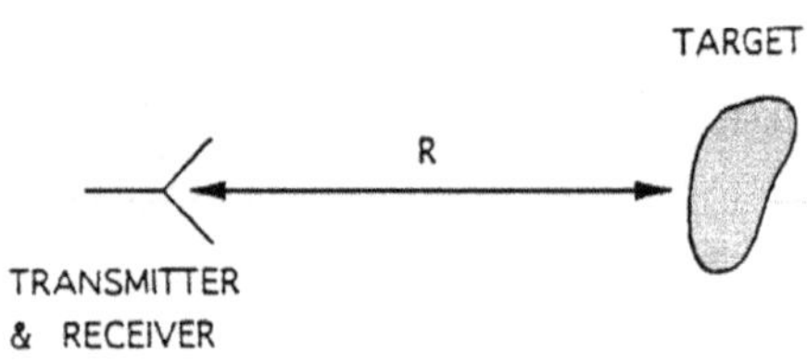

MONOSTATIC RADAR

Bistatic cross section is defined as the scattering cross section in any specified direction other than back toward the source. In most applications, the transmitter and receiver are located on the same platform and frequently share the same antenna. In this case the radar is monostatic. Monostatic cross section or back scattering cross section is defined as the scattering cross section in the direction toward the source.

RADAR CROSS SECTION, CONTINUED

RADAR EQUATION, CONTINUED

The radar equation is usually written as

$$P_R = \frac{P_T \cdot D_T \cdot D_R \cdot \sigma \cdot \lambda^2}{(4\pi)^3 \cdot R^4}$$

or

$$R_{max} = \left[\frac{P_T \cdot D_T \cdot D_R \cdot \lambda^2 \cdot \sigma}{(4\pi)^3 \cdot P_{min}} \right]^{\frac{1}{4}}$$

where

R = distance between monostatic radar and target
P_T = radar transmitting power
P_R = radar receive power
D_R = transmitting antenna directivity
D_T = receiving antenna directivity
σ = target "cross section" or radar cross section
λ = wavelength

For a given scattering object, upon which a plane wave is incident, that portion of the scattering cross section corresponding to a specified polarization component of the scattered wave is considered the RCS.

SCATTERING CROSS SECTION

The RCS for a scattering object, upon which a plane wave is incident, is considered to be that portion of the scattering cross section corresponding to a specified polarization component of the scattering wave.

The RCS of a target can be simply stated as the projected area of an equivalent isotropic reflector that returns the same power per unit solid angle as the target returns; it is expressed by

$$\frac{\sigma \cdot P_i}{4\pi} = P_S \cdot R^2$$

where

P_i = power density incident on target
P_S = power density scattered by target
$\frac{\sigma \cdot P_i}{4\pi}$ = power scattered in 4π steradians solid angle
$P_S \cdot R^2$ = power per unit solid angle reflected to the receiver

Because RCS is a far-field quantity, RCS can then be expressed as

$$\sigma = \lim_{R \to \infty} \left(4\pi R^2 \cdot \frac{P_S}{P_i} \right)$$

Power density can be expressed in terms of electric or magnetic fields, and RCS can be expressed as

RADAR CROSS SECTION, CONTINUED

SCATTERING CROSS SECTION, CONTINUED

$$\sigma = \lim_{R \to \infty} 4\pi \cdot R^2 \cdot \left| \frac{E_S}{E_i} \right|^2 \quad \text{or} \quad \sigma = \lim_{R \to \infty} 4\pi \cdot R^2 \cdot \left| \frac{H_S}{H_i} \right|^2$$

Because E_i, H_i are fixed and E_S, H_S vary as $1/R$ in the far field, σ has a limit as $R \to \infty$. The unit for RCS σ is area, typically 1 m^2. Because RCS can have many orders of magnitude variation, it is usually measured on a logarithmic (decibel) scale that uses 1 m^2 as reference. Thus the name of the unit: dBsm or dBm^2.

$$\sigma_{\text{dBsm}} = 10 \cdot \log_{10}\left(\frac{\sigma}{1}\right)$$

As indicated in the figure below, typical values of RCS range from 40 dBsm (10,000 m^2) for ships and large bombers to –30 dBsm (0.001 m^2) for insects. Modern radars are capable of detecting flocks of birds and even swarms of insects. The radar's computer will examine all detections and can discard these targets on the basis of their velocity and trajectory. However, the fact that such low RCS targets are detectable can be explained by the radar equation, which states that detection range varies as the fourth root of σ.

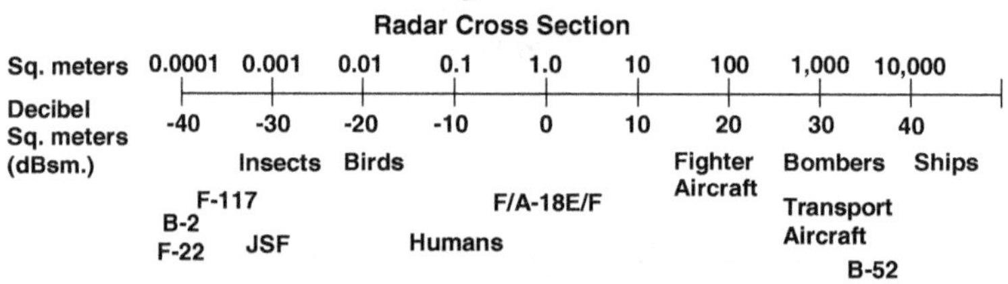

Typical values of RCS. (Source: *Aviation Week and Space Technology*, Feb. 5, 2001, and AIAA, Reston, VA. Reproduced with permission.)

RADAR FREQUENCY BANDS AND GENERAL USAGES

Band Designation	Frequency Range	General Usage
VHF	50–300 MHz	Very long range surveillance
UHF	300–1000 MHz	Very long range surveillance
L	1–2 GHz	Long range surveillance, enroute traffic control
S	2–4 GHz	Moderate range surveillance, terminal traffic control, long range weather
C	4–8 GHz	Long range tracking, airborne weather detection
X	8–12 GHz	Short range tracking, missile guidance, mapping, marine radar, airborne intercept
Ku	12–18 GHz	High resolution mapping, satellite altimetry
K	18–27 GHz	Little used (water vapor absortion)
Ka	27–40 GHz	Very high resolution mapping, airport surveillance
Millimeter	40–100 + GHz	Experimental

RADAR CROSS SECTION, CONTINUED

RADAR FREQUENCY BANDS AND GENERAL USAGES, CONTINUED

A few long-range ballistic missile defense radars operate in the 300 MHz (UHF) band, but most others use frequencies greater than 1 GHz (L band and above), as shown in the table. Low-frequency radars are capable of handling more power because the applied voltages can be higher without causing electrical breakdown. Finally, ambient noise is lowest in the 1–10 GHz band, and low-altitude atmospheric attenuation favors frequencies below 18 GHz.

SCATTERING MECHANISMS

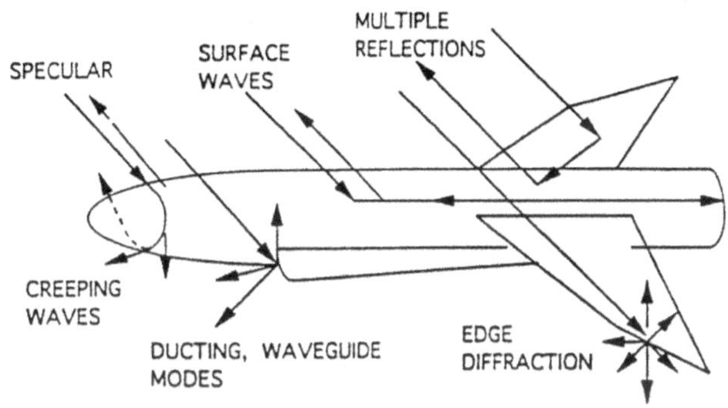

Most scattering levels relative to the peak RCS may be quite small but, away from the peaks, scattering due to these mechanisms can dominate. The scattering mechanisms are described in the following sections.

REFLECTION

This mechanism yields the highest RCS peaks, but these peaks are limited in number because Snell's law must be satisfied. A surface location where the angle of incidence is equal to the angle of reflection is called a specular point. When the specular points are those surface locations where the local normal point back to the illuminating radar, it is referred to as backscatter. Multiple reflections can occur when multiple surfaces are present. For instance, the incident plane wave could possibly reflect off the fuselage, hit a fin, and then return to the radar.

DIFFRACTION

Diffracted waves are those scattered from discontinuities such as edges and tips. The waves diffracted from these shapes are less intense than reflected waves, but they can emerge over a wide range of angles. However, a diffraction phenomenon called edge scattering may result in a large RCS in the specular direction. Edge specular points are where edge normals point back toward the illuminating radar. A leading edge spike occurs when an incident wave, having its electric field parallel to the leading edge, is diffracted from the edge. Trailing

SCATTERING MECHANISMS, CONTINUED

DIFFRACTION, CONTINUED

edge spike is the specular diffraction from an incident wave having its electric field perpendicular to the trailing edge.

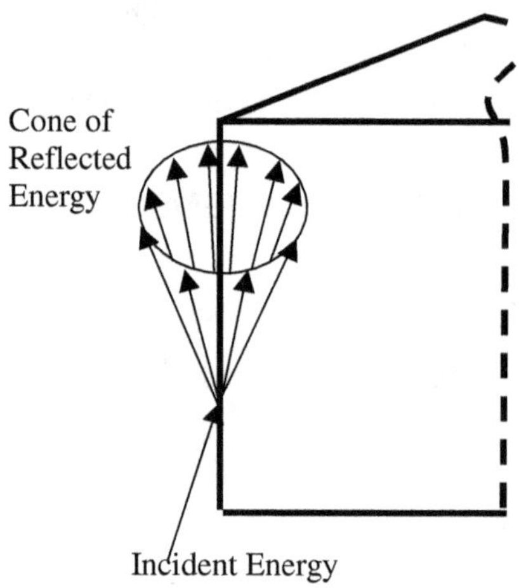

1. Angle of Incidence = Angle of Reflection
2. Bistatic RCS diffraction pattern is a cone.
3. For Backscatter RCS, the cone shaped pattern becomes a disk perpendicular to edge.

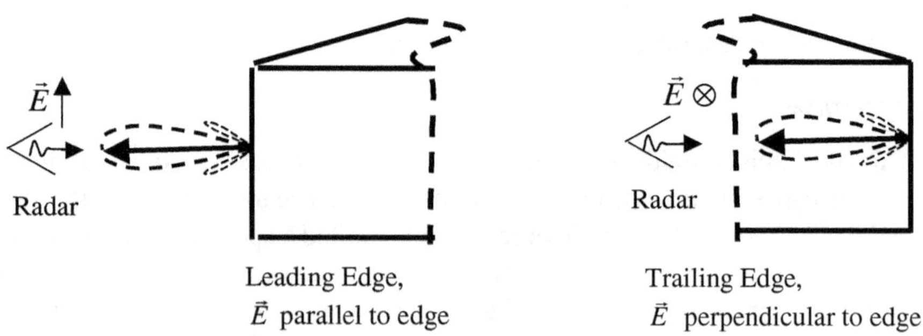

Leading and trailing edge spike.

SURFACE WAVES

The term "surface wave" refers to a wave traveling along the surface of a body and includes several types of waves. In general, the target acts as a transmission line guiding the wave along its surface. On curved bodies, the surface wave will continuously radiate in the direction tangential to the surface. If the surface is a smooth closed shape such as a sphere or cylinder (when viewed nearly broadside), the wave will circulate around the body many times. These are called creeping

SCATTERING MECHANISMS, CONTINUED

SURFACE WAVES, CONTINUED

waves because they appear to creep around the back of a curved body. Radiating surface waves on flat bodies are usually called leaky waves. Traveling waves appear on slender bodies and along edges and suffer little attenuation as they propagate. If the surface is terminated with a discontinuity such as an edge, the traveling wave will be reflected back toward its origin. Traveling wave RCS lobes can achieve surprisingly large levels.

A metallic sphere of radius *a* is commonly used as a calibration target for RCS measurement; its RCS vs circumference in wavelengths *ka* is illustrated below. This figure also illustrates three frequency regimes that are applicable to general targets. For small *ka* ($\lambda >> a$) the RCS of the sphere is $\sigma/\pi a^2 = 9 \cdot (ka)^4$. This is also known as the Rayleigh scattering law, which is applicable when λ is large in comparison with the target characteristic dimension. The oscillatory nature of the resonance (Mie) region results from the interference between creeping waves and specular reflected waves. At very large *ka* ($\lambda << a$), the scattering behavior is similar to optical scattering; i.e., $\sigma/\pi a^2 = 1$. In this regime, simple formulas can be derived for several canonical shapes assuming that the phase of the wavefront is constant across the impinged target area. Then term "constant phase region" refers to this approximation for RCS at very high frequency.

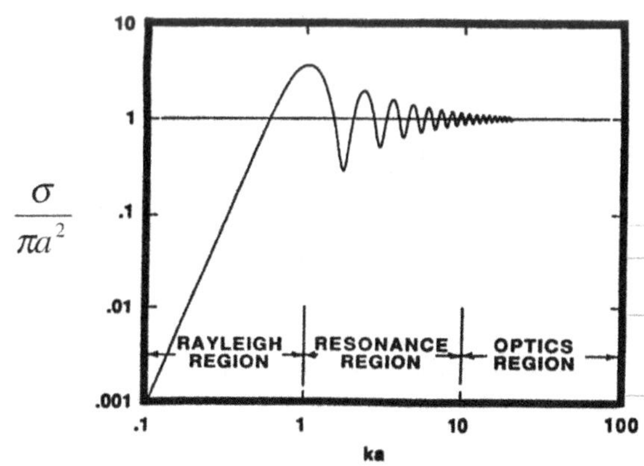

RCS of a metallic sphere with radius of *a*. (Source: *Radar Cross Section,* page 53, figure 3-5, by Knott, Schaeffer, and Tuley. Copyright © 1985, Artech House Publishers, Norwood, MA. Reproduced with permission of Artech House, www.artechhouse.com.)

DUCTING (WAVEGUIDE MODES)

Ducting occurs when a wave is trapped in a partially closed structure. An example is an air inlet cavity on a jet propelled vehicle. At high frequencies, where the wavelengths are small relative to the duct dimensions, the wave enters the cavity

SCATTERING MECHANISMS, CONTINUED

DUCTING (WAVEGUIDE MODES), CONTINUED

and many bounces can occur before a ray hits the engine face and bounces back toward the radar, as illustrated in the following figure. The ray can take many paths, and therefore, rays will emerge at most all angles. The result is a large, broad RCS lobe. In this situation, radar absorbing material (RAM) is effective in reducing the backscatter. At the resonant frequencies of the duct, waveguide modes can be excited by the incoming wave. At lower frequencies, where the wavelength is relatively larger than the duct dimensions, no energy can couple into the duct and most of the scattering occurs in the lip region of the duct.

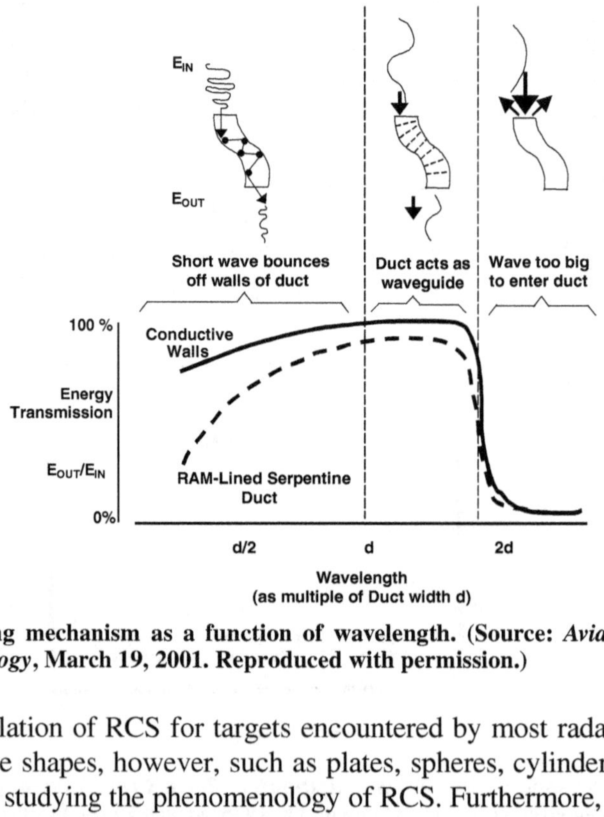

Duct scattering mechanism as a function of wavelength. (Source: *Aviation Week and Space Technology*, March 19, 2001. Reproduced with permission.)

The calculation of RCS for targets encountered by most radars is complicated. Simple shapes, however, such as plates, spheres, cylinders, and wires, are useful in studying the phenomenology of RCS. Furthermore, complex targets can be decomposed into primitives (basic geometrical shapes that can be assembled to form a more complex shape). As shown in the following figure, an aircraft can be decomposed into cylinders, plates, cones, and hemispheres. A collection of basic shapes will give a rough order of magnitude RCS estimate that can be used during the initial design stages of a platform. The locations and levels of the largest RCS lobes (spikes) are of most concern at this stage of the design process. The accuracy of the RCS calculation at other angles will depend on how the interactions between the various shapes are handled.

SCATTERING MECHANISMS, CONTINUED

DUCTING (WAVEGUIDE MODES), CONTINUED

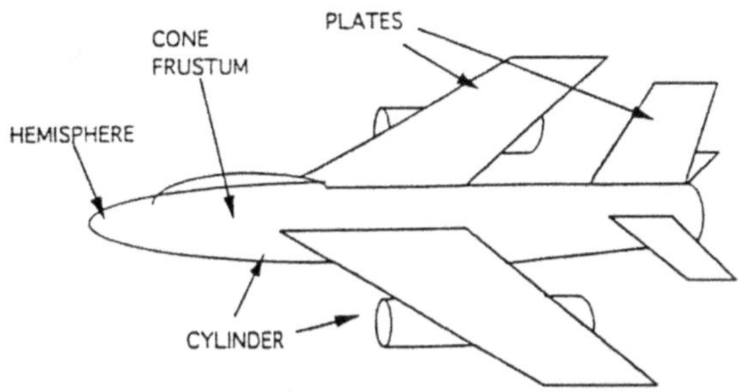

Aircraft represented by geometric components.

RADAR CROSS SECTION ESTIMATION FORMULAS

TWO-DIMENSIONAL TO THREE-DIMENSIONAL RCS CONVERSION

In the optics regime, where $l \gg \lambda$, RCS of a three-dimensional (3-D) object with constant cross section and length of l can be approximated by using the two-dimensional (2-D) RCS solution and then converting from the 2-D RCS to 3-D RCS by the following formula:

$$\sigma_{3D} = \left(\frac{2 \cdot l^2}{\lambda}\right) . \sigma_{2D}$$

where

σ_{3D} = RCS of 3-D target
σ_{2D} = RCS of the 2-D cross section
l = length of object ($l \gg \lambda$)
λ = wavelength

SPHERE (OPTICS REGIME)

RADIUS = A $\boldsymbol{\sigma = \pi \cdot a^2}$

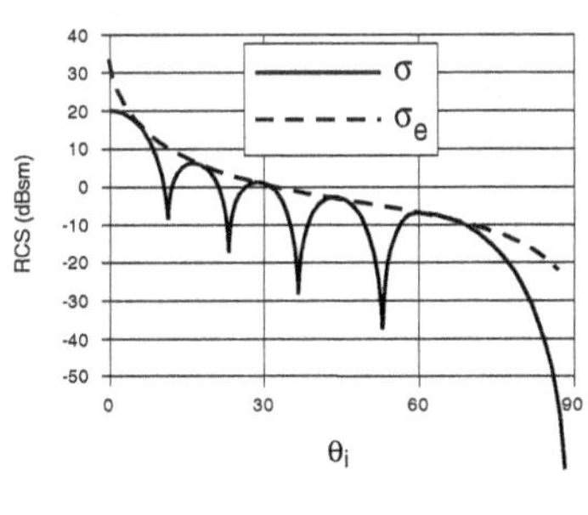

$$\sigma = \pi \cdot a^2$$

$$\sigma = k \cdot a \cdot l^2 \left(\cos\theta_i \frac{\sin(k \cdot l \cdot \sin\theta_i)}{k \cdot l \cdot \sin\theta_i} \right)^2$$

$$\sigma_e = \frac{a}{k \cdot \tan^2 \theta_i}$$

At broadside or constant phase region:

$$\sigma = k \cdot a \cdot l^2$$

RADAR CROSS SECTION ESTIMATION FORMULAS, CONTINUED

SPHERE (OPTICS REGIME), CONTINUED

WEDGE

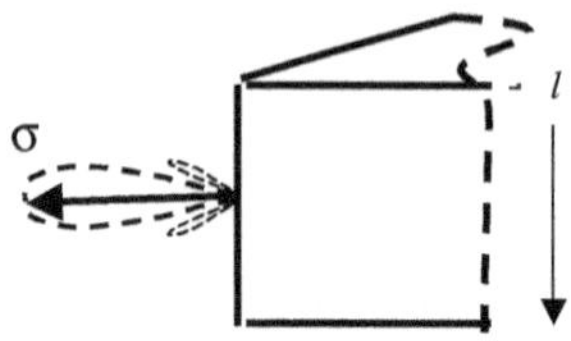

For a wedge of a length l, the edge-on broadside RCS is

$$\sigma = \frac{l^2}{\pi}$$

TRAVELING WAVE SCATTERING

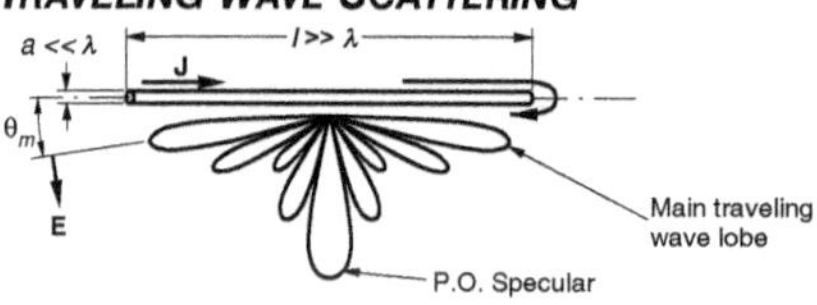

Traveling wave on long thin wire:

1) Traveling wave guided by the wire
2) Wave reflected at aft end
3) Surface attenuation and aft end termination reduces the traveling wave lobes
4) $\theta_m \approx 49.3 \cdot \sqrt{\lambda / l}$

RADAR CROSS SECTION DESIGN GUIDELINES

A benefit of RCS reduction includes the ability to carry out covert missions. Stealth greatly enhances system survivability and thus increases the probability of successful missions. Reduced radar detection range also allows the pilot to take less evasive maneuvers to avoid being detected by the radar. Because the target signature is small, the electronic countermeasure effectiveness will be increased.

However, for any system or platform, RCS reduction is often achieved at the expense of other performance measures. In the case of aircraft, increased stealth has resulted in decreased aerodynamic performance and increased complexity. Complextiy, of course, translates into increased cost. The trend in the design of new platforms has been to integrate all of the engineering disciplines into a common process. Thermal, mechanical, aerodynamic, and RCS analyses are carried out in parallel, using a common database. Changes made to the structure or materials for the purpose of decreasing RCS are automatically recorded in the database for use in all other design analyses. This approach is called concurrent engineering.

Reducing the most intense sources of scattering is the easiest, and the payoff in decibels is greatest. For example, just tilting a surface or realigning an edge only a few degrees can reduce the RCS presented to a monostatic radar by 20 or 30 dB. Further reduction is much more difficult because second-order scattering mechanisms (multiple reflections, diffractions, surface waves, etc.) become important. Therefore, it can be more costly to drop the RCS the next 5 dB than it was for the first 30 dB. With this point in mind, it is evident that the guidelines for designing a low-RCS vehicle will not be as extensive as those for an ultralow-RCS vehicle. The guidelines for both vehicles have several basic points in common, however, and these are summarized as follows:

RADAR CROSS SECTION DESIGN GUIDELINES, CONTINUED

1) Design for specific threats when possible to minimize cost. Keep in mind the threat radar frequency, whether it is monostatic or bistatic, and the target aspect angle that will be presented to the radar.
2) Orient large flat surfaces and align edges away from high priority quiet zones.
3) Use radar absorbing materials (RAM) or coatings to reduce specular/traveling wave reflections.
4) Maintain tight tolerances on large surfaces and materials.
5) Treat trailing edges to avoid traveling wave and edge lobes.
6) Avoid corner reflectors (dihedrals or trihedrals).
7) Do not expose cavity inlets, especially the engine front face, to the incoming radar. Use a mesh cover or locate the inlets/engine front face out of view of the radar.
8) Shield high-gain antennas from out-of-band threats.
9) Avoid discontinuities in geometry and material to minimize diffraction and traveling wave radiation.

As an example of the application of these points, consider the following figures. The first shows a sketch of a typical fighter aircraft with the scattering sources labeled. Note that size of RCS depends on aspect angle and the relative magnitude stated in this figure is for maximum RCS from the aircraft. The second shows a more stealthy design and describes its RCS reduction measures that follow many of the guidelines listed above.

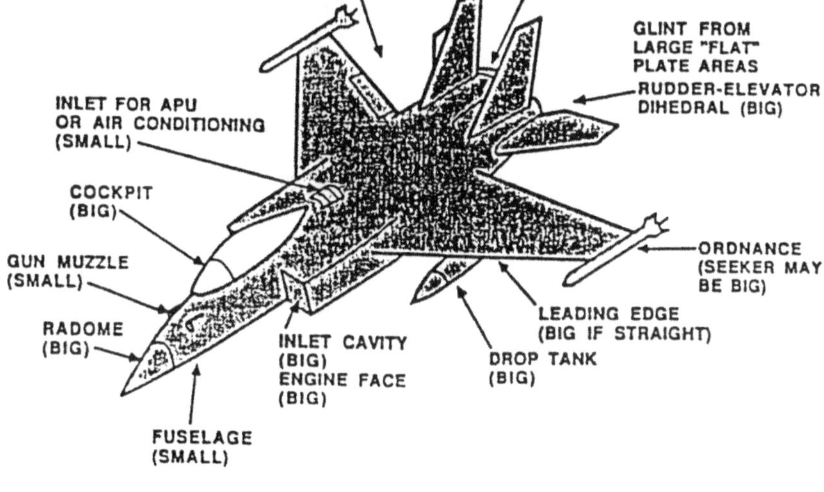

Standard jet fighter aircraft design with scattering sources indicated. (Source: *Radar Cross Section Lectures,* by A. E. Fuhs. Copyright © 1984, AIAA, New York. All rights reserved. Reprinted with permission of AIAA.)

RADAR CROSS SECTION DESIGN GUIDELINES, CONTINUED

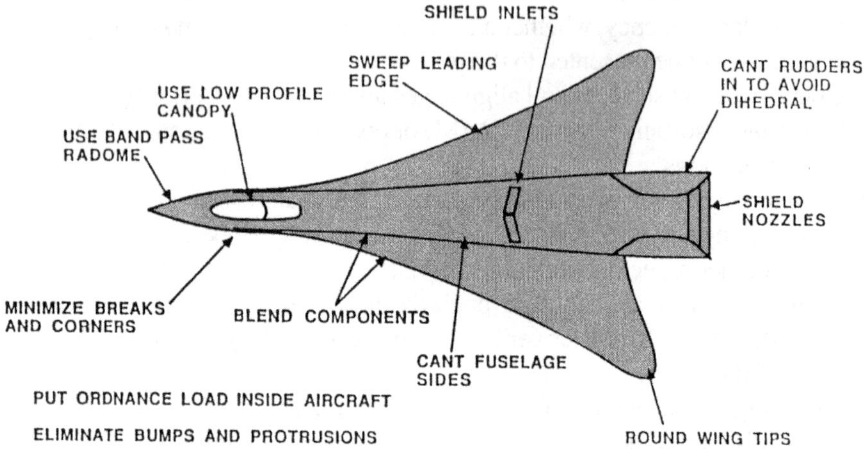

Stealthy jet fighter aircraft design with RCS reduction guidelines incorporated. (Source: *Radar Cross Section Lectures*, by A.E. Fuhs. Copyright © 1984, AIAA, New York. All rights reserved. Reprinted with permission AIAA.)

Section 9

Aircraft and Helicopter Design

VEHICLE DEFINITIONS

GEOMETRY

The following figures and formulas provide an introduction on geometric relationships concerning vehicle physical dimensions. These dimensions are used throughout this section.

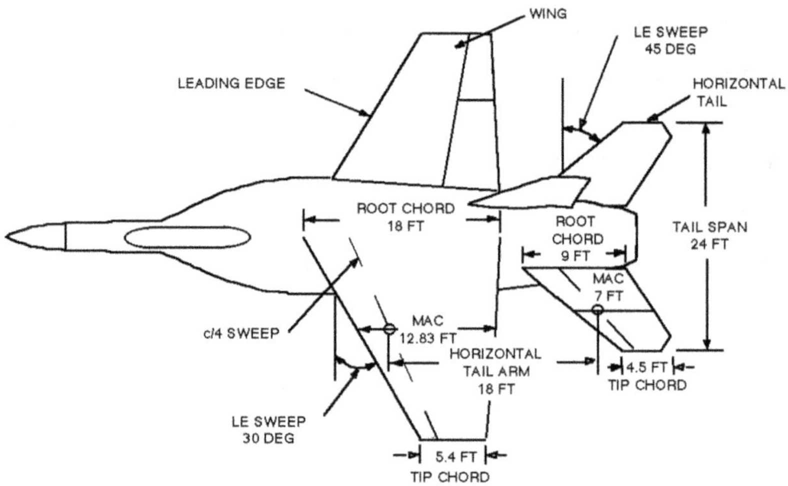

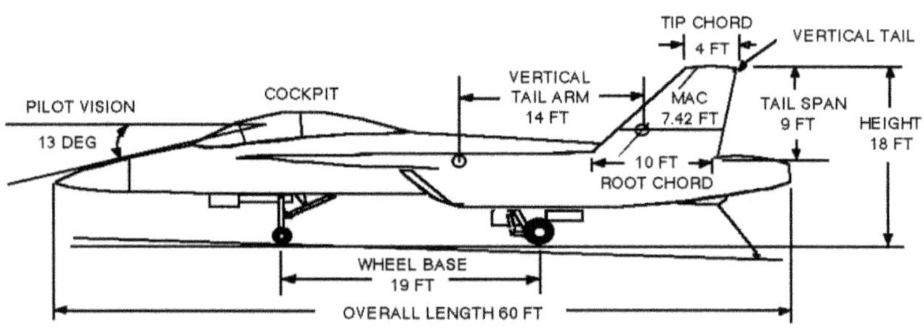

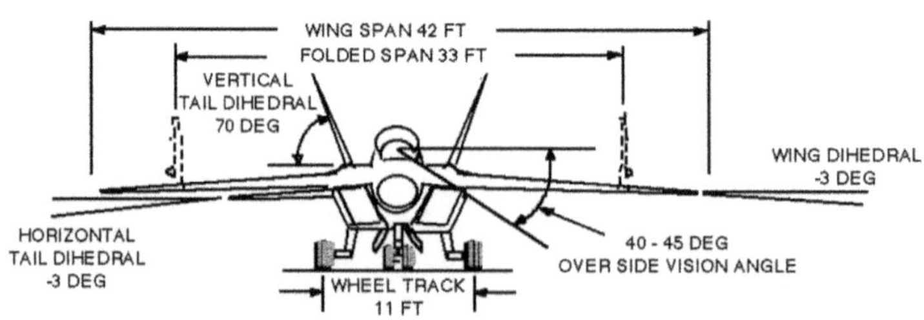

VEHICLE DEFINITIONS, CONTINUED

GEOMETRY, CONTINUED

Geometry	Unit	Wing	H-Tail	V-Tail
LE sweep angle	deg	30.0	45.0	45.0
c/4 sweep angle	deg	23.1	42.2	33.6
Reference area	m^2 (ft^2)	45.7 (491.4)	15.1 (162)	5.9 (63) each
Projected span	m (ft)	12.8 (42)	7.3 (24)	2.7 (9)
m.a.c.	ft	12.83	7	7.42
Aspect ratio	AR	3.59	3.55	1.28
Taper ratio	λ	0.3	0.5	0.4
Thickness ratio	t/c	0.05	0.04	0.03
Dihedral	Γ	– 3 deg	– 3 deg	70 deg
Airfoil	——	MOD NACA 65A	MOD NACA 65A	MOD NACA 65A
Tail volume	$\overline{V}$	n/a	0.462 ($\overline{V}_H$)	0.28 (V_V)

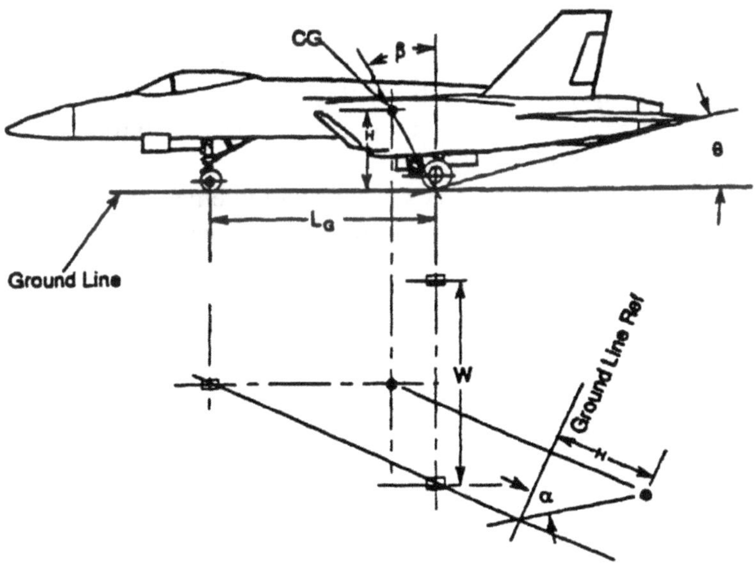

where
L_G = wheel base
W = wheel track
H = height c.g. to ground reference
β = tip back angle
θ = tail down angle
α = turn over angle
Gear in normal static position
Use most aft c.g. for β, keep $\beta > \theta$
Use most forward c.g. for α
Use landing weight c.g. for θ

VEHICLE DEFINITIONS, CONTINUED

GEOMETRY, CONTINUED

The following definitions and equations apply to trapezoidal planforms, as illustrated here.

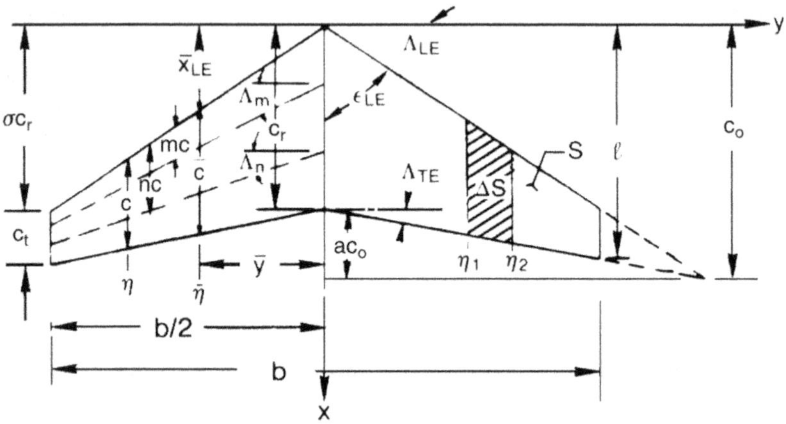

c_o = overall length of zero-taper-ratio planform having same leading- and trailing-edge sweep as subject planform

σ = ratio of chordwise position of leading edge at tip to the root chord length = $(b/2)\tan\Lambda_{LE}(1/c_r)$

η_1, η_2 = span stations of boundary of arbitrary increment of wing area

Λ_m, Λ_n = sweep angles of arbitrary chordwise locations

m, n = nondimensional chordwise stations in terms of % c

GENERAL

$$\eta = \frac{y}{b/2}$$

$$\lambda = c_t/c_r$$

$$c = c_r[1 - \eta(1-\lambda)]$$

$$\tan\Lambda = 1/\tan\epsilon$$

$$x_{LE} = (b/2)\eta\tan\Lambda_{LE}$$

$$\ell = \frac{b}{2}\tan\Lambda_{LE} + c_t = c_r\frac{1-a\lambda}{1-a}$$

AREA

$$S = \frac{b^2}{AR} = \frac{b}{2}c_r(1+\lambda) = \frac{b}{2}c_0(1-a)(1-\lambda)$$

$$= \frac{c_0^2(1-a)(1-\lambda^2)}{\tan\Lambda_{LE}}$$

$$\Delta S = \frac{b}{2}c_r[2-(1-\lambda)(\eta_1-\eta_2)]\frac{\eta_2-\eta_1}{2}$$

VEHICLE DEFINITIONS, CONTINUED

GEOMETRY, CONTINUED

ASPECT RATIO

$$\mathrm{AR} = \frac{b^2}{S} = \frac{2b}{c_r(1+\lambda)} = \frac{4(1-\lambda)}{(1-a)(1+\lambda)\tan\Lambda_{\mathrm{LE}}}$$

CUTOUT FACTOR

$$a = \frac{\tan\Lambda_{\mathrm{TE}}}{\tan\Lambda_{\mathrm{LE}}} = 1 - \frac{c_r(1-\lambda)}{(b/2)\tan\Lambda_{\mathrm{LE}}} = 1 - \frac{4(1-\lambda)}{\mathrm{AR}(1+\lambda)\tan\Lambda_{\mathrm{LE}}}$$

SWEEP ANGLES

$$\tan\Lambda_{\mathrm{LE}} = \frac{1}{a}\tan\Lambda_{\mathrm{TE}} = \frac{c_r(1-\lambda)}{(b/2)(1-a)} = \frac{4(1-\lambda)}{\mathrm{AR}(1+\lambda)(1-a)}$$

$$= \frac{c_0(1-\lambda)}{b/2} = \frac{\mathrm{AR}(1+\lambda)\tan\Lambda_{c/4} + (1-\lambda)}{\mathrm{AR}(1+\lambda)} = \frac{4\tan\Lambda_{c/4}}{3+a}$$

$$\tan\Lambda_m = \tan\Lambda_{\mathrm{LE}}[1-(1-a)m]$$

$$\tan\Lambda_m = \tan\Lambda_{\mathrm{LE}} - \frac{4m}{\mathrm{AR}}\left(\frac{1-\lambda}{1+\lambda}\right)$$

$$\cos\Lambda_m = (\tan\Lambda_{\mathrm{LE}})^{-1}\left\{\left(\frac{1}{\tan\Lambda_{\mathrm{LE}}}\right)^2 + [1-(1-a)m]^2\right\}^{-\frac{1}{2}}$$

MEAN AERODYNAMIC CHORD (m.a.c.)

$$\bar{c} = \frac{2}{S}\int_0^{b/2} c^2\,\mathrm{dy} = \frac{2}{3}c_r\left(1+\frac{\lambda^2}{1+\lambda}\right)$$

$$= \frac{2}{3}c_0(1-a)\left(1+\frac{\lambda^2}{1+\lambda}\right)$$

$$= \frac{4}{3}\left(\frac{S}{\mathrm{AR}}\right)^{\frac{1}{2}}\left[1-\frac{\lambda}{(1+\lambda)^2}\right]$$

$$= \frac{2}{3}\left(c_r + c_t - \frac{c_r \times c_t}{c_r + c_t}\right)$$

$$\bar{\eta} = \frac{2}{S}\int_0^{b/2} c y\,\mathrm{dy} = \frac{1-(\bar{c}/c_r)}{1-\lambda} = \frac{1}{3}\left(\frac{1+2\lambda}{1+\lambda}\right)$$

$$\bar{x}_{\mathrm{LE}} = \bar{y}\tan\Lambda_{\mathrm{LE}}$$

VEHICLE DEFINITIONS, CONTINUED

GEOMETRY, CONTINUED

ROOT CHORD

$$c_r = \frac{S}{(b/2)(1+\lambda)} = \frac{4(b/2)}{\mathrm{AR}(1+\lambda)}$$

LOCATION OF TIP L.E. IN 90 ROOT CHORD

$$\sigma = \frac{\mathrm{AR}}{4}(1+\lambda)\tan\Lambda_{\mathrm{LE}}$$

FORCE-VELOCITY

AIRPLANE AXIS SYSTEM

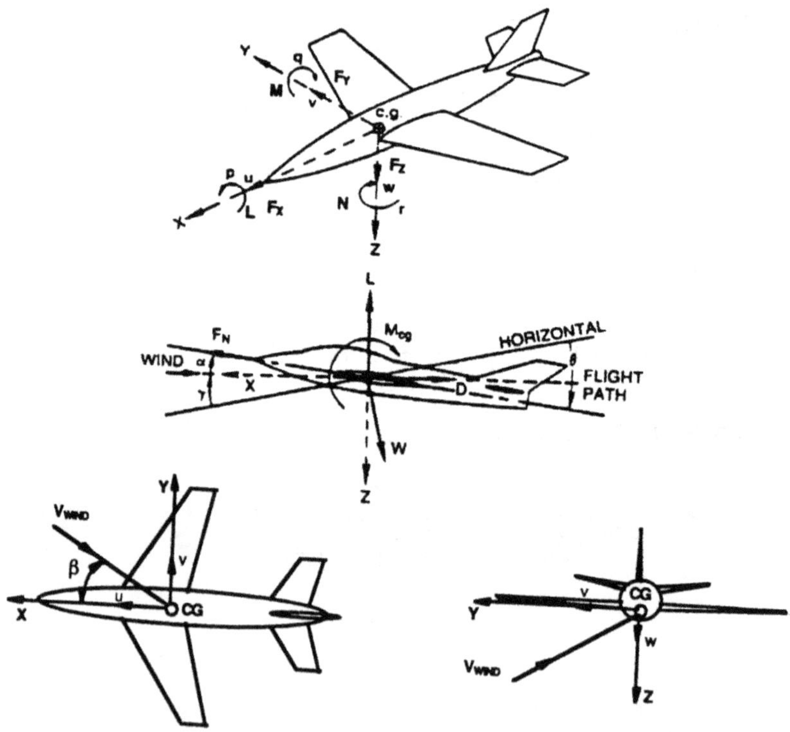

Axis	Force Along	Moment About	Linear Velocity	Attitude Angle	Ang. Vel.	Inertia	Wind Relative Attitude Angle
X	F_x	L	u	ϕ	p	I_x	σ
Y	F_y	M	v	θ	q	I_y	α
Z	F_z	N	w	ψ	r	I_z	β

AERODYNAMICS

BASIC AERODYNAMIC RELATIONSHIPS

AR = aspect ratio = b^2/S

C_D = drag coefficient = $D/qS = C_{DO} + C_{Di}$

C_{Di} = induced drag coefficient = $C_L^2/(\pi ARe)$

C_L = lift coefficient = L/qS

C_ℓ = rolling-moment coefficient = rolling moment / qbS

C_m = pitching-moment coefficient = pitching moment / qcS

C_n = yawing-moment coefficient = yawing moment / qbS

C_y = side-force coefficient = side force / qS

D = drag = $C_D qS$

d = equivalent body diameter = $\sqrt{4A_{\text{MAX}}/\pi}$

$\overline{\text{FR}}$ = fineness ratio = ℓ/d

L = lift = $C_L qS$

M = Mach number = V/a

P = planform shape parameter = $S/b\ell$

q = dynamic pressure = $\frac{1}{2}(\rho V^2) = \frac{1}{2}(\rho a^2 M^2)$

R_n = Reynolds number = $V\ell\rho/\mu$

R_o = $d/2$ = equivalent body radius

$(t/c)_{\text{RMS}}$ = root-mean-square thickness ratio

$$(t/c)_{\text{RMS}} = \left[\frac{1}{b/2-r}\int_r^{b/2}(t/c)^2\,\mathrm{dy}\right]^{\frac{1}{2}}$$

V = true airspeed = $V_e/\sigma^{1/2}$

$\bar{x}$ = chordwise location from apex to c/z (equivalent to chordwise location of centroid of area)

$$\bar{x} = \frac{2}{S}\int_o^{b/2} c\left(x+\frac{c}{2}\right)\mathrm{dy}$$

$\bar{x}_{\text{LE}}$ = chordwise location of leading edge of m.a.c.

$$\bar{x}_{\text{LE}} = x - \frac{\bar{c}}{2}$$

$\bar{y}$ = spanwise location of $\bar{c}$ (equivalent to spanwise location of centroid of area)

$$\bar{y} - \frac{2}{S}\int_o^{b/2} cy\,\mathrm{dy}$$

β = $\sqrt{M^2-1}$ (supersonic), $\sqrt{1-M^2}$ (subsonic)

ε = complement to wing sweep angle = 90 deg$-\Lambda_{\text{LE}}$

η = nondimensional span station = $y/(b/2)$

λ = taper ratio, tip-to-root chord = c_t/c_r

σ = air density ratio = ρ/ρ_o

υ = kinematic viscosity = μ/ρ

AERODYNAMICS, CONTINUED

SYMBOLS

a	= speed of sound
A_{MAX}	= maximum cross-sectional area
a.c.	= aerodynamic center
b	= wing span[a]
c	= chord[a]
$\overline{c}$	= mean aerodynamic chord (m.a.c.)
C_{DO}	= drag coefficient at zero lift
c.g.	= center of gravity
c.p.	= center of pressure
c_r	= root chord[a]
c_t	= tip chord[a]
d	= diameter
$(dA/dx)AFT$	= slope of aft end of configuration distribution curve
e	= Oswald (wing) efficiency factor
g	= acceleration due to gravity
i	= angle of incidence
K_{BODY}	= body wave-drag factor
K_{LE}	= wing shape factor
ℓ	= characteristic length[a]
S, S_{REF}	= reference area[a]
S_{EXP}	= exposed planform area
T	= temperature
t	= airfoil maximum thickness at span station y
t/c	= airfoil thickness ratio (parallel to axis of symmetry)
V_e	= equivalent velocity
x	= general chordwise location, parallel to plane of symmetry[a]
y	= general spanwise location, perpendicular to plane of symmetry[a]
α	= angle of attack, chord plane to relative wind
α_{Lo}	= angle of attack for zero lift
Γ	= dihedral angle
δ	= pressure ratio (p/p_o)
γ	= ratio of specific heats
Λ	= sweep-back angle
Λ_{LE}	= wing leading-edge sweep angle[a]
Λ_{TE}	= wing trailing-edge sweep angle[a]
Λ_{HL}	= flap-hinge-line sweep angle
μ	= coefficient of absolute viscosity
ρ	= density
R	= specific gas constant
p	= pressure

[a]Defined in figure appearing on page 9–4.

AERODYNAMICS, CONTINUED

SPEED OF SOUND VS TEMPERATURE

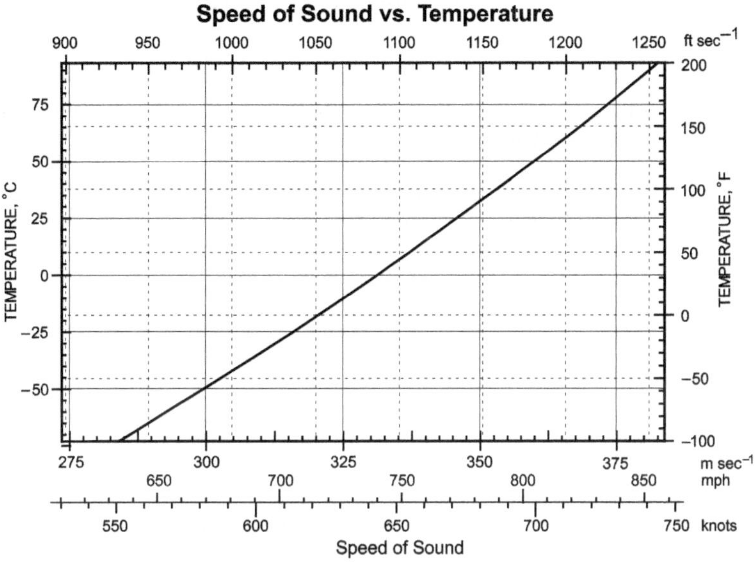

DYNAMIC PRESSURE (q) VS MACH NUMBER

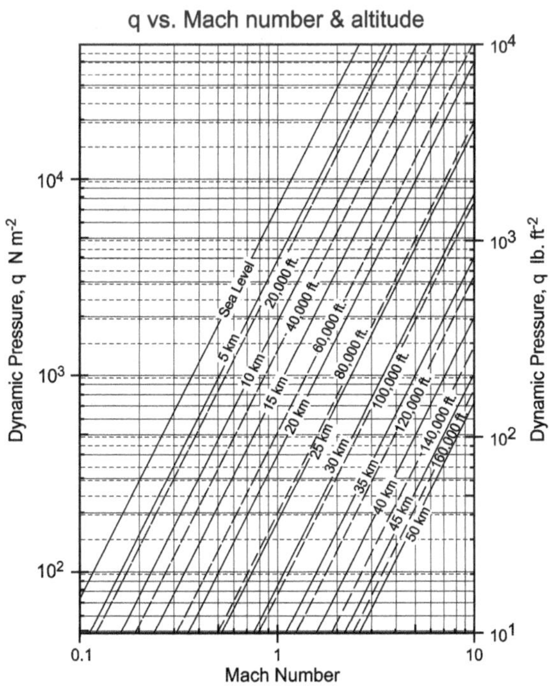

AERODYNAMICS, CONTINUED

STANDARD ATMOSPHERE

Standard atmosphere is a hypothetical vertical distribution of atmospheric temperature, pressure, and density which, by international or national agreement, is taken to be the representative of the atmosphere for the purpose of altimeter calculations, aircraft design, performance calculations, etc. The data presented in the following tables is based on the U.S. Standard Atmosphere–1976.

Altitude	Temperature	Pressure	Density	Kinematic Viscosity					Speed of Sound	
h, km	K	p, mm Hg	ρ, kg/m^3	$\nu \times 10^5$ m^2/s	δ, (p/p_o)	σ, (ρ/ρ_o)	$\sigma^{1/2}$	$q/M^2 \times 10^{-3}$	a, m/s	$a\sigma^{1/2}$
0	288.2	760.0	1.225	1.464	1.0000	1.0000	1.0000	70.93	340.3	340.3
1	281.7	674.1	1.112	1.584	0.8870	0.9075	0.9526	62.91	336.4	320.5
2	275.2	596.3	1.006	1.717	0.7846	0.8216	0.9064	55.65	332.5	301.4
3	268.7	525.9	0.9091	1.865	0.6919	0.7421	0.8615	49.08	328.6	283.1
4	262.2	462.3	0.8191	2.029	0.6083	0.6687	0.8177	43.15	324.6	265.4
5	255.7	405.2	0.7361	2.212	0.5331	0.6009	0.7752	37.81	320.5	248.5
6	249.2	353.9	0.6597	2.416	0.4656	0.5385	0.7338	33.03	316.4	232.2
7	242.7	308.0	0.5895	2.646	0.4052	0.4812	0.6937	28.74	312.3	216.6
8	236.2	267.0	0.5252	2.903	0.3513	0.4287	0.6548	24.92	308.1	201.7
9	229.7	230.6	0.4663	3.194	0.3034	0.3807	0.6170	21.52	303.8	187.4
10	223.2	198.3	0.4127	3.522	0.2609	0.3369	0.5804	18.51	299.5	173.8
11	216.7	169.8	0.3639	3.895	0.2234	0.2971	0.5450	15.84	295.1	160.8
12	216.7	145.0	0.3108	4.560	0.1908	0.2537	0.5037	13.53	295.1	148.6
13	216.7	123.8	0.2655	5.339	0.1629	0.2167	0.4655	11.56	295.1	137.4
14	216.7	105.8	0.2268	6.251	0.1392	0.1851	0.4302	9.871	295.1	127.0
15	216.7	90.34	0.1937	7.318	0.1189	0.1581	0.3976	8.431	295.1	117.3
16	216.7	77.16	0.1654	8.568	0.1015	0.1350	0.3675	7.201	295.1	108.4
17	216.7	65.91	0.1413	10.03	0.0867	0.1153	0.3396	6.151	295.1	100.2
18	216.7	56.29	0.1207	11.75	0.0741	0.0985	0.3139	5.253	295.1	92.6
19	216.7	48.08	0.1031	13.75	0.0633	0.0841	0.2901	4.487	295.1	85.6
20	216.7	41.07	0.0880	16.10	0.0540	0.0719	0.2681	3.832	295.1	79.1
21	217.7	35.09	0.0749	19.00	0.0462	0.0611	0.2472	3.275	295.7	73.1
22	218.7	30.00	0.0637	22.42	0.0395	0.0520	0.2281	2.800	296.4	67.6
23	219.7	25.67	0.0543	26.42	0.0338	0.0443	0.2105	2.396	297.1	62.5
24	220.7	21.98	0.0463	31.12	0.0289	0.0378	0.1943	2.051	297.8	57.9
25	221.7	18.83	0.0395	36.62	0.0248	0.0322	0.1795	1.758	298.5	53.6
26	222.7	16.15	0.0337	43.07	0.0212	0.0275	0.1658	1.507	299.1	49.6
27	223.7	13.86	0.0288	50.61	0.0182	0.0235	0.1533	1.293	299.8	45.9
28	224.7	11.90	0.0246	59.43	0.0157	0.0201	0.1417	1.110	300.5	42.6
29	225.7	10.22	0.0210	69.73	0.0135	0.0172	0.1311	0.9541	301.1	39.5
30	226.7	8.790	0.0180	81.77	0.0116	0.0147	0.1213	0.8203	301.8	36.6

(continued)

p_o = standard pressure at sea level, 760 mm Hg, 1.013×10^5 N/m^2 or Pa
ρ_o = standard density at sea level, 1.225 kg/m^3
δ = pressure ratio
σ = density ratio
a = speed of sound

AERODYNAMICS, CONTINUED

STANDARD ATMOSPHERE, CONTINUED

Altitude h, ft	Temperature °R	Pressure p, in Hg	Density $\rho \times 10^4$ slug/ft^3	Kinematic Viscocity $\nu \times 10^4$ ft^2/s	δ, (p/p_o)	σ, (ρ/ρ_0)	$\sigma^{1/2}$	q/M^2	Speed of Sound a, knots	$a\sigma^{1/2}$
0	518.7	29.92	23.77	1.576	1.0000	1.0000	1.0000	1482	661.5	661.5
1000	515.1	28.86	23.08	1.614	0.9644	0.9710	0.9854	1429	659.2	649.6
2000	511.5	27.82	22.41	1.653	0.9298	0.9428	0.9710	1378	656.9	637.8
3000	508.0	26.82	21.75	1.694	0.8962	0.9150	0.9566	1328	654.6	626.2
4000	504.4	25.84	21.11	1.735	0.8637	0.8881	0.9424	1280	652.3	614.7
5000	500.8	24.90	20.48	1.778	0.8320	0.8616	0.9282	1233	650.0	603.4
6000	497.3	23.98	19.87	1.823	0.8014	0.8359	0.9143	1188	647.7	592.2
7000	493.7	23.09	19.27	1.868	0.7716	0.8107	0.9004	1143	645.4	581.1
8000	490.1	22.22	18.68	1.916	0.7428	0.7859	0.8865	1100	643.0	570.0
9000	486.6	21.39	18.11	1.965	0.7148	0.7619	0.8729	1059	640.7	559.2
10,000	483.0	20.58	17.55	2.016	0.6877	0.7383	0.8593	1019	638.3	548.5
11,000	479.4	19.79	17.01	2.067	0.6614	0.7156	0.8459	980.2	636.0	538.0
12,000	475.9	19.03	16.48	2.121	0.6360	0.6933	0.8327	942.6	633.6	527.6
13,000	472.3	18.29	15.96	2.176	0.6113	0.6714	0.8194	906.0	631.2	517.2
14,000	468.7	17.58	15.45	2.234	0.5875	0.6500	0.8062	870.4	628.8	507.0
15,000	465.2	16.89	14.96	2.293	0.5643	0.6294	0.7933	836.4	626.4	497.0
16,000	461.6	16.22	14.48	2.354	0.5420	0.6092	0.7805	803.3	624.0	487.1
17,000	458.0	15.57	14.01	2.418	0.5203	0.5894	0.7677	771.3	621.6	477.2
18,000	454.5	14.94	13.55	2.484	0.4994	0.5700	0.7550	740.1	619.2	467.5
19,000	450.9	14.34	13.1	2.553	0.4791	0.5511	0.7424	709.9	616.8	457.9
20,000	447.3	13.75	12.67	2.622	0.4595	0.5330	0.7301	681.2	614.3	448.5
21,000	443.8	13.18	12.24	2.697	0.4406	0.5149	0.7176	652.8	611.9	439.1
22,000	440.2	12.64	11.83	2.771	0.4223	0.4977	0.7055	625.9	609.4	429.9
23,000	436.6	12.11	11.43	2.849	0.4046	0.4809	0.6934	599.8	606.9	420.9
24,000	433.1	11.60	11.03	2.932	0.3876	0.4640	0.6812	574.1	604.4	411.7
25,000	429.5	11.10	10.65	3.016	0.3711	0.4480	0.6694	549.8	601.9	402.9
26,000	425.9	10.63	10.28	3.103	0.3552	0.4325	0.6576	526.3	599.4	394.2
27,000	422.4	10.17	9.920	3.193	0.3398	0.4173	0.6460	503.6	596.9	385.6
28,000	418.8	9.725	9.570	3.286	0.3250	0.4026	0.6345	481.7	594.4	377.2
29,000	415.3	9.298	9.230	3.383	0.3107	0.3883	0.6231	460.6	591.9	368.8
30,000	411.7	8.885	8.890	3.487	0.2970	0.3740	0.6116	439.9	589.3	360.4
30,000	411.7	8.885	8.890	3.487	0.2970	0.3740	0.6116	439.9	589.3	360.4
31,000	408.1	8.488	8.570	3.591	0.2837	0.3605	0.6004	420.4	586.8	352.3
31,000	408.1	8.488	8.570	3.591	0.2837	0.3605	0.6004	420.4	586.8	352.3
32,000	404.6	8.106	8.260	3.698	0.2709	0.3475	0.5895	401.6	584.2	344.4
33,000	401.0	7.737	7.950	3.813	0.2586	0.3345	0.5783	383.1	581.6	336.4
34,000	397.4	7.382	7.650	3.933	0.2467	0.3218	0.5673	365.4	579.0	328.5
35,000	393.9	7.041	7.370	4.051	0.2353	0.3101	0.5568	348.9	576.4	321.0
36,000	390.3	6.712	7.090	4.178	0.2243	0.2983	0.5461	332.6	573.8	313.4
37,000	390.0	6.397	6.760	4.379	0.2138	0.2844	0.5333	316.8	573.6	305.9
38,000	390.0	6.097	6.440	4.597	0.2038	0.2709	0.5205	301.8	573.6	298.5
39,000	390.0	5.811	6.140	4.821	0.1942	0.2583	0.5082	287.8	573.6	291.5
40,000	390.0	5.538	5.850	5.060	0.1851	0.2461	0.4961	274.2	573.6	284.5
45,000	390.0	4.355	4.600	6.435	0.1455	0.1935	0.4399	215.6	573.6	252.3
50,000	390.0	3.425	3.620	8.177	0.1145	0.1523	0.3902	169.7	573.6	223.8
55,000	390.0	2.693	2.850	10.39	0.0900	0.1199	0.3463	133.6	573.6	198.6
60,000	390.0	2.118	2.240	13.22	0.0708	0.0942	0.3070	105.0	573.6	176.1
65,000	390.0	1.665	1.760	16.82	0.0557	0.0740	0.2721	82.49	573.6	156.1
70,000	392.4	1.310	1.380	21.56	0.0438	0.0581	0.2409	65.08	575.3	138.6
75,000	395.1	1.033	1.080	27.72	0.0345	0.0454	0.2132	51.29	577.3	123.1
80,000	397.9	0.815	0.840	35.85	0.0273	0.0353	0.1880	40.17	579.3	108.9
85,000	400.6	0.645	0.660	45.89	0.0216	0.0278	0.1666	31.78	581.3	96.87
90,000	403.3	0.511	0.520	58.59	0.0171	0.0219	0.1479	25.21	583.3	86.28
95,000	406.1	0.405	0.410	74.74	0.0135	0.0172	0.1313	20.01	585.3	76.87
100,000	408.8	0.322	0.320	96.30	0.0108	0.0135	0.1160	15.72	587.3	68.14

p_o = standard pressure at sea level, 29.92 in Hg, 14.70 lb/in.^{2}, ρ_o = standard density at sea level, 23.77 × 10^{-4} slugs/ft^3, δ = pressure ratio, σ = density ratio, α = speed of sound

AERODYNAMICS, CONTINUED

AIRSPEED RELATIONSHIPS

IAS—indicated airspeed (read from cockpit instrumentation, includes cockpit-instrument error correction)

CAS—calibrated airspeed (indicated airspeed corrected for airspeed-instrumentation position error)

EAS—equivalent airspeed (calibrated airspeed corrected for compressibility effects)

TAS—true airspeed (equivalent airspeed corrected for change in atmospheric density)

$$\text{TAS} = \frac{\text{EAS}}{\sqrt{\sigma}}$$

Mach number:

$$M = \frac{V_a}{a} = V_a \Big/ \sqrt{\gamma g R T}$$

where

V_a = true airspeed

a = sonic velocity

γ = specific heat ratio

g = gravitational constant

R = gas constant

T = ambient temperature

The following are equivalent at 4.6 km (15,000 ft), 30°C (86°F) day:

M = 0.428
TAS = 149 m/s (290 kn)
CAS = 111 m/s (215 kn)
EAS = 110 m/s (213 kn)

Change in velocity with change in air density, ρ, at constant power:

$$V_2 = V_1 \sqrt[3]{\frac{\rho_1}{\rho_2}} \quad \text{(approximate)}$$

Change in velocity with change in power at constant air density:

$$V_2 = V_1 \sqrt[3]{\frac{\rho_2}{\rho_1}} \quad \text{(approximate)}$$

Aerodynamics, Continued

Airspeed Conversion Charts

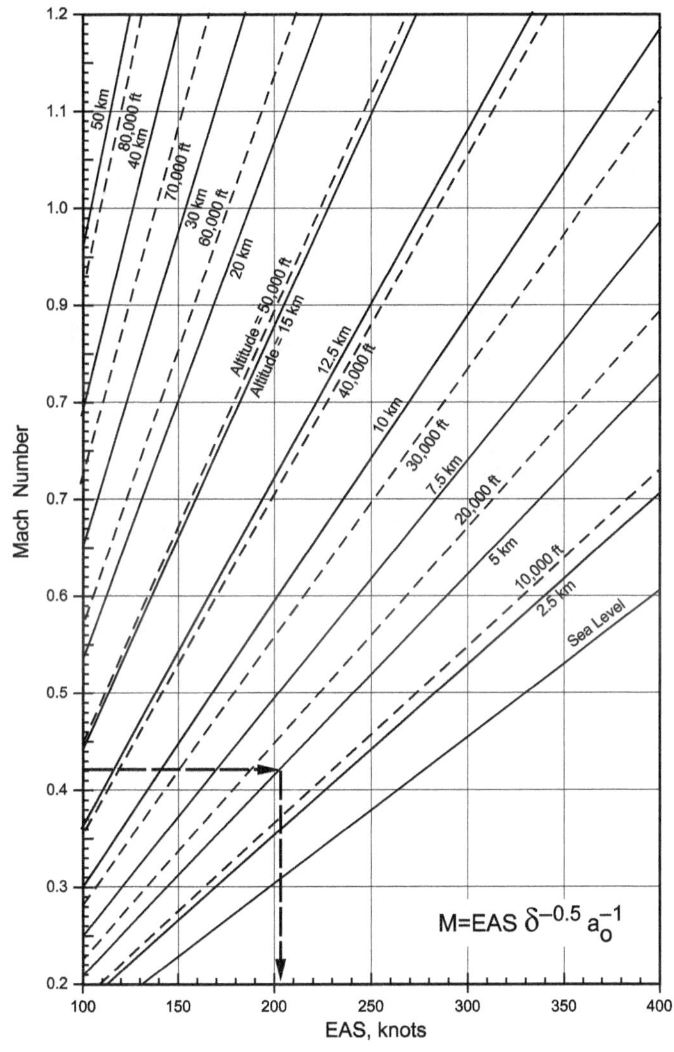

AERODYNAMICS, CONTINUED

AIRSPEED CONVERSION CHARTS, CONTINUED

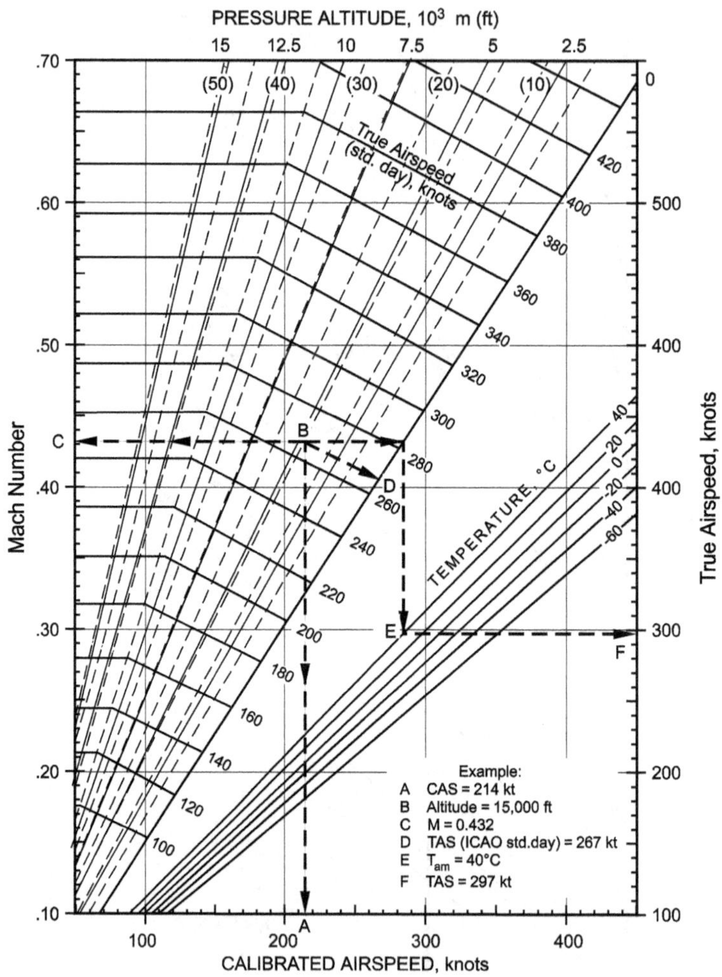

AERODYNAMICS, CONTINUED

AIRSPEED CONVERSION CHARTS, CONTINUED

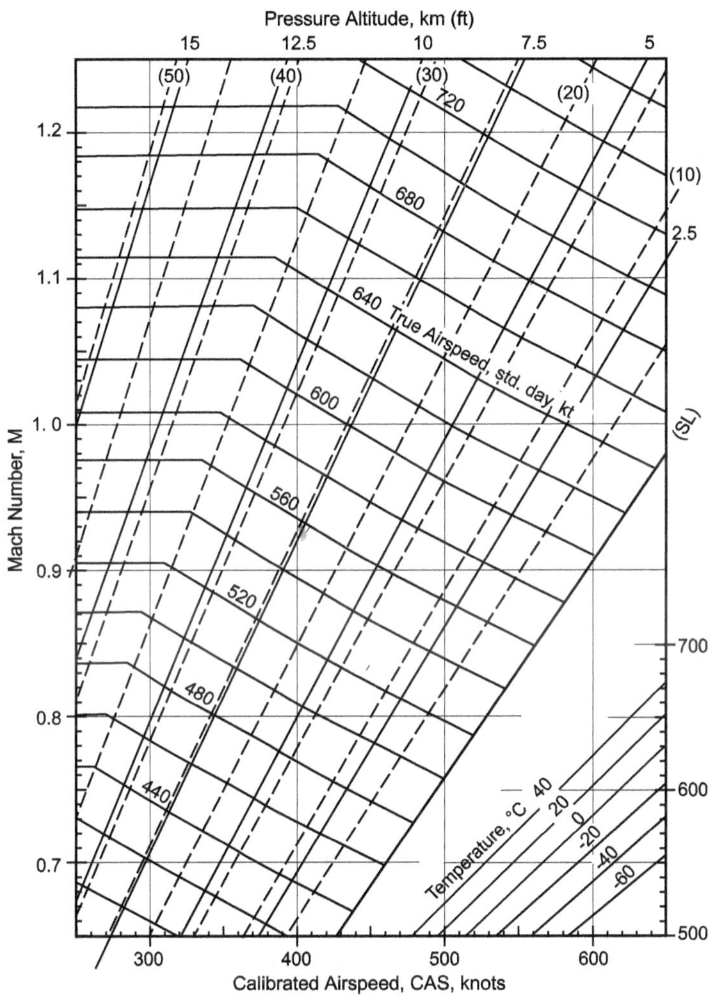

AERODYNAMICS, CONTINUED

MINIMUM DRAG

SUBSONIC

The basic minimum drag of an aerodynamic vehicle consists of not only friction drag but also drag due to the pressure forces acting on the vehicle.

The following equations present a method of predicting minimum drag:

$$\text{Minimum drag} = C_{D_{\min}} qS$$

$$C_{D_{\min}} = \frac{\sum \left(C_{f_{\text{comp}}} \times A_{w_{\text{comp}}} \right)}{S} + C_{D_{\text{camber}}} + C_{D_{\text{base}}} + C_{D_{\text{misc}}}$$

where the component skin-friction coefficients are evaluated according to the following equations:

$A_{w_{\text{comp}}}$ = component wetted area

c = lifting surface exposed streamwise m.a.c.

$C_{f_{\text{comp}}}$ = component drag coefficient

$C_{f_{\text{wing}}} = C_{f_{FP}} \left[1 + L(t/c) + 100(t/c)^4 \right] R_{\text{LS}}$

$C_{f_{\text{fuselage}}} = C_{f_{\text{FP}}} \left[1 + 1.3(\text{FR})^{-1.5} + 44(\text{FR})^{-3} \right] R_{\text{fus}}$

$C_{f_{\text{nacelle}}} = C_{f_{\text{FP}}} \times Q[1 + 0.35(\text{FR})^{-1}$

$C_{f_{\text{canopy}}} = C_{f_{\text{FP}}} \left[1 + 1.3(\text{FR})^{-1.5} + 44(\text{FR})^{-3} \right]$

$C_{f_{\text{horiz \& vert tails (one piece)}}} = C_{f_{\text{FP}}} \left[1 + L(t/c) + 100(t/c)^4 \right] R_{\text{LS}}$

$C_{f_{\text{horiz \& vert tails (hinged)}}} = (1.1)\, C_{f_{\text{FP}}} \left[1 + L(t/c) + 100\,(t/c)^4 \right] R_{\text{LS}}$

$C_{f_{ext.\,store}} = C_{f_{FP}} \times Q \left[1 + 1.3(\text{FR})^{-1.5} + 44(\text{FR})^{-3} \right]$

$C_{D_{\text{camber}}} = 0.7 \left(\Delta\, C_L^2 \right) \left(S_{\text{EXP}} / S \right)$ (do not use for conical camber)

$C_{D_{\text{base}}}$ = base drag: a good estimate can be obtained by using a base pressure coefficient of $C_p = -0.1$. (More detailed discussion of base drag may be found in Hoerner's *Fluid Dynamic Drag*.)

$C_{D_{\min}}$ = minimum drag coefficient

$C_{D_{\text{misc}}}$ = miscellaneous drag coefficient

$C_{f_{\text{FP}}}$ = White–Christoph's flat-plate turbulent-skin-friction coefficient based on Mach number and Reynolds number (in which characteristic length of lifting surface equals exposed m.a.c.)

$\overline{\text{FR}}$ = fineness ratio

= length/diameter (for closed bodies of circular cross section)

$= \text{length} / \sqrt{(\text{width})(\text{height})}$ (for closed bodies of irregular cross section and for nacelles)

$= \text{length} \Big/ a\sqrt{1 + \frac{1}{2}\left(1 - \frac{a}{b}\right)\left(\frac{a}{b}\right)^2}$ (for closed bodies of elliptic cross section, where a = minor axis and b = major axis)

AERODYNAMICS, CONTINUED

MINIMUM DRAG, CONTINUED

SUBSONIC, CONTINUED

L = thickness location parameter
= 1.2 for $(t/c)_{max}$ located at $x \geq 0.3c$
= 2.0 for $(t/c)_{max}$ located at $x < 0.3c$

m.a.c. = mean aerodynamic chord

$\Lambda_{max_{t/c}}$ = sweep of lifting-surface at maximum thickness t/c

q = dynamic pressure

Q = interference factor
= 1.0 for nacelles and external stores mounted out of the local velocity field of the wing
= 1.25 for external stores mounted symmetrically on the wing tip
= 1.3 for nacelles and external stores if mounted in moderate proximity of the wing
= 1.5 for nacelles and external stores mounted flush to the wing (The same variation of the interference factor applies in the case of a nacelle or external store strut-mounted to or flush-mounted on the fuselage.)

R_{LS} = lifting surface factor (see Lifting Surface Correction graph)

R_{fus} = fuselage correction factor (see Fuselage Corrections graph), $R_{e_{fus}}$ based on length

S = wing gross area

S_{EXP} = exposed wing area

t = maximum thickness of section at exposed streamwise m.a.c.

Example

Calculation of uncambered-wing drag coefficient for subsonic case.

Reference wing area, $S = 92.9$ m^2 (1000 ft^2)
Wetted area, $A_w = 150.5$ m^2 (1620 ft^2)
Velocity, $V = 169.5$ m/s (556 ft/s) = 0.54 M
Altitude, $H = 6700$ m (22,000 ft)
Mean chord, $\bar{c} = 3.74$ m (12.28 ft)
Sweep at maximum thickness t/c = 11% at 35% chord = 24 deg
Reynolds number, $R_e = V\bar{c}/\nu = (169.5)(3.74)/2.577 \times 10^{-5} = 24.60 \times 10^6$
Thickness location parameter, $L = 1.2$
Lifting surface factor, $R_{LS} = 1.13$ (see Lifting Surface Correction graph)
Basic skin-friction coefficient, $C_{f_{FP}} = 0.00255$ (see Turbulent Skin Friction Coefficient graph)
Wing skin-friction factor, $C_f = 0.0033$
Wing minimum-drag coefficient, $C_d = 0.00535$

AERODYNAMICS, CONTINUED

MINIMUM DRAG, CONTINUED

SUBSONIC-COMPONENT CORRECTION FACTORS

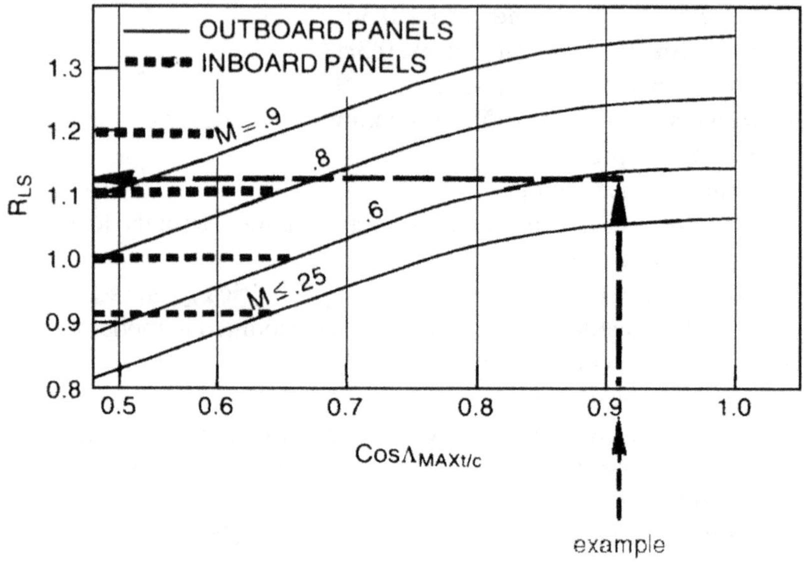

Lifting surface correction.

Apply ratio A_{wet}/S_{ref} value for the fuselage plus attached items (to respective sets of curves, dashed or solid).

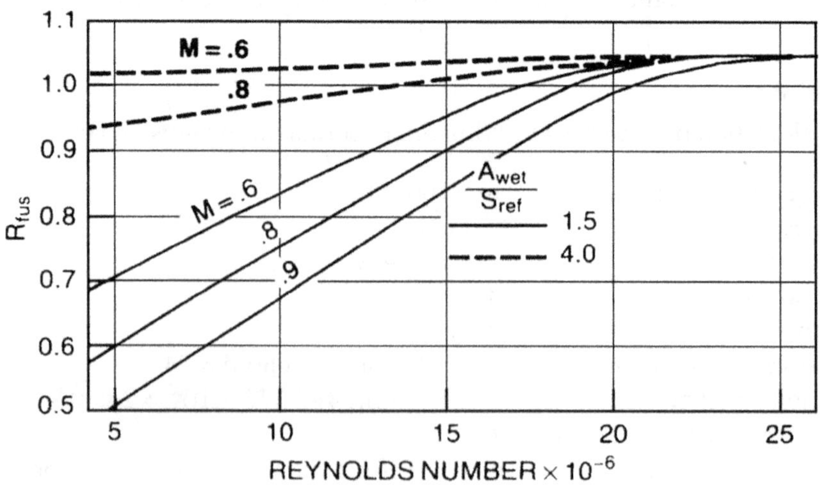

Fuselage corrections.

AERODYNAMICS, CONTINUED

MINIMUM DRAG, CONTINUED

SUBSONIC-COMPONENT CORRECTION FACTORS, CONTINUED

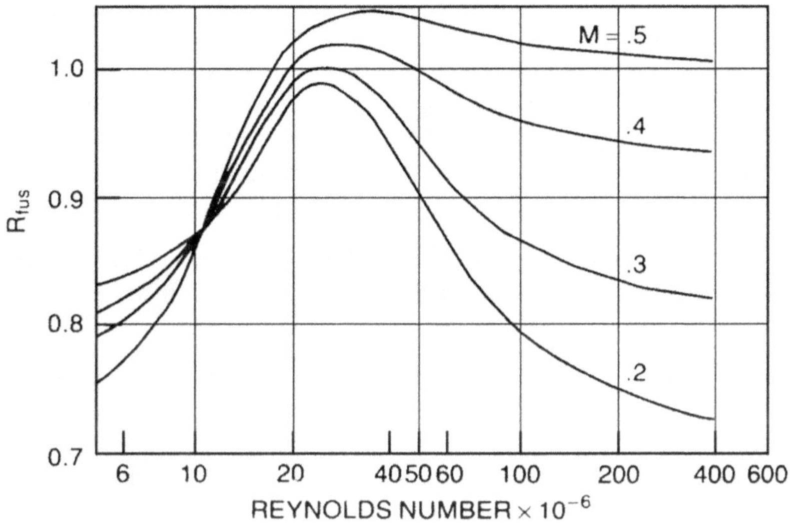

For Mach ≤ 0.5.

Turbulent Skin-Friction Coefficient

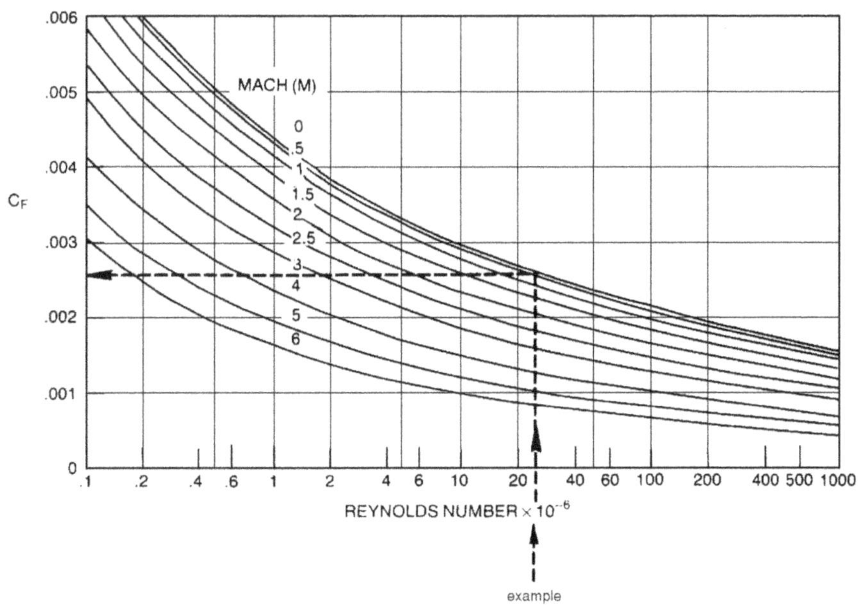

Insulated flat plate (White–Christoph).

AERODYNAMICS, CONTINUED

MINIMUM DRAG, CONTINUED

SUPERSONIC

Wave drag. At supersonic speeds, the pressure drag is associated with the shockwave pattern about the vehicle and is called wave drag. "Area-rule" techniques are generally employed for determining the wave drag of a configuration. Due to the lengthy calculations involved in solving the wave-drag equation, digital computers are used exclusively; accuracy primarily depends on the methods employed for geometrical manipulation. The NASA Harris area-rule program is used extensively in wave-drag calculations.

Skin friction. Supersonic skin-friction drag is calculated for each component utilizing flat-plate skin-friction coefficients (see Insulated Flat Plate graph).

Wave drag—wing. As previously discussed, the calculation of wave drag for most configurations requires the resources of a digital computer. As a first approximation for wave drag of an uncambered, untwisted, conventional trapezoidal wing with sharp-nosed airfoil section, the following can be used:

$$C_{D_{\text{wave wing}}} = \frac{K_{\text{LE}}\left(t/c\right)^2}{\beta} \quad (\text{for } \beta \tan\varepsilon \geq 1), \text{ where } \beta^2 = M^2 - 1$$

$$C_{D_{\text{wave wing}}} = K_{\text{LE}} \tan\varepsilon \left(t/c\right)^2 \quad (\text{for } \beta \tan\varepsilon < 1)$$

$$(\text{for } \varepsilon = 90 \text{ deg} - \Lambda_{\text{LE}})$$

where the shape factor K_{LE} is shown in the following table:

Configuration	K_{LE}
Single wedge	1
Symmetrical double wedge	4
Double wedge with maximum thickness at x/c	$\frac{(c/x)}{(1-x/c)}$
Biconvex section	5.3
Streamline foil with x/c = 50%	5.5
Round-nose foil with x/c = 30%	6.0
Slender elliptical airfoil section	6.5

AERODYNAMICS, CONTINUED

MINIMUM DRAG, CONTINUED

SUPERSONIC, CONTINUED

Wave drag—body. A first approximation for a body can be obtained from the preliminary wave-drag evaluation graph here for $M = 1.2$ using the expression:

$$C_{D_{\text{wave body}}} = \frac{A_{\text{MAX}}}{S_{\text{REF}}} \frac{K_{\text{BODY}}}{\overline{\text{FR}}^2}$$

Example

Calculate the wing wave-drag coefficient for the following conditions:

Mach number = 1.2
Airfoil = 6% thick symmetrical double wedge ($K_{\text{LE}} = 4.0$)
$\beta = 0.663$
$\varepsilon = 90$ deg
$\beta \tan \varepsilon = \infty$

$$C_{D_{\text{wave wing}}} = 0.0217$$

Example

Calculate the body wave-drag coefficient for the following conditions:

Mach number = 1.2
Fuselage = 0.91 m (3 ft) diameter, 9.1 m (30 ft) length ($\overline{\text{FR}} = 10$)
$dA/dx = 6.10$ m (20 ft)
Reference wing area = 6.22 m^2 (67 ft^2)
$K_{\text{BODY}} = 18$ (see Preliminary Wave-Drag Evaluation graph)

$$C_{D\,\text{wave body}} = 0.0190$$

PRELIMINARY WAVE-DRAG EVALUATION

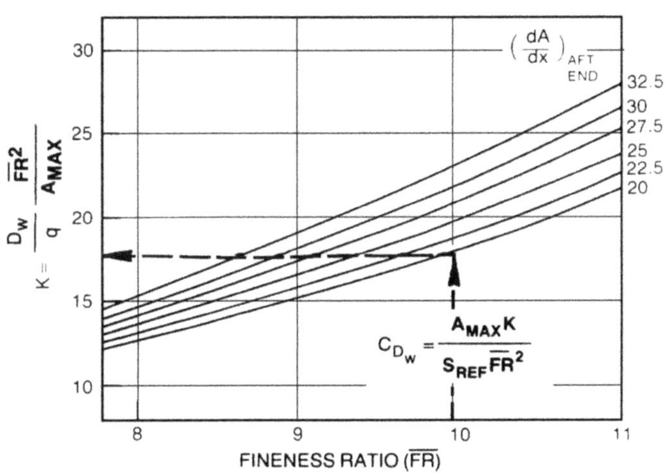

AERODYNAMICS, CONTINUED

INDUCED DRAG

SUBSONIC

The drag due to lift, or induced drag, reflects lift-producing circulation. Potential flow theory shows that the relationship to drag is a function of lift squared. Hence, the basic polar is parabolic. The parabolic polar is shifted from the origin by camber, wing incidence, minimum drag, etc. and deviates from its parabolic shape at higher lifts when flow separation exists.

$$C_{D_i} = KC_L^2$$

where C_L is total lift, including camber effects, and K = the parabolic drag constant. For plain wings below the parabolic polar break lift coefficient,

$$K = \frac{1}{\pi \text{AR} e}$$

The value of e, the wing efficiency factor, accounts for non-ellipticity of the lift distribution and the variation of viscous drag with α; typical values of e for high-subsonic jets are 0.75–0.85. The higher the wing sweep angle, the lower the e factor.

For wings with sharp leading edges, the drag due to lift can be approximated by $C_{D_i} = 0.95 C_L \tan\alpha$ (α = wing angle of attack).

Example

Induced drag at a lift coefficient of 0.8 for a vehicle with an effective Oswald efficiency factor e of 0.80 and an aspect ratio of 8.5 will be

$$C_{D_i} = \frac{C_L^2}{\pi \text{AR} e} = \frac{(0.8)^2}{\pi(8.5)(0.80)} = 0.030$$

SUPERSONIC

At supersonic speeds, the wave drag due to lift increases as $\sqrt{M^2 - 1}$ and is a function of planform geometry. For preliminary design evaluation, the following graphs provide sufficient accuracy. At supersonic speeds, the polars generally show no tendency to depart from a parabolic shape. Thus, there is no corresponding polar break as at subsonic speeds.

Example

For straight tapered planform with sharp leading edges,

Mach number = 1.2	Wing area, S = 6.22 m^2 (67 ft^2)	$P = S/b\ell = 1.0$
Aspect ratio = 1.5	Length = 2.04 m (6.69 ft)	$\beta = 0.663$
Span, b = 3.05 m (10.02 ft)	Taper ratio = 1.0	$\beta(b/2\ell) = 0.497$

$$\pi \text{AR} \frac{C_{Di}}{C_L^2}\left(\frac{P}{1+P}\right) = 0.6 \quad C_{Di} = 0.255 C_L^2$$

AERODYNAMICS, CONTINUED

INDUCED DRAG, CONTINUED

SUPERSONIC DRAG DUE TO LIFT (EMPIRICAL)

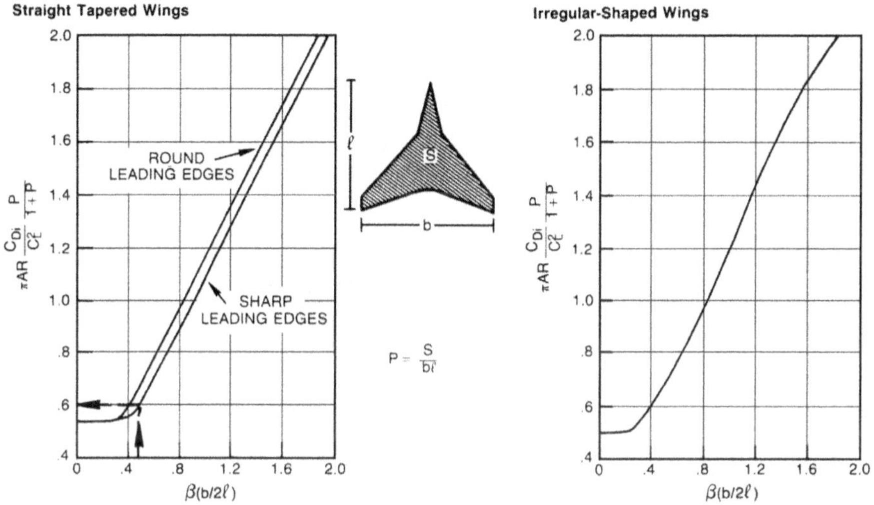

CRITICAL MACH NUMBER

Subsonic drag evaluation terminates at a Mach number where the onset of shock formations on a configuration causes a sudden increase in the drag level—the so-called "critical Mach number." The following graphs show simple working curves for it.

Example

Find critical Mach number for $C_L = 0.4$.

$t/c = 0.068$

$C_{L_{\text{DESIGN}}} = 0.2$

Aspect ratio = 3

$\Lambda_{c/4} = 45°$

$M_{\text{CR}_{\text{CL}=0}} = 0.895$ (see Critical Mach Number graph)

$M_{\text{CR}_{\text{CL}=0.4}} / M_{\text{CR}_{\text{CL}=0}} = 0.97$ (see Lift Effect on Critical Mach Number graph)

$$M_{\text{CR}_{\text{CL}=0.4}} = (0.97)(0.895) = 0.868$$

AERODYNAMICS, CONTINUED

CRITICAL MACH NUMBER, CONTINUED

CRITICAL MACH NUMBER CHART

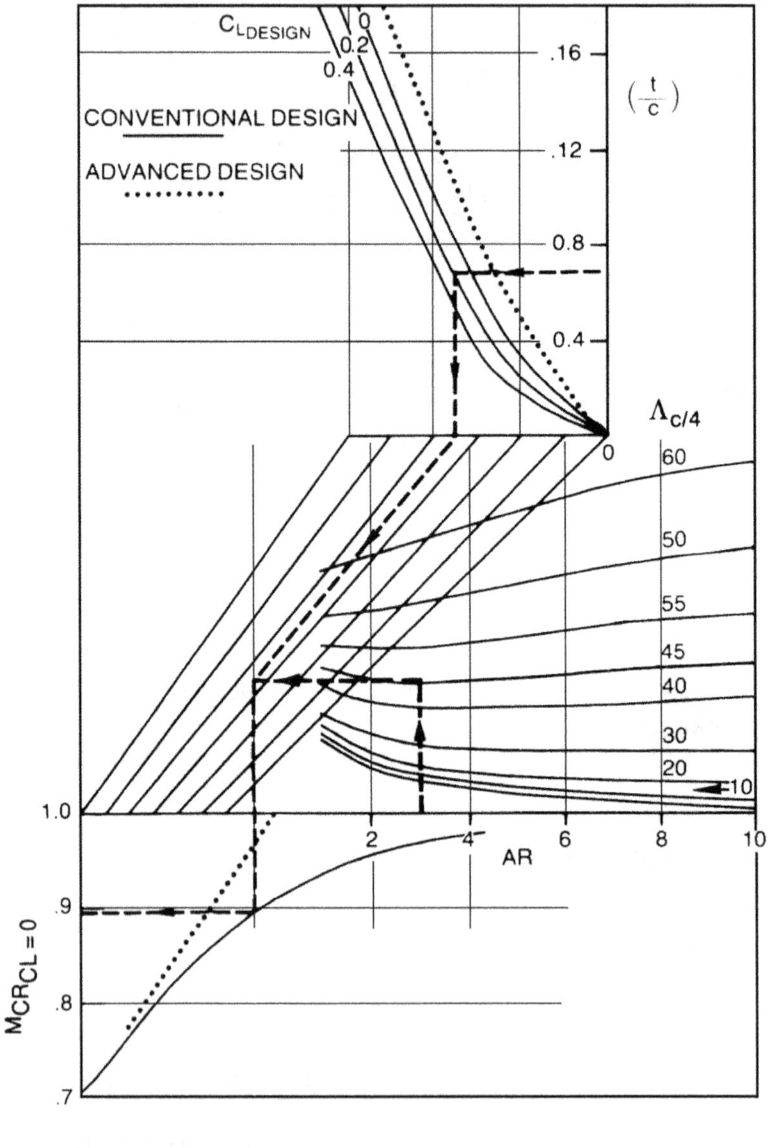

Aerodynamics, Continued

Critical Mach Number, Continued

Lift Effect on Critical Mach Number (For Conventional Airfoils)

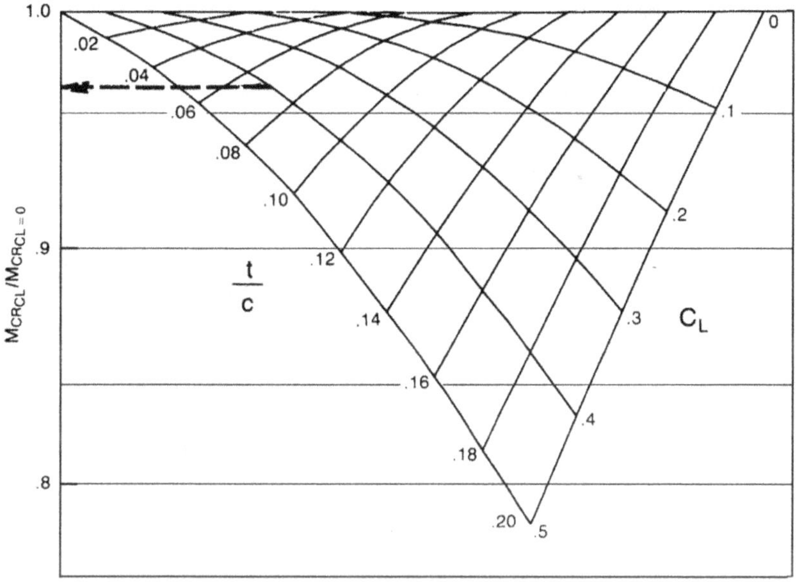

Drag Rise

Following the critical Mach number, the drag level increases abruptly. This phenomenon is associated with strong shocks occurring on the wing or body, causing flow separation. To estimate the drag rise increment at these conditions, Hoerner, in *Fluid Dynamic Drag*, gives the following empirical relation:

$$\Delta C_{D_{\text{Rise}}} = \left(K/10^3\right)\left[10\Delta M \Big/ \left(\frac{1}{\cos\Lambda_{\text{LE}}} - M_{\text{CR}}\right)\right]^n$$

where

K = 0.35 for 6-series airfoils in open tunnels
= 0.40 for airfoil sections with $t/c \approx 6\%$
= 0.50 for thicker airfoils and for 6-series airfoils

$\Delta M = M - M_{\text{CR}}$

$n = 3\Big/\left(1 + \frac{1}{\text{AR}}\right)$

AERODYNAMICS, CONTINUED

DRAG RISE, CONTINUED

Example

Determine the drag rise increment for the following:

ΔM	$= 0.05$
Aspect ratio	$= 3.0$
t/c	$= 0.068$ ($K = 0.40$)
Λ_{LE}	$= 50°$
M_{CR}	$= 0.895$

$$\Delta C_{D_{Rise}} = \frac{0.4}{1000}\left[\frac{10(0.05)}{\cos(50°)^{-1} - 0.895}\right]^{\frac{3}{1+1/3}} = 0.0002136$$

AERODYNAMIC CENTER

The prediction of wing-alone aerodynamic center (a.c.) may be made from curves presented in the following graphs, which show a.c. location as a fraction of the wing root chord. These curves are based on planform characteristics only and are most applicable to low-aspect-ratio wings. The characteristics of high-aspect-ratio wings are primarily determined by two-dimensional section characteristics of the wing.

The aerodynamic center of the wing is the dominant term in the determination of the aerodynamic center of the aircraft, also called the neutral point. The horizontal tail is second to the wing in determination of the neutral point. The fuselage, nacelles, and propulsion system also contribute to the neutral point. For an aircraft to be statically stable, the neutral point must be aft of the center of gravity. The distance that the neutral point is behind the center of gravity, in lengths of $\bar{c}$, is called the static margin, K_n. Measuring the location of the center of gravity (X) and the neutral point (X_{ac}) from the reference datum, such as the nose or firewall, the static margin can be determined by

$$K_n = \frac{X_{ac} - X}{\bar{c}} = \frac{C_{m_\alpha}}{C_{L_\alpha}}$$

Example

Determine the location of the aerodynamic center of a wing under the following conditions:

Mach number	$= 1.2$ ($\beta = 0.663$)
Λ_{LE}	$= 45$ deg
Aspect ratio	$= 2.0$
Taper ratio (λ)	$= 0.02$

$$\frac{X_{ac}}{C_r} = 0.51 \quad \text{(see Location of Wing Aerodynamic Center graph)}$$

AERODYNAMICS, CONTINUED

AERODYNAMIC CENTER, CONTINUED

LOCATION OF WING AERODYNAMIC CENTER

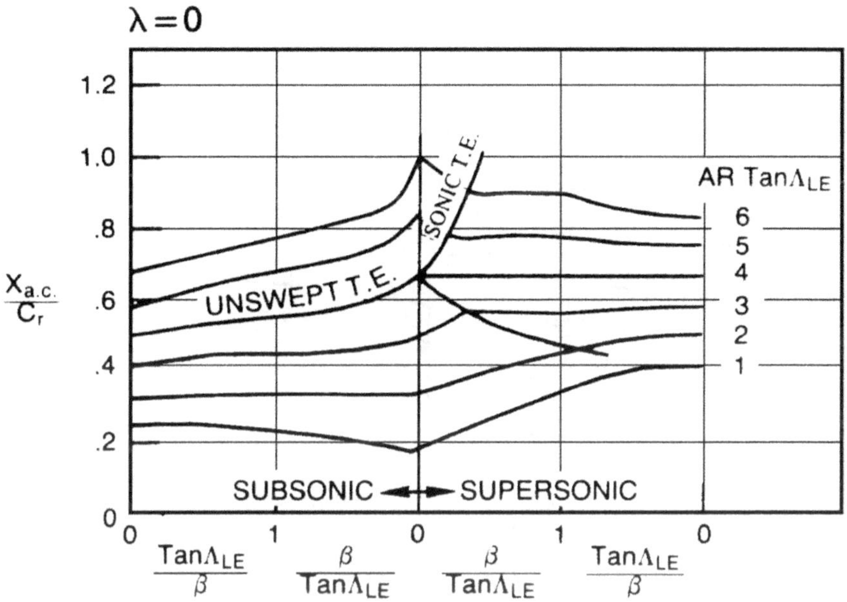

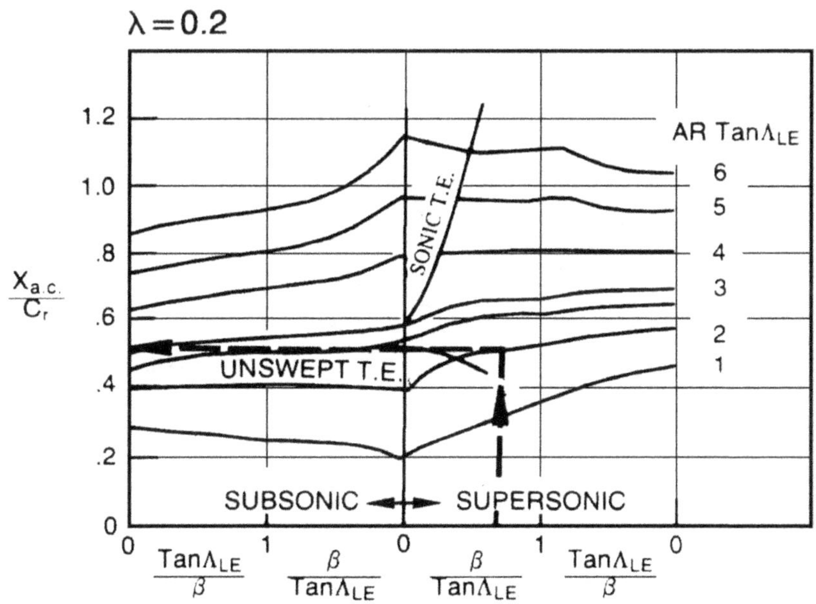

AERODYNAMICS, CONTINUED

AERODYNAMIC CENTER, CONTINUED

LOCATION OF WING AERODYNAMIC CENTER, CONTINUED

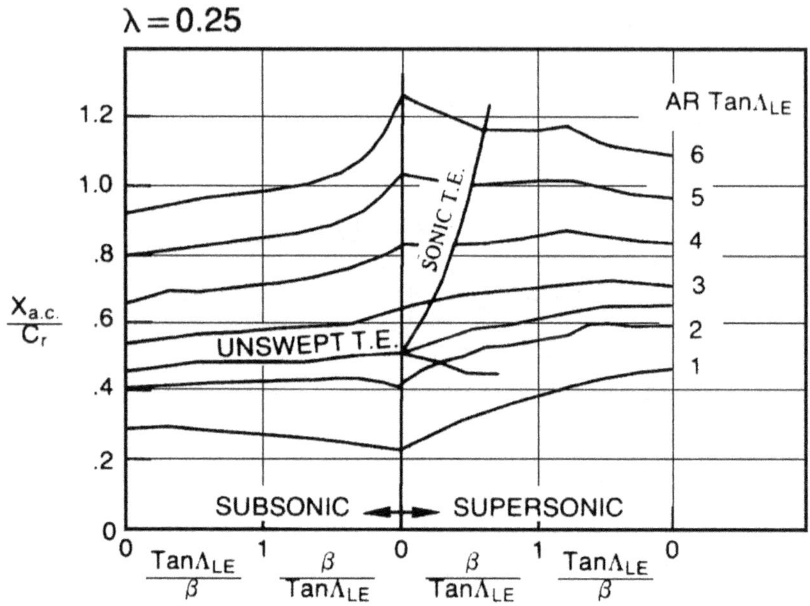

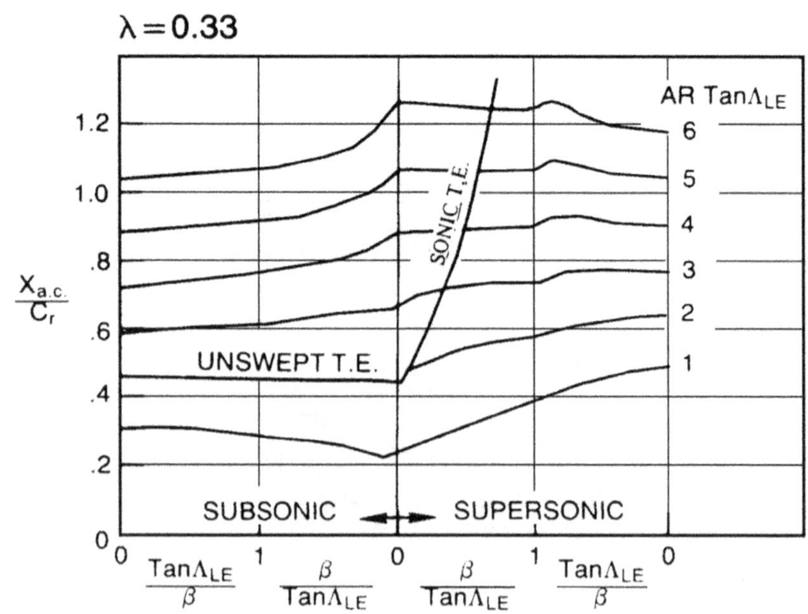

AERODYNAMICS, CONTINUED

AERODYNAMIC CENTER, CONTINUED

LOCATION OF WING AERODYNAMIC CENTER, CONTINUED

$\lambda = 0.5$

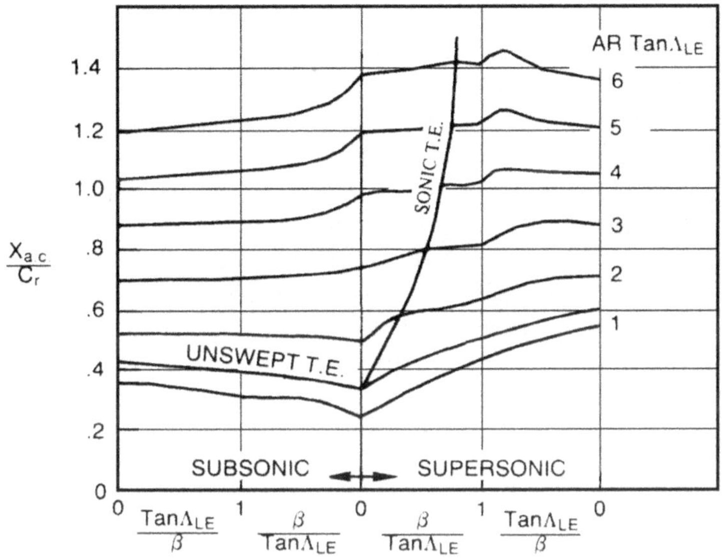

$\lambda = 1.0$

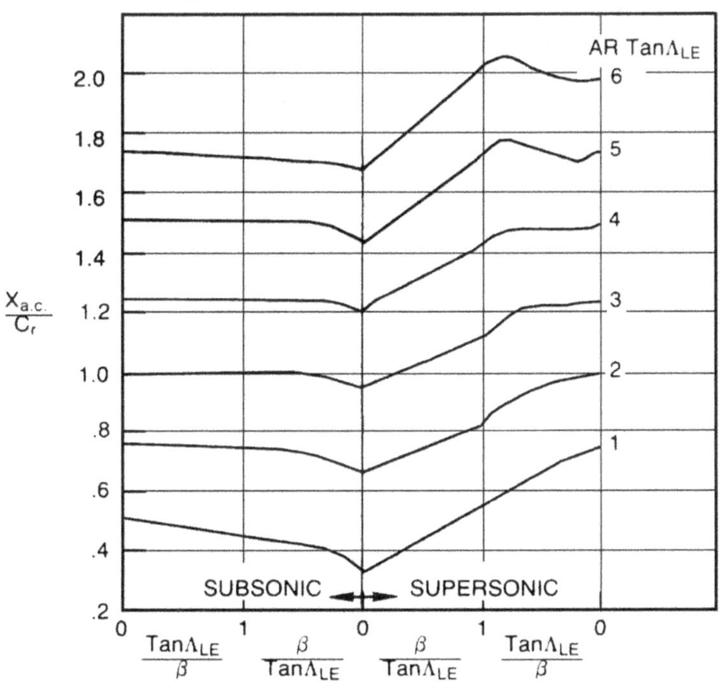

AERODYNAMICS, CONTINUED

MAXIMUM LIFT COEFFICIENT (C_{Lmax})

Determination of maximum lift depends exclusively on the viscous phenomena controlling the flow over the wing. As wing incidence increases, flow separates from the surface, and lifting pressures cease to be generated. These phenomena depend on the wing geometry: sweep, aspect ratio, taper ratio, thickness ratio, and airfoil section. The thickness ratio has a decidedly marked effect through the influence of the leading-edge radius. Being a viscous phenomenon, a strong effect of Reynolds number is apparent. As a consequence, the precise determination of C_{Lmax} for an arbitrary wing has never been satisfactorily accomplished. The *USAF Stability and Control Handbook* (WADD-TR-60-261: Lib. 57211) does contain such a method. However, correlations indicate errors exceeding 25%. This method works with section lift data and applies corrections for finite airplane effects. Therefore, predictions must depend on existing test data and experience.

HIGH-LIFT DEVICES

MAXIMUM LIFT INCREMENT DUE TO FLAPS

High-lift devices are designed for certain specialized functions. Generally they are used to increase the wing camber or in some other way to control the flow over the wing, for example, to prevent flow separation. Wing flaps increase the camber at the wing trailing edge, thus inducing a higher lift due to increased circulation at the same angle of attack as the plain wing. Plain-flap effectiveness can be determined very accurately. Other devices customarily employed are slats, slots, and special leading-edge modification. Evaluation of these devices depends on the application of NASA results. British report Aero 2185 and NACA Technical Note 3911 contain prediction curves and techniques for these devices.

At supersonic speeds, high-lift devices are generally not required for flow stabilization. However, flap-type controls may be used to trim the airplane pitching moments.

The determination of maximum lift increment due to trailing-edge flap deflection uses the method of NACA TN 3911. The maximum lift increment is given by

$$\Delta C_{L\max} = \Delta C_{\ell\max} \frac{C_{L_\alpha}}{C_{\ell_\alpha}} K_c \Delta K_b$$

where $\Delta C_{\ell\max}$ = increment of section lift coefficient due to flap deflection (see Princeton Report No. 349).

AERODYNAMICS, CONTINUED

HIGH-LIFT DEVICES, CONTINUED

MAXIMUM LIFT INCREMENT DUE TO FLAPS, CONTINUED

$$C_{L_\alpha}/C_{\ell_\alpha} = \frac{\text{AR}}{\frac{a_0}{\pi} + \left[\left(\frac{a_0}{\pi}\right)^2 + \left(\frac{\text{AR}}{\cos\Lambda_{C/2}}\right)^2\right]^{\frac{1}{2}}}$$

where

a_0 = section lift-curve slope, per radian
K_b = flap span factor (see Flap Span Factor graph)
K_c = flap chord factor (see Flap Chord Factor graph)

Care should be exercised in use of all prediction techniques for $\Delta C_{L\max}$ due to flap deflection, because flap effectiveness is modified to a large extent by the ability of the wing-leading-edge devices to maintain attached flow.

Example

Determine $\Delta C_{L\max}$ due to flap deflection, with the following wing characteristics:

Aspect ratio = 4.0
Section lift-curve slope = 5.73 per radian
Semichord sweep = 20 deg
Taper ratio = 0.2

$$C_{L_\alpha}/C_{\ell_\alpha} = \frac{4.0}{\frac{5.73}{\pi} + \left[\left(\frac{5.73}{\pi}\right)^2 + \left(\frac{4.0}{\cos(20)}\right)^2\right]^{\frac{1}{2}}} = 0.6197$$

For plain flaps with 50 deg deflection, 16% chord,

$\Delta C_{\ell\max}$ = 0.80 for a typical plain flap
Inboard span, η_I = 0.15 $K_{bi} = 0.22$ (see Flap Span Factor graph)
Outboard span, η_o = 0.65 $K_{bo} = 0.80$ (see Flap Span Factor graph)
$\Delta K_b = 0.8 - 0.22 = 0.58$
$(\alpha\delta)C_\ell$ = 0.5 (see Flap Chord Factor graph)
K_c = 1.1 (see Flap Chord Factor graph)

$$C_{L\max} = (0.8)(0.6197)(1.1)(0.58) = 0.3163$$

AERODYNAMICS, CONTINUED

HIGH-LIFT DEVICES, CONTINUED

FLAP CHORD FACTOR, K_c

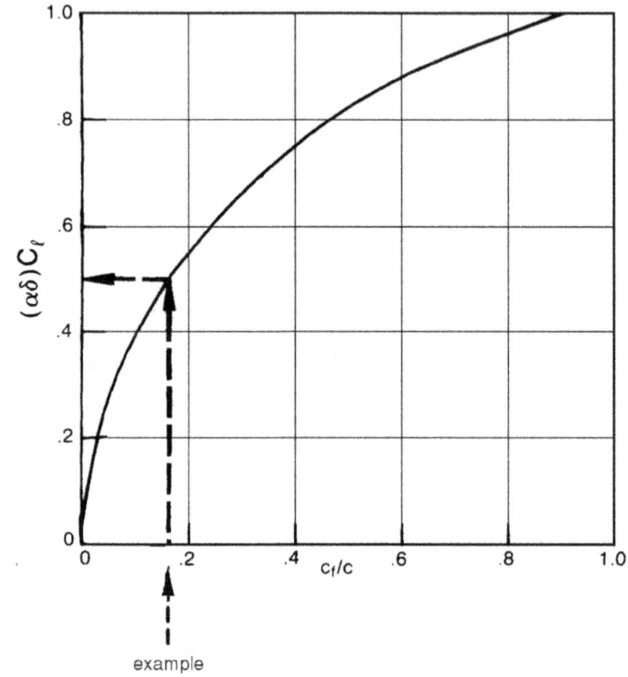

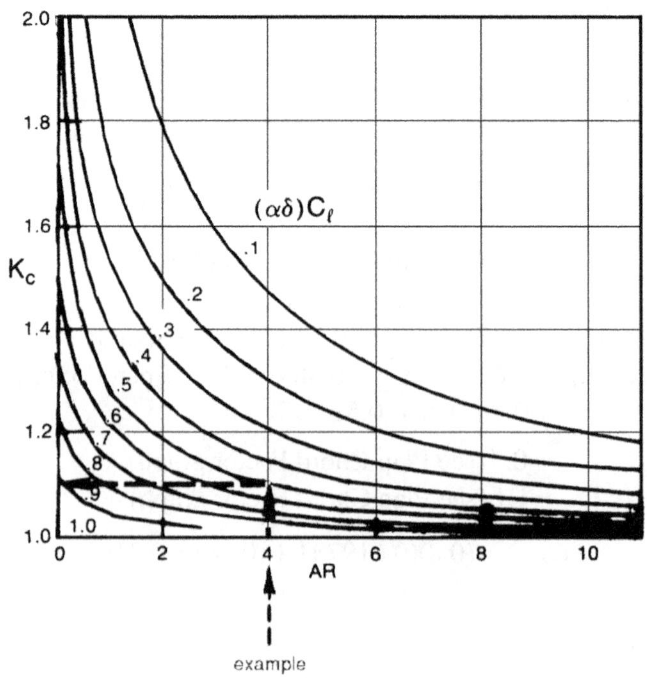

AERODYNAMICS, CONTINUED

HIGH-LIFT DEVICES, CONTINUED

FLAP SPAN FACTOR, K_b

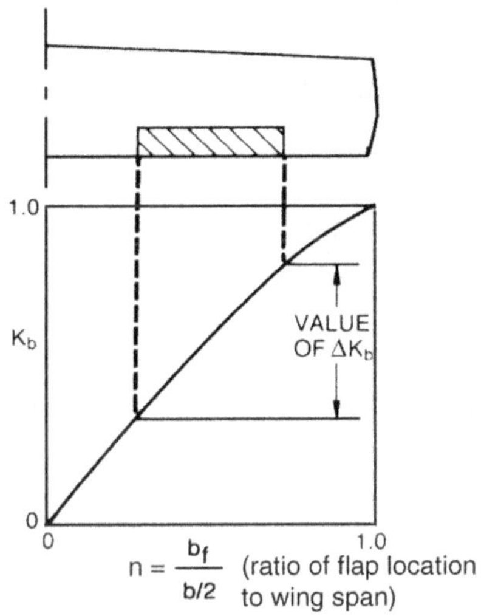

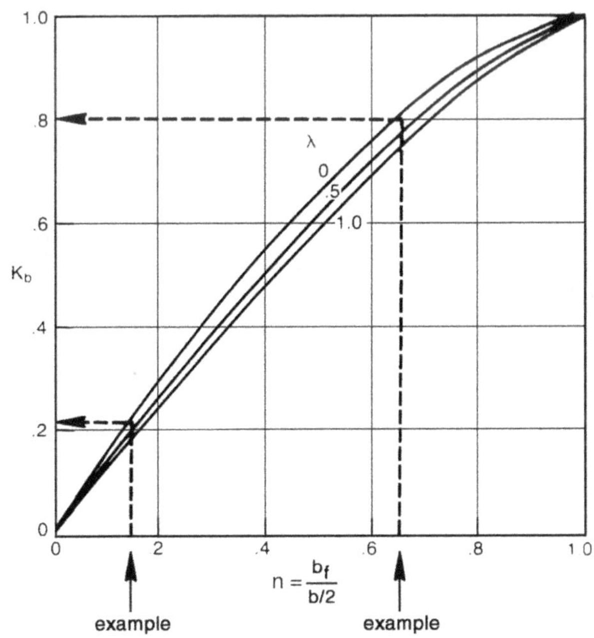

AERODYNAMICS, CONTINUED

HIGH-LIFT DEVICES, CONTINUED

EFFECTS OF FLAP DEFLECTION ON ZERO-LIFT ANGLE OF ATTACK

The set of Plain-Flap Effectiveness graphs may be used with the following expression to obtain subsonic, plain-flap effectiveness. Correction factors for body effect, partial span, and flap-leading-edge gap are shown.

$$\Delta\alpha_{LO} = (\partial\alpha/\partial\delta)\cos\Lambda_{HL}\,\delta[1 - (a + b\beta)|\delta|]$$

Example

Determine $\Delta\alpha_{LO}$ due to plain-flap deflection, with the following wing characteristics:

Aspect ratio = 4.0
Sweep at flap hinge line = 10 deg
Sweep at wing leading edge = 32 deg

For inboard plain flaps with 25 deg deflection, 15% flap chord,

Flap gap ratio, GAP/$\bar{c}_f$ = 0.002
Inboard span, η_I = 0.15
Outboard span, η_o = 0.65
Ratio of body diameter to wing span, $2R_0/b$ = 0.15
Approach Mach number = 0.1
$\beta = \sqrt{1-M^2} = 0.995$

Flap effectiveness = $-\partial\alpha/\partial\delta = 0.49$ (see Plain-Flap Effectiveness graph)

Sweep factor:

$$-a = 0.039 \qquad b = 0.034$$ (see Sweep factor graph)

Fuselage factor:

$$\frac{(\Delta\alpha_{Lo})_{WB}}{(\Delta\alpha_{Lo})_{W}} = 0.93$$ (see Wing–body graph)

Flap span factor:

$$S_{F(PS)}/S_{F(FS)} = \eta_0 - \eta_1 = 0.65 - 0.15 = 0.50$$

$$\frac{(\Delta\alpha_{Lo})_{PS}}{(\Delta\alpha_{Lo})_{FS}} = 0.55$$ (see Flap span graph)

Flap gap factor:

$$\frac{(\Delta\alpha_{Lo})_{gap}}{(\Delta\alpha_{Lo})_{sealed}} = 0.96$$ (see Flap Gap graph)

$$\Delta\alpha_{LO} = (-0.49)\cos(10\text{ deg})(25)[1 - (-0.039 + 0.034(0.995))(25)] \times (0.93)(0.55)(0.96) = -6.69\text{ deg}$$

AERODYNAMICS, CONTINUED

HIGH-LIFT DEVICES, CONTINUED

PLAIN-FLAP EFFECTIVENESS

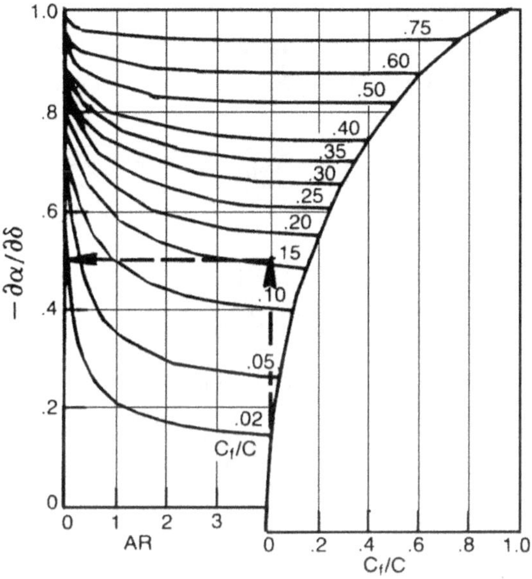

Full-span effectiveness.

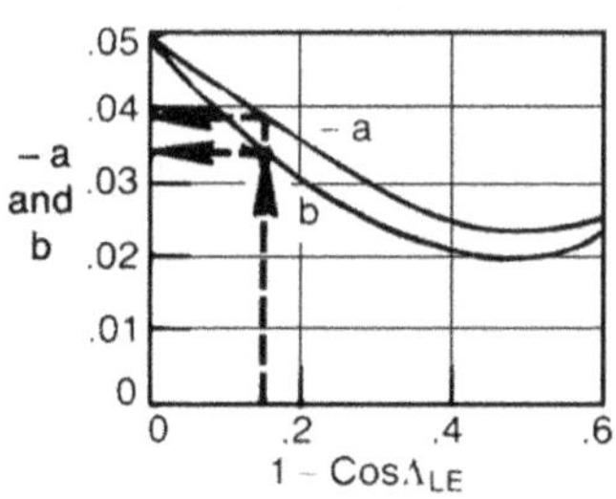

Sweep factor.

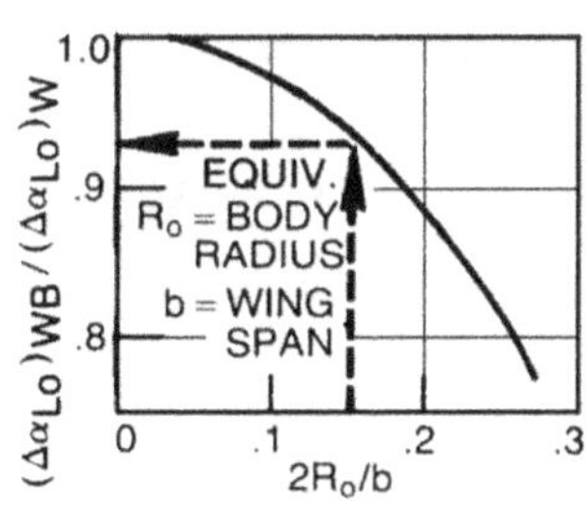

Wing-body.

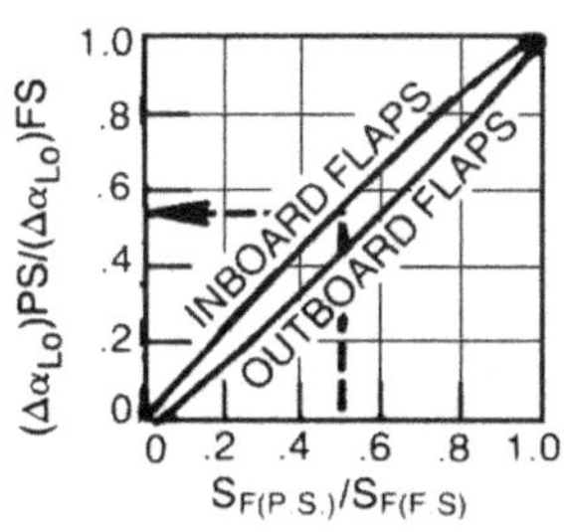

Flap span.

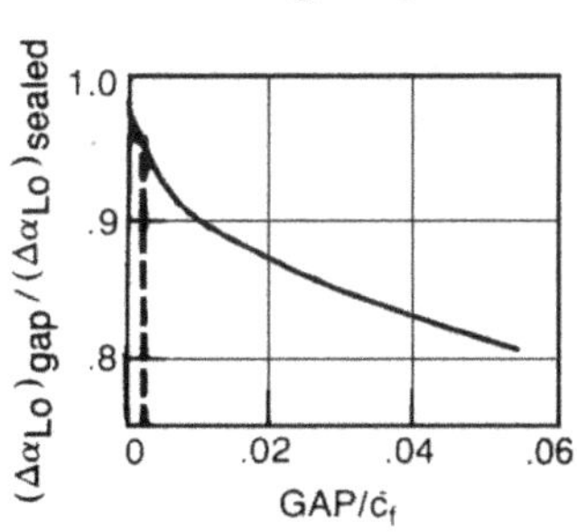

Flap gap.

AERODYNAMICS, CONTINUED

HIGH-LIFT DEVICES, CONTINUED

INDUCED DRAG DUE TO FLAPS DEPLOYED

Subsonic. Trailing-edge flaps on a wing give it a camber for improving lift. Elevons in a similar manner camber the wing, although they are utilized for control at cruise lift values. Drag due to lift ratio can be expressed as follows:

$$\frac{C_{Di\delta}}{C_{Di\delta}=0} = (1-F)\left(\frac{\alpha-\alpha_{Lo_\delta}}{\alpha-\alpha_{Lo}}\right)^2 + F\left(\frac{\alpha-\alpha_{Lo_\delta}}{\alpha-\alpha_{Lo}}\right)$$

where $F = \sin\Lambda_{c/4}(0.295 + 0.066\alpha - 0.00165\,\alpha^2)$.

The following graph shows working plots of these relationships. Zero-lift angle of attack with flaps deflected is obtained from the preceding section.

Supersonic. At the supersonic speeds, flaps as such are not likely to be used. However, elevons still are required on tailless configurations for pitch control. The subsonic data listed above may be used.

Example

Find induced-drag ratio for flaps deployed for the following wing characteristics: angle of attack at zero lift α_{LO} of –6 deg and quarter-chord sweep $\Lambda_{c/4}$ = 26 deg. For a plain flap with 25 deg deflection, and with angle of attack at zero-lift with flaps deflected α_{Lo_δ} of –10 deg, the induced-drag ratio at 10 deg angle of attack will be as follows.

$$\frac{F}{\sin\Lambda_{c/4}} = 0.79 \text{ (see Drag-Due-to-Lift Ratio with Flap Deflection graph)}$$

$$\frac{\alpha-\alpha_{Lo_\delta}}{\alpha-\alpha_{Lo}} = \frac{10-(-10)}{10-(-6)} = 1.25$$

$$F = (0.79)\sin(26\text{ deg}) = 0.3463$$

$$\frac{C_{Di\delta}}{C_{Di\delta}=0} = 1.454 \text{ (see Drag-Due-to-Lift Ratio with Flap Deflection graph)}$$

AERODYNAMICS, CONTINUED

HIGH-LIFT DEVICES, CONTINUED

DRAG-DUE-TO-LIFT RATIO WITH FLAP DEFLECTION

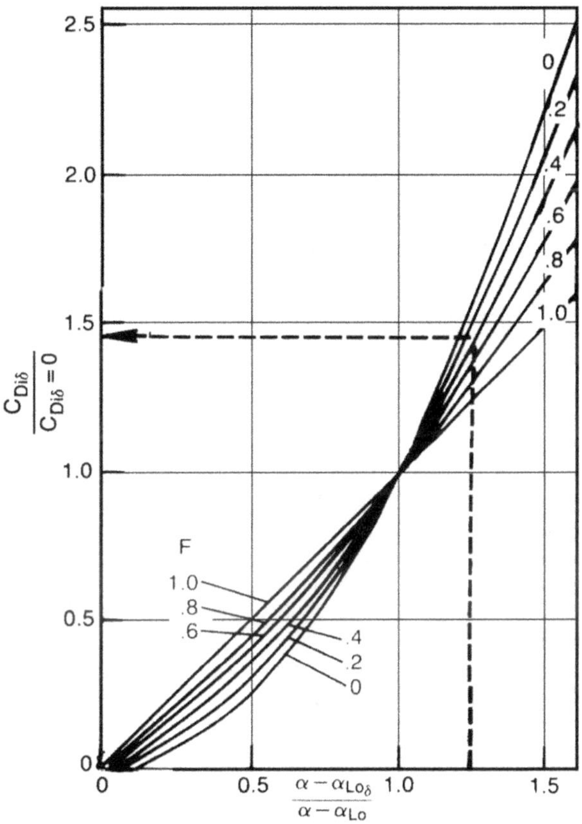

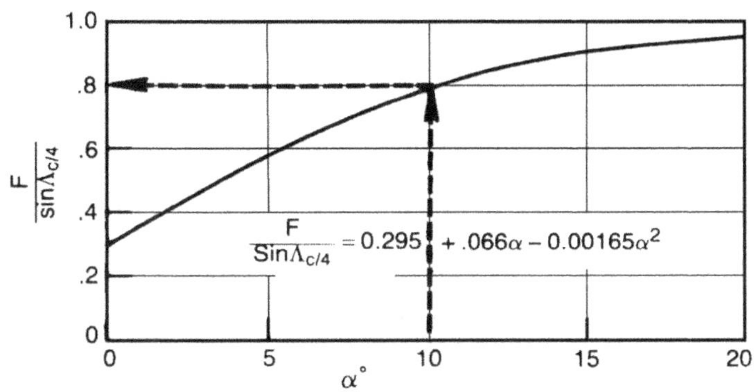

$$\frac{F}{\sin\Lambda_{c/4}} = 0.295 + .066\alpha - 0.00165\alpha^2$$

PERFORMANCE

NOMENCLATURE

A = margin above stall speed (typically 1.2 for takeoff and 1.15 for landing)
a = speed of sound at altitude
AR = aspect ratio
C_D = drag coefficient
C_L = lift coefficient
D = $(C_D)(q)(S)$ = aircraft drag
e = Oswald efficiency factor (wing)
F_{pl} = planform factor for takeoff calculations
F = thrust
F/D = ratio of thrust to drag
ΔFuel = fuel increment
g = acceleration due to gravity
h = altitude
h_{obs} = height of obstacle
K = $1/(\pi)(\text{AR})(e)$ = parabolic drag polar factor
k = $(L/D)(V/\text{SFC})$ = range constant
L = $(C_L)(q)(S)$ = aircraft lift
L/D = ratio of lift to drag
M = Mach number
M_{DD} = drag divergence Mach number
n = load factor (lift/weight)
p = atmospheric pressure at altitude
P_S = specific excess power
q = $\frac{1}{2}\rho V^2$ = dynamic pressure
R = range
R/C = rate of climb
R/D = rate of descent
S = reference wing area
S_{air} = air run distance
S_{gnd} = ground roll distance
S_{land} = total landing distance
S_{obs} = distance to clear obstacle
SFC = W_f/F = specific fuel consumption
S_{TO} = total takeoff distance
SR = V/W_f = specific range
ΔS = distance increment
Δt = time increment
V = velocity, true airspeed
V_{obs} = velocity at obstacle

PERFORMANCE, CONTINUED

NOMENCLATURE, CONTINUED

V_{stall} = $(2W/S\rho C_{L\max})^{1/2}$ = stall speed
V_{TD} = $A \times V_{stall}$ = landing velocity
ΔV = velocity increment
W = weight
W_f = fuel flow
γ = flight-path angle
μ = coefficient of rolling friction
μ_{BRK} = coefficient of braking friction
ρ = air density at an altitude
$\dot{\psi}$ = rate of turn

TAKEOFF

Takeoff-distance calculations treat ground roll and the distance to clear an obstacle. Obstacle requirements differ for commercial (10.7m (35 ft)) and military (15.2m (50 ft)) aircraft.

TAKEOFF GROUND ROLL

$$S_{gnd} = -\frac{W/S}{g\rho(C_D - \mu C_L)} \ln\left[1 - \frac{A^2(C_D - \mu C_L)}{((F/W) - \mu)C_{L_{\max}}}\right]$$

The stall margin A typically is 1.2.

TOTAL TAKEOFF DISTANCE

$$S_{TO} = (S_{gnd})(F_{pl})$$

The factor to clear an obstacle depends greatly on available excess thrust, flight path, and pilot technique. The following typical factors characterize planforms in ability to clear a 15.2m (50 ft) obstacle:

Planform	F_{pl}
Straight wing	1.15
Swept wing	1.36
Delta wing	1.58

PERFORMANCE, CONTINUED

TAKEOFF, CONTINUED

TOTAL TAKEOFF DISTANCE, CONTINUED

Example

Takeoff distance for a straight-wing aircraft with the following characteristics:

W = 98,288 N (22,096 lb)
S = 24.3 m^2(262 ft^2)
C_L = 0.46
C_D = 0.3538
S_{gnd} = 382.5 m (1255 ft)
$C_{L_{max}}$ = 1.8
F = 85,806 N (19,290 lb)
μ = 0.025
A = 1.2
ρ = 1.225 kg m^{-3} (0.0023769 lb-s^2-ft^{-4})
F_{pl} = 1.15

$$S_{gnd} = -\frac{98{,}288/24.3}{(9.807)(1.225)(0.3538-(0.025)(0.46))} \times \ln\left[1-\frac{(1.2)^2(0.3538-(0.025)(0.46))}{((85{,}806/98{,}288)-0.025)(1.8)}\right]$$

$$= 382.6 \text{ m}$$

$$S_{T0} = (382.6)(1.15) = 440 \text{ m (1444 ft)}$$

TAKEOFF FUEL ALLOWANCE

For brake release to initial climb speed,

$$\text{Fuel} = \frac{W_1 V_1}{g(F_1 - D_1)}\left(\frac{W_{f_0} + W_{f_1}}{2}\right)$$

where

W_1 = weight at start of climb
V_1 = initial climb speed, m/s (ft/s)
g = 9.807 m/s^2 (32.174 ft/s^2)
F_1 = maximum power thrust at initial climb speed
D_1 = drag at 1-g flight condition, initial climb speed
W_{f_0} = maximum power fuel flow at brake release, kg/s (lb/s)
W_{f_1} = maximum power fuel flow at initial climb speed, kg/s (lb/s)

PERFORMANCE, CONTINUED

TAKEOFF, CONTINUED

CLIMB

Time, fuel, and distance to climb from one altitude (h_1) to another (h_2) can be calculated in increments and then summed. By using this technique, specific climb speed schedules—i.e., constant Mach number climb and maximum rate of climb—can be depicted.

RATE OF CLIMB

For small angles, the rate of climb can be determined from

$$R/C = (F - D)V/W\left(1 + \frac{V}{g}\cdot\frac{\mathrm{d}V}{\mathrm{d}h}\right)$$

where $V/g \cdot \mathrm{d}V/\mathrm{d}h$ is the correction term for flight acceleration.

The following table gives acceleration corrections for typical flight procedures:

Altitude	Procedure	$\frac{V}{g}\cdot\frac{dV}{dh}$
11 km (36,089 ft) or less	Constant CAS	0.5668 M^2
	Constant Mach	–0.1332 M^2
Over 11 km (36,089 ft)	Constant CAS	0.7 M^2
	Constant Mach	Zero

For low subsonic climb speeds, the acceleration term is usually neglected.

$$R/C = (F - D)V/W$$

FLIGHT-PATH GRADIENT

$$\gamma = \sin^{-1}\left(\frac{F - D}{W}\right)$$

TIME TO CLIMB

$$\Delta t = \frac{2(h_2 - h_1)}{(R/C)_1 + (R/C)_2}$$

DISTANCE TO CLIMB

$$\Delta S = V(\Delta t)$$

PERFORMANCE, CONTINUED

CLIMB, CONTINUED

FUEL TO CLIMB

$$\Delta\text{Fuel} = W_f(\Delta t)$$

Sum increments for total.

ACCELERATION

The time, fuel, and distance for acceleration at a constant altitude from one speed to another can be calculated in increments and then summed up. By using this technique, specific functions can be simulated (e.g., engine power spool-up).

TIME-TO-ACCELERATE INCREMENT

$$\Delta t = \frac{\Delta V}{g\left(\frac{F-D}{W}\right)}$$

DISTANCE-TO-ACCELERATE INCREMENT

$$\Delta S = V(\Delta T)$$

FUEL-TO-ACCELERATE INCREMENT

$$\Delta\text{Fuel} = Wf(\Delta t)$$

Sum increments to yield total time, fuel, and distance to accelerate.

CRUISE

The basic cruise distance can be determined by using the Breguet range equation for jet aircraft, as follows.

CRUISE RANGE

$$R = L/D(V/\text{SFC})\ln(W_0/W_1)$$

where subscripts "0" and "1" stand for initial and final weight, respectively.

CRUISE FUEL

$$\text{Fuel} = W_0 - W_1 = W_f(e^{R/k} - 1)$$

where k, the range constant, equals $L/D(V/\text{SFC})$.

PERFORMANCE, CONTINUED

CRUISE, CONTINUED

DASH RANGE

For flight at constant Mach number and constant altitude, the following equation gives dash range:

$$R = \left(\frac{V}{\text{SFC}}\right)\left(\frac{1}{F}\right)(\text{Fuel})$$

For large excursions of weight, speed, and altitude during cruise, it is recommended that the range calculations be divided into increments and summed up for the total.

Example

Find cruise distance for an aircraft with the following characteristics:

W_0 = 7167 kg (15,800 lb)
W_1 = 6622 kg (14,600 lb)
V = 138 m/s (268 Kn)
SFC = 3.5×10^{-4} kg/kg-s (1.26 lb/h/lb)
L/D = 9.7

$$R = 9.7\,(138/3.5 \times 10^{-4})\ln(7167/6622) = 302 \text{ km } (163 \text{ n mile})$$

CRUISE SPEEDS

Cruise-speed schedules for subsonic flight can be determined by the following expressions.

OPTIMUM MACH NUMBER (M_{DD}), OPTIMUM-ALTITUDE CRUISE

First calculate the pressure at altitude.

$$P = \frac{W}{0.7\left(M_{DD}^2\right)\left(C_{L_{DD}}\right)S}$$

Then enter value from Cruise-Altitude Determination graph for cruise altitude.

PERFORMANCE, CONTINUED

CRUISE, CONTINUED

OPTIMUM MACH NUMBER, CONSTANT-ALTITUDE CRUISE

Optimum occurs at maximum $M\,(L/D)$.

$$M = \sqrt{\frac{W/S}{0.7P}\sqrt{\frac{3K}{C_{D_{\min}}}}}$$

CONSTANT MACH NUMBER, OPTIMUM-ALTITUDE CRUISE

Derive optimum altitude, as above, except M_{DD} and $C_{L_{DD}}$ are replaced with values for the specified cruise conditions.

CRUISE-ALTITUDE DETERMINATION

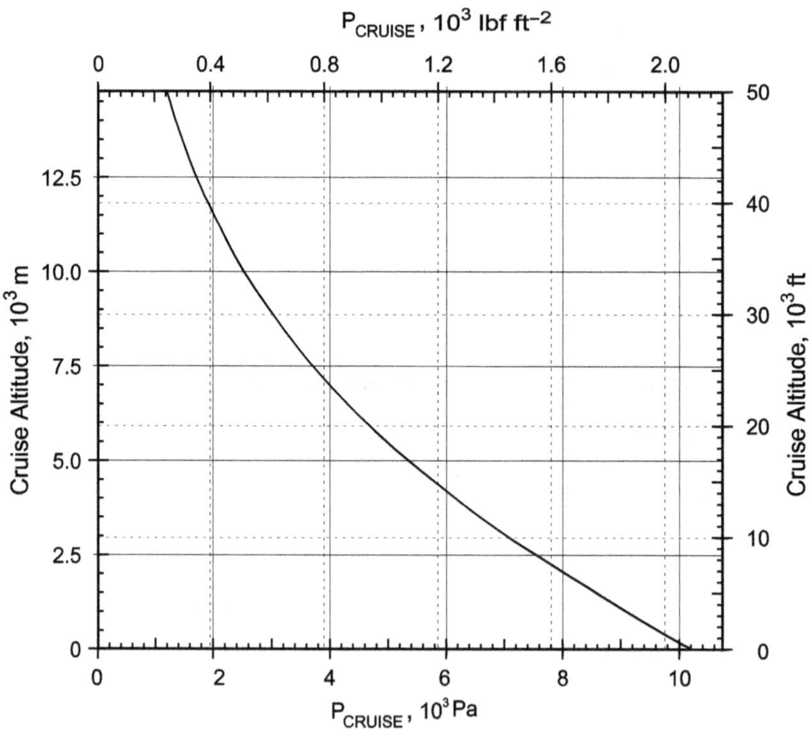

PERFORMANCE, CONTINUED

LOITER

Loiter performance is based on conditions at $(L/D)_{max}$, because maximum endurance is of primary concern.

LOITER SPEED

$$M = \sqrt{\frac{W/S}{0.7(P)(C_L)_{(L/D)_{max}}}}$$

where

$$C_{L(L/D)_{max}} = \sqrt{\frac{C_{D_{min}}}{K}}$$

LOITER TIME

$$t = L/D_{max}\left(\frac{1}{SFC}\right)\ln\left(\frac{W_0}{W_1}\right)$$

LOITER FUEL

$$\text{Fuel} = W_f e^y \quad \text{where } y = \left\{\frac{t(SFC)}{L/D_{max} - 1}\right\}$$

Example

Find time to loiter at altitude for aircraft with the following characteristics:

Initial weight, W_0	= 5023 kg (11,074 lb)
Final weight, W_1	= 4536 kg (10,000 lb)
L/D_{max}	= 10.2
SFC	= 6.04×10^{-4} kg/kg-s (2.175 lb/h/lb)
t	$= 10.2(1/6.04 \times 10^{-4})\ln(5023/4536) = 1722\text{s}\ (0.478\text{h})$

MANEUVER

The measure of maneuverability of a vehicle can be expressed in terms of sustained and instantaneous performance. For sustained performance, thrust available from the engines must equal the drag of the vehicle (i.e., specific excess power equals zero).

PERFORMANCE, CONTINUED

MANEUVER, CONTINUED

SPECIFIC EXCESS POWER

$$P_S = \left(\frac{F - D}{W}\right)V$$

TURN RADIUS

$$\text{TR} = \frac{V^2}{g\sqrt{n^2 - 1}}$$

TURN RATE

$$\dot{\psi} = \frac{V}{\text{TR}} = \frac{g}{V}\sqrt{n^2 - 1}$$

FUEL REQUIRED FOR N SUSTAINED TURNS

$$\text{Fuel} = \frac{(\text{SFC})(F)(N)(\text{fac})}{\dot{\psi}}$$

where fac = 360 if $\dot{\psi}$ is in degrees per second; fac = 2π if $\dot{\psi}$ is in radians per second.

Example

Determine turn radius for the following vehicle:

$V = 186$ m/s (610 fps)

$n = 1.3\ g$

$$\text{TR} = \left(\frac{186}{9.807\sqrt{1.3^2 - 1}}\right) = 4247\,\text{m}\ (13{,}923\ \text{ft})$$

LANDING

Landing distance calculations cover distance from obstacle height to touchdown and ground roll from touchdown to a complete stop.

PERFORMANCE, CONTINUED

LANDING, CONTINUED

APPROACH DISTANCE

$$S_{\text{air}} = \left(\frac{V_{\text{obs}}^2 - V_{\text{TD}}^2}{2g} + h_{\text{obs}} \right) (L/D)$$

where V_{obs} = speed at obstacle, V_{TD} = speed at touchdown, h_{obs} = obstacle height, and L/D = lift-to-drag ratio.

LANDING GROUND-ROLL

$$S_{\text{gnd}} = \frac{(W/S)}{g\rho\,(C_D - \mu_{\text{BRK}} C_L)} \ln \left[1 - \frac{A^2 (C_D - \mu_{\text{BRK}} C_L)}{((F/W) - \mu_{\text{BRK}}) C_{L_{\max}}} \right]$$

This equation, of course, also describes takeoff roll, except being positive to account for deceleration from touchdown to zero velocity.

Normally the distance would require a two-second delay to cover the time required to achieve full braking. Commercial requirements may also dictate conservative factors be applied to the calculated distances.

Example

Find landing distance for an aircraft with the following characteristics.

W	= 75,620 N (17,000 lb)	S_{gnd}	= 375 m (1229 ft)
S	= 24.3 m^2 (262 ft^2)	S_{land}	= $S_{\text{air}} + S_{\text{gnd}}$ = 502 m (1647 ft)
C_L	= 0.46	μ_{BRK}	= 0.6
C_D	= 0.352	A	= 1.15
$C_{L_{\max}}$	= 1.44	V_{obs}	= 68.9 m/s (226 ft/s)
ρ	= 1.225 kg m^{-3} (0.0023769 lb-s^2-ft^{-4})	V_{TD}	= 61.0 m/s (200 ft/s)
F	= 0.0 N (lb)	L/D	= 1.88
S_{air}	= 127 m (418 ft)	h_{obs}	= 15.2 m (50 ft)

AIRCRAFT COMBAT SURVIVABILITY

Aircraft combat survivability (ACS) is the capability of an aircraft to avoid or withstand a man-made hostile environment. This environment is made up of several effects that influence ACS, each having an associated probability of occurrence or of causing significant damage. A summary of these effects and probabilities with additional considerations follows.

SINGLE SHOT SCENARIO

NOMENCLATURE

P_A = Probability that a threat weapon is active, searching for, and ready to encounter the aircraft that enter into its defended area

$P_{D/A}$ = Conditional probability that the aircraft is detected, given that the threat is active

$P_{L/D}$ = Conditional probability that the aircraft is tracked, a fire control solution is obtained, and a missile is launched or a gun is fired at the aircraft, given that the threat was active and detected the aircraft

$P_{I/L}$ = Conditional probability that the threat propagator intercepts the aircraft, given that the threat propagator was launched or fired at the aircraft

$P_{H/I}$ = Conditional probability that the threat propagator hits the aircraft, given that the propagator has intercepted the aircraft

$P_{K/H}$ = Conditional probability that the aircraft is killed, given a hit by the propagator

$P_{F/I}$ = Conditional probability that the proximity fuze on the propagator detonates a high-explosive (HE) warhead, given that the propagator has intercepted the aircraft

$P_{K/F}$ = Conditional probability that an aircraft is killed, given a proximity-fuzed warhead detonation

AIRCRAFT COMBAT SURVIVABILITY, CONTINUED

PROBABILITY EQUATIONS

$P_H = P_A\, P_{D/A}\, P_{L/D}\, P_{I/L}\, P_{H/I}$	Probability the aircraft is hit in the scenario
$P_K = P_H\, P_{K/H}$	Probability the aircraft is killed in the scenario (by a hit)
$P_F = P_A\, P_{D/A}\, P_{L/D}\, P_{I/L}\, P_{F/I}$	Probability the HE warhead on the propagator fuzes in the scenario
$P_K = P_F\, P_{K/F}$	Probability the aircraft is killed in the scenario (by a proximity-fuzed detonation)
$P_{E/A} = P_{D/A}\, P_{L/D}$	Probability the aircraft is engaged by an active weapon
$P_E = P_A\, P_{E/A}$	Probability the aircraft is engaged in the scenario
$P_{E\ (single\ shot)} = P_{SS} = P_A\, P_{E/A}$	Probability a single shot is fired at the aircraft in the scenario
$P_{H/SS} = P_{I/L}\, P_{H/I}$	Probability the aircraft is hit, given a single shot
$P_{F/SS} = P_{I/L}\, P_{F/I}$	Probability of fuzing, given a single shot
$P_{K/SS} = P_{SSK} = P_{H/SS}\, P_{K/H}$	Probability the aircraft is killed (by a hit), given a single shot
$P_{K/SS} = P_{SSK} = P_{F/SS}\, P_{K/F}$	Probability the aircraft is killed (by a proximity-fuzed detonation), given a single shot
$P_K = P_{SS}\, P_{K/SS}$	Probability the aircraft is killed in a single-shot scenario
$P_S = 1 - P_K$	Probability the aircraft survives the single-shot scenario

MULTIPLE, INDEPENDENT SHOT SCENARIO

NOMENCLATURE

P_E = Probability the aircraft is engaged in the scenario with one or more shots

P_{Ki} = Probability the "i-th" shot kills the aircraft, given a shot at a live aircraft

P_{Si} = Probability the "i-th" shot does not kill the aircraft, given a shot at a live aircraft

AIRCRAFT COMBAT SURVIVABILITY, CONTINUED

PROBABILITY EQUATIONS

$P_K^{(i)} = P_{S1}P_{S2}...P_{Si-1}P_{Ki}$	Probability the aircraft is killed by the "i-th" shot in the scenario
$P_{K/E} = P_K^{(1)} + P_K^{(2)} ... + P_K^{(N)}$	Probability the aircraft is killed in the engagement consisting of N shots
$P_{S/E} = P_{S1}P_{S2}...P_{SN} = (1-P_{K1})(1-P_{K2})...(1-P_{KN})$	Probability the aircraft survives the N (different) shot engagements
$P_{S/E} = (1-P_{K/SS})^N = 1-P_{K/E}$	Probability the aircraft survives the N (identical) shot engagements

POTENTIAL COMPONENT AND SYSTEM KILL MODES

Component/System	Kill Mode
Fuel	Fuel supply depletion
	Fire/explosion
	In-tank/ullage
	Void space
	Hydrodynamic ram
Structural	Fracture/removal
	Pressure overload
	Thermal weakening
	Delamination/fiber buckling
	Connection failure
Avionics	Mechanical damage
	Fire/overheat
Armament	Fire/explosion
Propulsion	Air inlet flow distortion
	Engine failure
	Fuel ingestion
	Foreign object damage
	Fan/compressor damage
	Combustor damage
	Turbine damage
	Exhaust duct or after-burner damage
	Engine fire
	Engine subsystem or control failure
	Loss of lubrication
	Engine controls and accessories failure
Power train and rotor blade or propeller	Mechanical/structural damage
	Loss of lubrication

(continued)

AIRCRAFT COMBAT SURVIVABILITY, CONTINUED

POTENTIAL COMPONENT AND SYSTEM KILL MODES, CONTINUED

Component/System	Kill Mode
Flight control	Disruption of the control signal path
	Loss of pilot
	Loss of control lines
	Computer failure
	Sensor damage
	Loss of control power
	Hydraulic failure
	Electrical failure
	Actuator damage
	Damage to control surfaces/hinges
	Hydraulic fluid fire
Electrical power	Severing/grounding
	Mechanical damage
	Overheating
Crew	Injury/death
	Life support failure

BLAST EFFECTS

Aircraft can be vulnerable to blast environments generated by threats carrying explosives. Parameters of interest in vulnerability assessments include the peak overpressure (P_o), duration of the positive phase (Δ), and total impulse of the positive phase (I), which is the area under the overpressure-time history for the positive phase. The following procedure is used for estimating peak overpressure P_o and total impulse I for any charge size and altitude. The method makes use of the following empirical expressions derived from blast curves to relate peak overpressure (P^*) and positive-phase duration (Δ^*) for a 1 lb charge of TNT at sea level for a distance (z^*) from the 1 lb charge at sea level. For these empirical equations, P^* is in psi, Δ^* is in msec, and z^* is in feet. To calculate overpressure and total impulse at a range z from a charge of size W at an altitude h:

$$z^* = \frac{z}{\alpha\beta}$$

$$\alpha = W^{1/3} \qquad \text{where } W = \text{lbs of TNT}$$

$$\beta = \left[\frac{\text{ambient pressure at sea level (psi)}}{\text{ambient pressure at altitude, } h \text{ (psi)}}\right]^{1/3}$$

AIRCRAFT COMBAT SURVIVABILITY, CONTINUED

BLAST EFFECTS, CONTINUED

$$\gamma = \left[\frac{\text{ambient temperature at sea level (K)}}{\text{ambient temperature at altitude, } h\text{ (K)}} \right]^{1/2}$$

$$P^* = (1050)\, z^{*-2.2} \qquad 2 \text{ ft} \le z^* \le 10 \text{ ft}$$
$$P^* = (355)\, z^{*-1.73} \qquad 10 \text{ ft} \le z^* \le 20 \text{ ft}$$

$$\Delta^* = -0.41 + 2.3 \log (z^*) \qquad 2 \text{ ft} \le z^* \le 6 \text{ ft}$$
$$\Delta^* = -1.27 + 3.44 \log (z^*) \qquad 6 \text{ ft} \le z^* \le 20 \text{ ft}$$

$$I^* \cong 0.37\, (P^*)\, (\Delta^*)$$

$$P_o = \frac{P^*}{\beta^3} \qquad I = \frac{I^* \alpha \gamma}{\beta^2}$$

In some cases, it is desirable to know the range at which a particular overpressure or total impulse will be experienced. In such cases, the method outlined above can be used parametrically to generate tables or plots giving P_o and I as functions of z for the charge size and altitude of interest. The resulting tables or plots can then be used to determine the range at which a particular value of overpressure or total impulse occurs.

PENETRATOR EFFECTS

The probability an aircraft is killed by a single random hit on the aircraft is given by

$$P_{K/H} = A_V / A_P$$

where A_V is the vulnerable area and A_P is the presented area of the aircraft. Both A_V and A_P are calculated from the perspective of an observer looking in the direction of the flight of the penetrator. For non-redundant components with no component overlap

$$A_V = \sum_{i=1}^{N} A_{vi}$$

AIRCRAFT COMBAT SURVIVABILITY, CONTINUED

PENETRATOR EFFECTS, CONTINUED

where N is the total number of components and A_{vi} is the vulnerable area of the "*i*-th" component. Methods to calculate total vulnerable area for cases involving non-redundant components with overlap or redundant components with or without overlap are more complicated.

Calculating vulnerable areas often requires the ability to determine if a threat can penetrate far enough into the aircraft to interact with a vulnerable critical component. There are many analytical methods or computational tools available to do this. Presented below are the equations for one such method, called Project THOR. The equations and coefficients below are empirically derived using test data in the units indicated.

V_{50} = velocity at which the probability of penetration equals 0.5 (ft/s)
h = thickness of target plate (in.)
A_f = average impact area of fragment (in.^{2})
W_f = striking weight of fragment (grains)
θ = angle of obliquity ($\theta = 0°$ for a normal impact, $\theta = 90°$ for a fragment traveling parallel to the face of the target plate)
V_r = residual velocity of the fragment after penetrating the target plate (ft/s)
V = striking velocity of the fragment (ft/s)
W_r = residual weight of the fragment after penetrating the target plate (grains)
$c, \alpha, \beta, \gamma, \lambda$ = THOR coefficients

$$V_{50} = 10^c \left[\left(hA_f \right) \right]^{\alpha} W_f^{\beta} (\sec\theta)^{\gamma}$$

$$V_r = V - 10^c \left[\left(hA_f \right) \right]^{\alpha} W_f^{\beta} (\sec\theta)^{\gamma} V^{\lambda}$$

$$W_r = W - 10^c \left[\left(hA_f \right) \right]^{\alpha} W_f^{\beta} (\sec\theta)^{\gamma} V^{\lambda}$$

The following THOR coefficients are provided as an example for Aluminum Alloy 2024-T3. THOR coefficients are valid for fragment weights between 5 and 240 grains, striking velocities between 1200 and 11,000 ft/s, target thicknesses between 0.02 and 2.00 inches, and obliquity angles between 0 and 80 degrees.

AIRCRAFT COMBAT SURVIVABILITY, CONTINUED

PENETRATOR EFFECTS, CONTINUED

THOR COEFFICIENTS FOR ALUMINUM ALLOY 2024-T3

Equation for	c	α	β	γ	λ
V_{50}	6.185	0.903	−0.941	1.098	N/A
V_r	7.047	1.029	−1.072	1.251	−0.139
W_r	−6.663	0.227	0.694	−0.361	1.901

THREAT CHARACTERIZATION

Weights and diameters for several surface-to-air missile warheads are shown below. The primary piece of information needed to characterize blast environments is the equivalent weight of the explosive in the warhead. The equivalent explosive weight is usually expressed in pounds of TNT and accounts for the type and amount of explosive in the warhead, as well as the effects of the casing on the equivalent yield of the warhead.

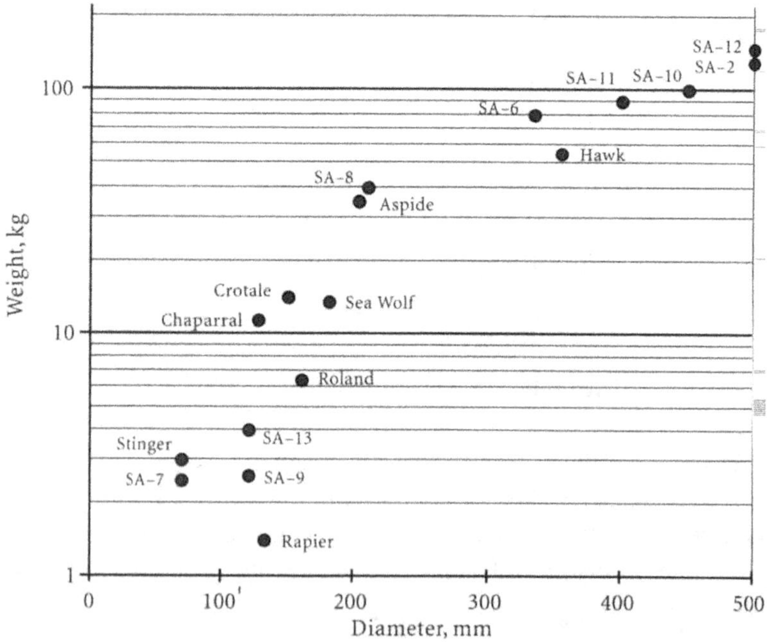

AIRCRAFT COMBAT SURVIVABILITY, CONTINUED

THREAT CHARACTERIZATION, CONTINUED

Information needed to characterize fragment environments generated by a cased warhead typically includes the mass distribution of the resulting fragment spray, as well as the velocities associated with the fragment spray. Velocity information of interest includes initial velocities and directions for fragments in the fragment spray, the effect of warhead velocity on fragment velocity, and the effect of aerodynamic drag on the decay in fragment velocity as a function of distance traveled by the fragment. Parameters for three generic warheads are shown below. The angles α_1 and α_2 refer to polar angles measured relative to the longitudinal axis of the warhead, with $\alpha = 0°$ corresponding to the nose of the warhead and $\alpha = 180°$ corresponding to the tail of the warhead. Fragments are assumed to be uniformly distributed between the polar angles α_1 and α_2 (inclusive).

FRAGMENT PARAMETERS FOR THREE GENERIC WARHEADS

Parameter	Small Warhead	Medium Warhead	Large Warhead
α_1, deg	75	80	85
α_2, deg	105	100	95
Frag weight, g (grains)	2.59 (40)	6.48 (100)	5.18 (80)
Number of frags	2600	3200	8300
Frag velocity, m s^{-1} (ft/s)	1524 (5000)	1823 (6000)	2743 (9000)

ADDITIONAL CONSIDERATIONS

The effects of penetrators, such as fragments or high-explosive incendiary (HEI) rounds, can be more devastating if they induce hydrodynamic ram in fluid-filled containers, such as fuel tanks, or if they cause flammable materials to ignite. Hydrodynamic ram refers to pressures generated in the fluid in a container as the penetrator imparts kinetic energy to the fluid. These pressures can cause additional damage to the walls of the container or cause fluid to leak out of holes in the container. If the penetration event also involves the detonation of an explosive round, the damage can be even more severe.

AIRCRAFT COMBAT SURVIVABILITY, CONTINUED

ADDITIONAL CONSIDERATIONS, CONTINUED

Combustion is an additional concern in situations in which an ignition source can encounter fuel vapors and in which oxygen has concentrations within limits referred to as flammability limits. Flammability diagrams for three common aviation fuels are provided below. The area under the curve for each type of fuel is the region of flammability. To the left of the curve, the fuel-air mixture is lean, and to the right of the curve, the mixture is rich, both conditions being less likely for combustion to occur.

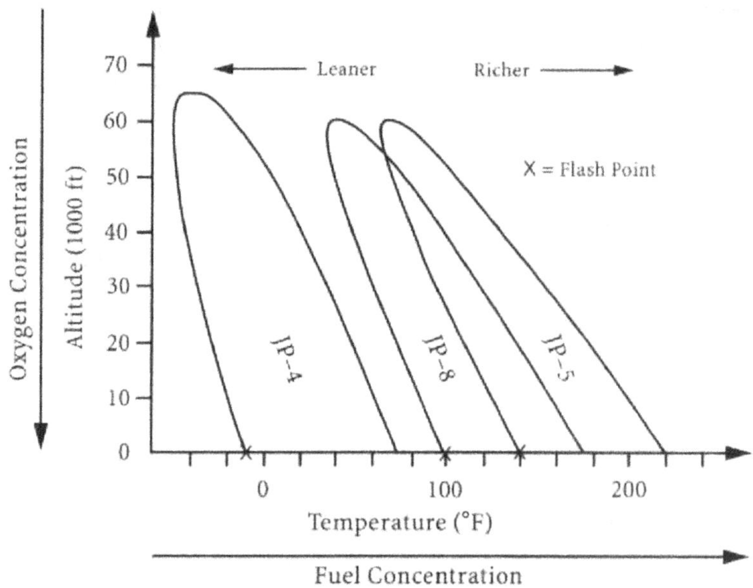

HELICOPTER DESIGN

GEOMETRY

POSITIVE SIGN AND VECTOR CONVENTIONS FOR FORCES ACTING ON THE HELICOPTER

Helicopter geometry is similar to the aircraft geometry shown on page 9–6, except for the addition of rotating blades. The following figure shows a typical side view of a helicopter with appropriate nomenclature. For a more detailed geometry breakdown, refer the helicopter references at the end of section.

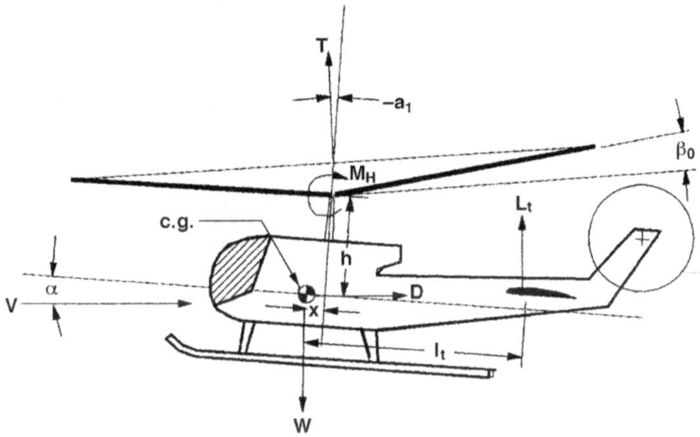

c.g. = center of gravity	l_t = distance, c.g. to horizontal tail
V = airflow	α = fuselage angle of incidence
x = distance, c.g. to T	$-a_1$ = mast angle
β_0 = coning angle	D = drag
M_H = hub moment	T = rotor thrust vector
L_T = lift from horizontal tail	DL = disc loading
	A_d = disc area

Source: *Engineering Design Handbook, Helicopter Engineering, Part One: Preliminary Design*, Headquarters, U.S. Army Materiel Command, Aug. 1974.

PRELIMINARY DESIGN PROCESS

Common Design Requirements	Design Constraints
Payload	Compliance with applicable safety standards
Range of endurance	Maximum disc loading
Critical hover or vertical climb condition	Choice of engine from list of approved engines
Maximum speed	Maximum physical size
Maximum maneuver load factor	Maximum noise level
	Minimum one-engine-out performance
	Minimum auto-rotate landing capability

HELICOPTER DESIGN, CONTINUED

PRELIMINARY DESIGN PROCESS, CONTINUED

OUTLINE OF TYPICAL DESIGN PROCESS STEPS

1) Guess at the gross weight (GW) and installed power on the basis of existing helicopters with performance similar to that desired.
2) Estimate the fuel required using a specific fuel consumption (SFC) of 0.5 for piston engines or 0.4 for turboshaft engines applied to the installed power. Estimate required mission time considering pre-takeoff warm-up, climb to cruise, cruise, descent, and operation at landing site times. Thirty minutes reserve fuel is a standard condition.

$$\text{Fuel} = \text{SFC} \times \text{hp}_{\text{installed}} \times \text{Mission time}$$

3) Calculate the useful load (UL).

$$\text{UL} = \text{crew} + \text{payload} + \text{fuel}$$

4) Assume a value of the ratio UL/GW based on existing helicopter and trends. Use the following figure for guidance:

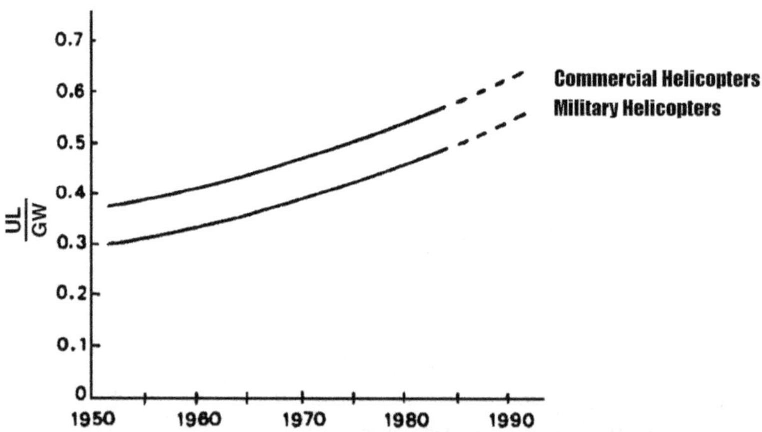

Historic trend of ratio of useful load to gross weight. (Source: *Military Helicopters—Design Technology,* page 30, by R. W. Prouty. Copyright © 1989, Jane's Defence Data, Coulsdon, Surrey, UK. Reproduced with permission from Jane's Information Group.)

5) Estimate gross weight as:

$$\text{GW} = \frac{\text{UL}}{\text{UL}/\text{GW}}$$

Compare this value with the original estimate (step 1). Modify the estimate of installed power and fuel if the two calculations are significantly different.

HELICOPTER DESIGN, CONTINUED

PRELIMINARY DESIGN PROCESS, CONTINUED

OUTLINE OF TYPICAL DESIGN PROCESS STEPS, CONTINUED

6) Assume a disc loading at the maximum allowable value, or at the highest deemed practical, and lay out the configuration based on the rotor radius corresponding to this disc loading and the estimated gross weight.

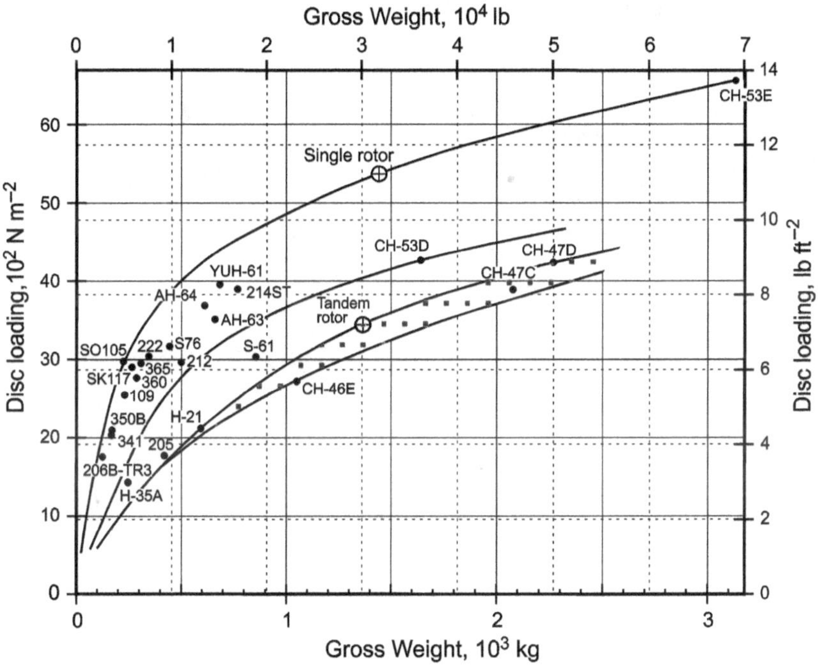

Disc loading trends. (Source: "Computer-Aided Helicopter Design," by Rosenstein and Stanzione. AHS 37th Forum. Copyright © 1981, American Helicopter Society, Alexandria, VA. Reproduced with permission from AHS.)

7) Select a figure of merit (FM) target. The FM is a measure of the rotor's efficiency in hover, based on the minimum power required to hover. For a well designed, state-of-the-art rotor designed primarily for hover, assume FM = 0.8. For a rotor designed for improved performance at high speed (and less efficient in hover), assume FM = 0.6.

HELICOPTER DESIGN, CONTINUED

PRELIMINARY DESIGN PROCESS, CONTINUED

OUTLINE OF TYPICAL DESIGN PROCESS STEPS, CONTINUED

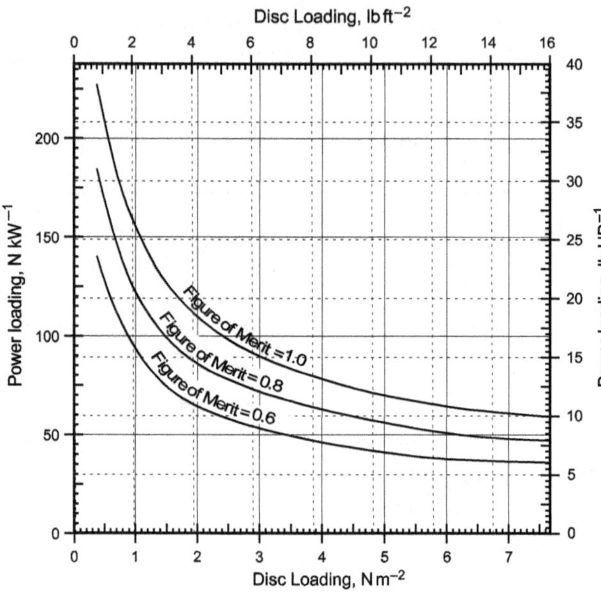

Main-rotor hover performance. (Source: *Military Helicopters—Design Technology,* page 23, by R. W. Prouty. Copyright © 1989, Jane's Defence Data, Coulsdon, Surrey, UK. Reproduced with permission from R. W. Prouty.)

HELICOPTER DESIGN, CONTINUED

PRELIMINARY DESIGN PROCESS, CONTINUED

OUTLINE OF TYPICAL DESIGN PROCESS STEPS, CONTINUED

8) Make first design decisions for main rotor tip speed, based on maximum speed.

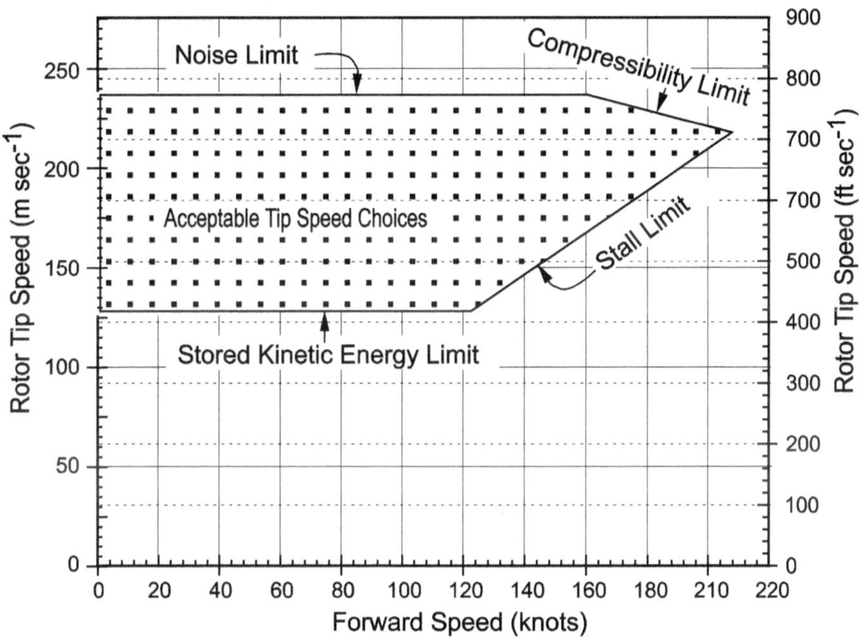

Constraints on choice of tip speeds. (Source: *Military Helicopters—Design Technology*, page 50, by R. W. Prouty. Copyright © 1989, Jane's Defence Data, Coulsdon, Surrey, UK. Reproduced with permission from R. W. Prouty.)

HELICOPTER DESIGN, CONTINUED

PRELIMINARY DESIGN PROCESS, CONTINUED

OUTLINE OF TYPICAL DESIGN PROCESS STEPS, CONTINUED

9) Make first estimates of drag in forward flight. Estimate *fe* [equivalent flat plate area, m^2 (ft^2)] by using existing helicopters designed for similar missions as a basis.

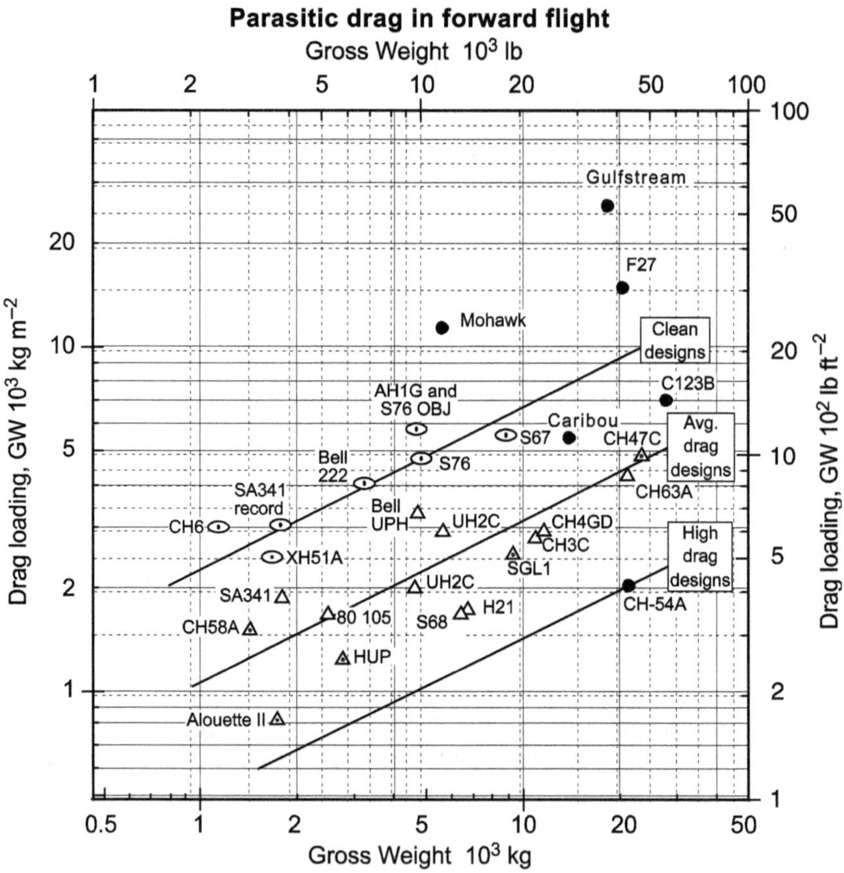

Helicopter and airplane drag state of the art. (Source: "Computer-Aided Helicopter Design," by Rosenstein and Stanzione, AHS 37th Forum. Copyright © 1981, American Helicopter Society, Alexandria, VA. Reproduced with permission from AHS.)

HELICOPTER DESIGN, CONTINUED

PRELIMINARY DESIGN PROCESS, CONTINUED

OUTLINE OF TYPICAL DESIGN PROCESS STEPS, CONTINUED

10) Calculate installed power required to satisfy vertical rate of climb and maximum speed requirements at specified altitude and temperature. Calculating the installed power properly requires additional information such as gross weight, maximum speed, altitude, and vertical rate of climb. For detailed discourse on installed power, refer to other sources such as *Helicopter Performance, Stability, and Control* by R.W. Prouty. Flat plate drag estimates are useful to make this calculation, and the following figure may be used.

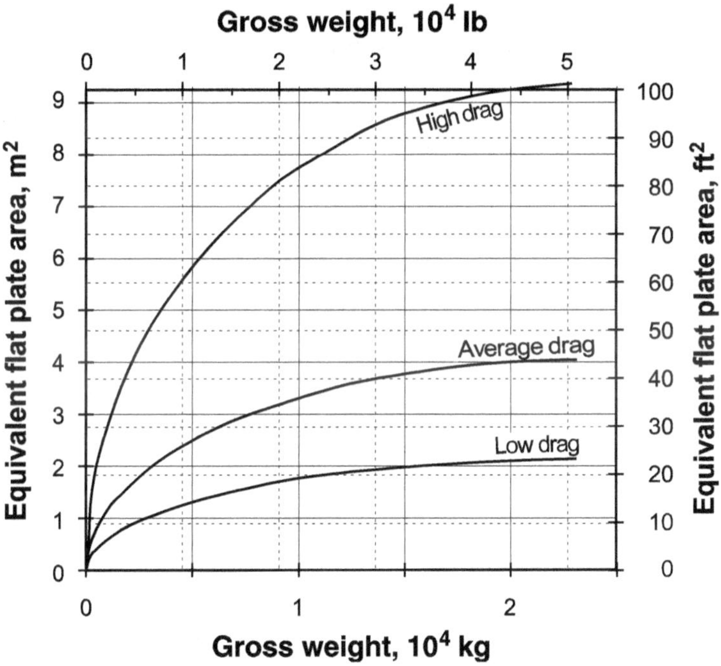

Statistical trend for equivalent flat plate area. (Source: *Military Helicopters—Design Technology,* page 30, by R.W. Prouty. Copyright © 1989, Jane's Defence Data, Coulsdon, Surrey, U.K. Reproduced with permission from Jane's Information Group.)

11) Select engine, or engines, that satisfy both requirements and constraints. Decrease disc loading if necessary to use approved engine if the vertical rate of climb is the critical flight condition. Use retracting landing gear and other drag reduction schemes to use approved engine if high speed is the critical flight condition.
12) Recalculate the fuel required based on the design mission and on known engine characteristics.

HELICOPTER DESIGN, CONTINUED

PRELIMINARY DESIGN PROCESS, CONTINUED

OUTLINE OF TYPICAL DESIGN PROCESS STEPS, CONTINUED

13) Calculate group weights based on statistical methods modified by suitable state-of-the-art assumptions. If the resulting gross weight is different from the gross weight currently being used, return to appropriate previous step.
14) Perform tradeoff studies with respect to disc loading, tip speed, solidity, twist, taper, type and number of engines, and so on to establish smallest allowable gross weight.
15) Continue with layout and structural design. Modify group weight statement as the design progresses.
16) Make detailed drag and vertical drag estimates based on drawings and model tests if possible.
17) Maintain close coordination between the team members to ensure that design decisions and design compromises are incorporated in the continual updating of the various related tasks.
18) The result.

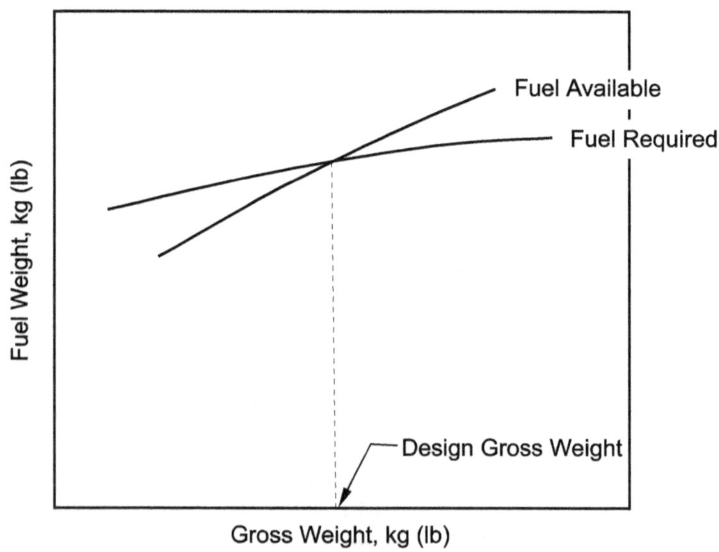

Result of preliminary design study. (Source: *Military Helicopters—Design Technology*, page 31, by R.W. Prouty. Copyright © 1989, Jane's Defence Data, Coulsdon. Surrey, UK. Reproduced with permission from Jane's Information Group.)

The gross weight that makes the fuel available equal to the fuel required is the design gross weight (dGW) as shown. As a fallout of this process, the difference in the slopes of the two lines of fuel weight vs gross weight yields the growth (GF)—the change in gross weight that is forced by a unit increase in the payload or the structural weight. The growth factor is

HELICOPTER DESIGN, CONTINUED

PRELIMINARY DESIGN PROCESS, CONTINUED

OUTLINE OF TYPICAL DESIGN PROCESS STEPS, CONTINUED

$$GF = \frac{1}{\frac{dF_{avail}}{dGW} - \frac{dF_{req}}{dGW}}$$

where the two slopes are taken at the design gross weight. The denominator is always less than unity; so the growth factor is always less than unity.

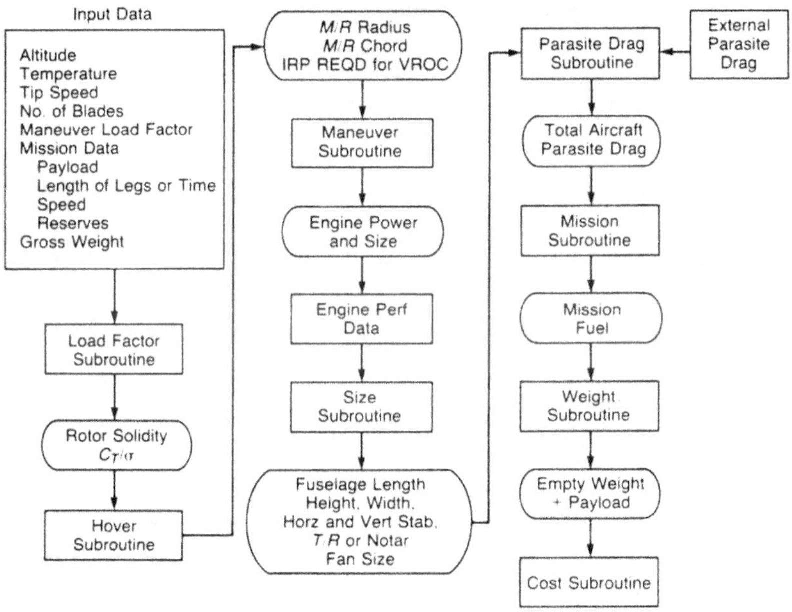

Block diagram of typical computer program for 13 initial steps of helicopter preliminary design. (Source: *Military Helicopters—Design Technology,* page 31, by R.W. Prouty. Copyright © 1989, Jane's Defence Data, Coulsdon, Surrey, UK. Reproduced with permission from Jane's Information Group.)

HELICOPTER DESIGN, CONTINUED

TAIL ROTOR DIAMETER SIZING TREND

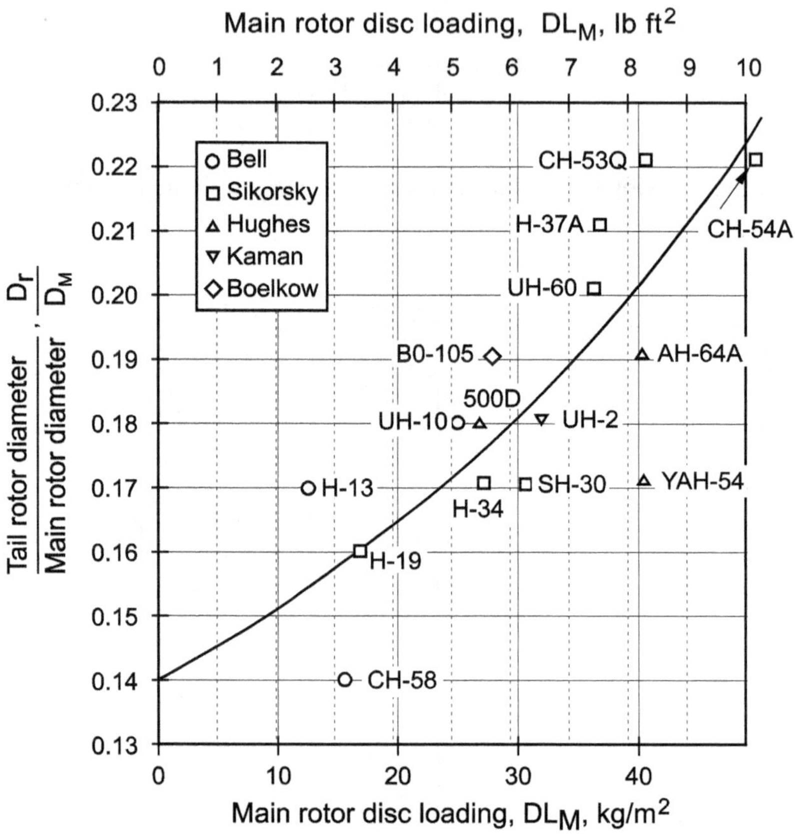

Source: Wiesner and Kohler, *Tail Rotor Design Guide,* U.S. Army Air Mobility Research and Development Laboratory, Fort Eustis, VA, USAAMRDAL TR 73–99, 1973.

HELICOPTER DESIGN, CONTINUED

PERFORMANCE

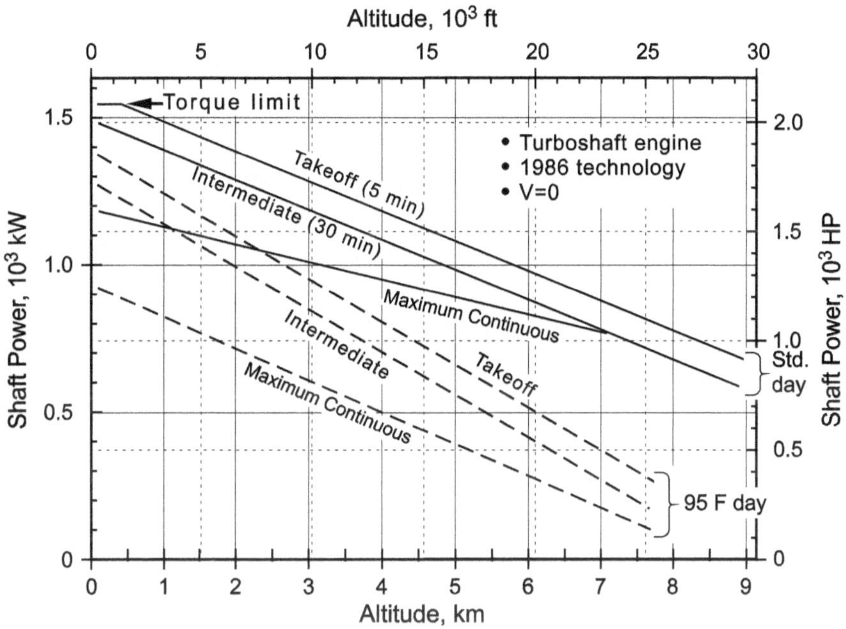

Typical uninstalled engine ratings as a function of altitude and temperature. (Source: *Helicoper Performance, Stability, and Control*, page 274, figure 4.1, by R.W. Prouty. Copyright © 1995, Krieger Publishing Company, Malabar, FL. Reproduced with permission of Krieger.)

HELICOPTER DESIGN, CONTINUED

PERFORMANCE, CONTINUED

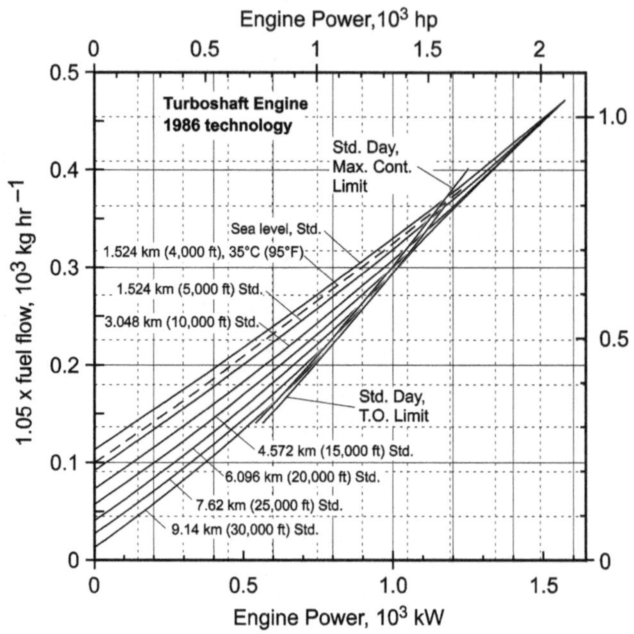

Typical engine fuel flow characteristics. (Source: *Helicopter Performance, Stability, and Control*, page 276, figure 4.3, by R. W. Prouty, Copyright © 1995, Krieger Publishing Company, Malabar, FL. Reproduced with permission of Krieger.)

HELICOPTER DESIGN, CONTINUED

PERFORMANCE, CONTINUED

BASIC FLIGHT LOADING CONDITIONS

Critical flight loading conditions normally considered in the design of a pure helicopter are defined as follows:

- Maximum speed (design limit speed V_H)
- Symmetrical dive and pullout at design limit speed V_{DL} and at 0.6 V_H, approximately the speed of maximum load factor capability
- Vertical takeoff (jump takeoff)
- Rolling pullout
- Yaw (pedal kicks)
- Autorotational maneuvers

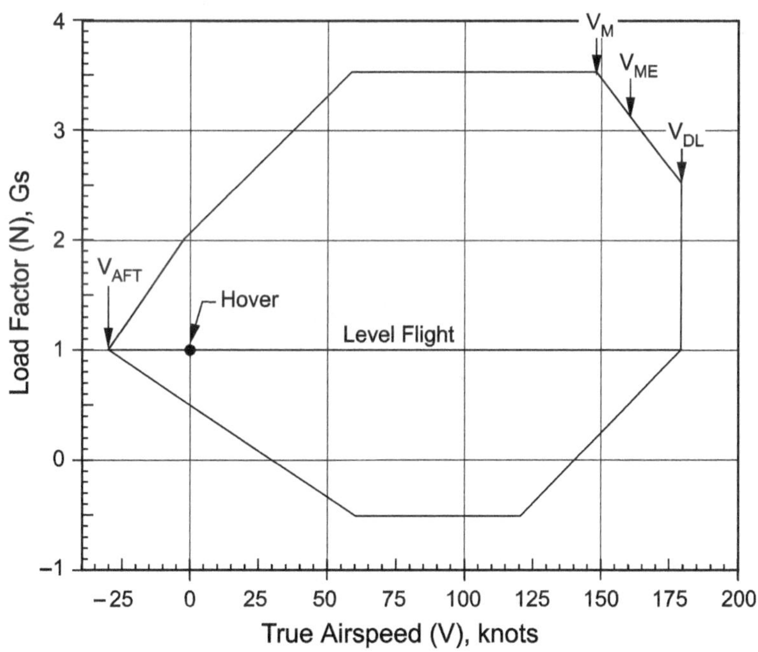

A typical design *V–N* diagram. (Source: *Helicopter Aerodynamics* (*Rotor & Wing International*), by R. W. Prouty. Copyright © 1995, PJS Publications. Reproduced with permission from Raymond W. Prouty.)

These, and other limits, are normally set by the customer or certifying authority and are depicted in the velocity-load (*V–N*) diagram. Other parameters usually defined in the *V–N* diagram are never-to-exceed velocity (V_{NE}) and maximum rearward velocity (V_{AFT}). The design structural envelope must satisfy the *V–N* diagram limits.

HELICOPTER DESIGN, CONTINUED

PERFORMANCE, CONTINUED

ROTOR THRUST CAPABILITIES

The maximum rotor thrust capabilities are shown below.

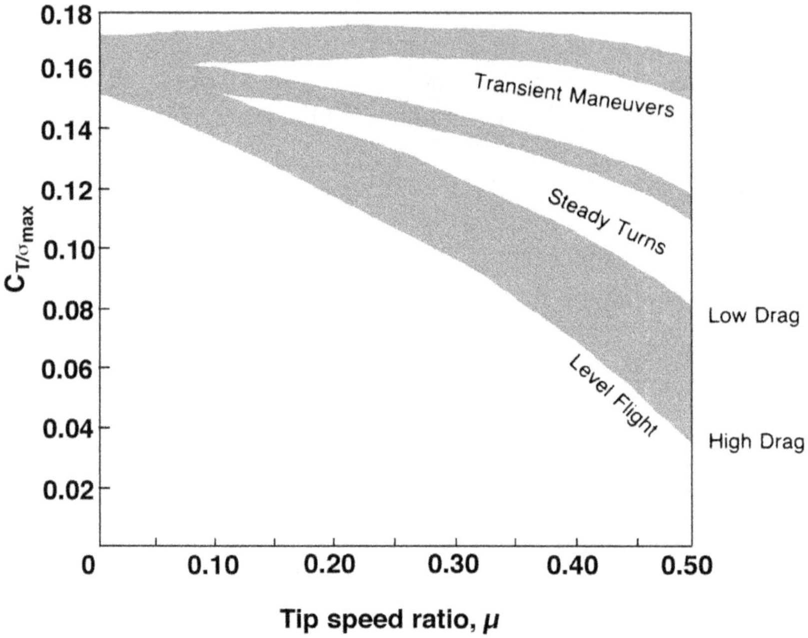

Source: *Helicopter Performance, Stability, and Control*, page 345, figure 5.2, by R. W. Prouty. Copyright © 1995, Krieger Publishing Company, Malabar, FL. Reproduced with permission of Krieger.

Rotor thrust capability

$$\frac{C_T}{\sigma} = \frac{\text{Rotor thrust}}{\text{Density of air} \times \text{blade area} \times (\text{tip speed})^2}$$

and tip speed ratio

$$\mu = \frac{\text{Forward speed of helicopter}}{\text{Tip speed}}$$

HELICOPTER DESIGN, CONTINUED

ONE-HOUR HELICOPTER DESIGN PROCESS

REQUIREMENTS

As an illustration of the procedure, it is helpful to use a specific example. This will be a small battlefield transport helicopter designed to meet the following performance requirements:

- Payload: five fully-equipped troops @ 103 kg = 515 kg (1135 lb)
- Crew: two people @ 91 kg = 182 kg (400 lb)
- Maximum speed at sea level: 370 km/h (200 kn) at the 30-min rating
- Cruise speed at sea level: at least 324 km/h (175 kn) at the maximum continuous engine rating
- Range: 556 km (300 n mile) at continuous engine rating with 30-min fuel reserve
- Vertical rate-of-climb: 2.3 m/s (450 ft/min) @ 1.2 km (4000 ft), 35°C (95°F), with 95% of 30-min rating
- Engines: two with sea level maximum continuous rating of 485 kW (650 hp) each, 30-min rating of 600 kW (800 hp) each

INITIAL GROSS WEIGHT ESTIMATING

Estimate the fuel required to do the mission. Assume a specific fuel consumption of 0.3 kg/kg-h (0.5 lb/hp-h). For cruise at continuous engine rating and 324 km/h (175 kn) for 556 km (300 n mile), this gives 645 kg (1422 lb) including reserve. When added to the payload and the crew weight, the resultant "useful load" is 1342 kg (2960 lb).

Estimate the gross weight using the Historic Trend of Useful Load to Gross Weight curve. Two lines are shown on the curve, one for commercial helicopters and a slightly lower one for combat helicopters, which are penalized by the necessity to carry things such as redundant components, armor protection, and self-sealing fuel tanks. For a design of the 1990's, the ratio for the combat helicopter has been chosen as 0.5, which means that the example helicopter should have a gross weight of about 2684 kg (5920 lb).

CHECK ON MAXIMUM FORWARD SPEED

Estimate the drag characteristics by using the statistical trend for equivalent plate area as shown in the Statistical Trend for Equivalent Flat Plate Area curve, which is based on a number of existing helicopters. Choosing to use the line labeled "Average Drag," the helicopter will have an equivalent flat plate area (fp) of 1.5 m^2 (16 ft^2). The maximum speed can be estimated using a fraction of the total installed power (P_{av}) to overcome parasitic drag with the following equation:

HELICOPTER DESIGN, CONTINUED

ONE-HOUR HELICOPTER DESIGN PROCESS, CONTINUED

CHECK ON MAXIMUM FORWARD SPEED, CONTINUED

For P_{av} in watts:

$$V_{\max} = 3.6 \left(\frac{2a\,P_{av}}{\rho f p} \right)^{1/3}, \text{ km/h}$$

where a = fraction of total installed power available to overcome parasitic drag, typically 0.50–0.70.

For P_{av} in hp:

$$V_{\max} = 4.85 \left(\frac{2a\,P_{av,\,\text{hp}}}{\rho f p} \right)^{1/3}, \text{ knots}$$

BIBLIOGRAPHY

Abbot, I. H., and Von Doenhoff, A. E., *Theory of Wing Sections*, Dover, New York, 1958.

AMCP 706-201, *Engineering Design Handbook, Helicopter Engineering, Part One: Preliminary Design*, Headquarters, U.S. Army Materiel Command, Aug. 1974.

Anon., "Jet Transport Performance Methods," The Boeing Co., Document D6-1460, 1967.

Bligh, J. A., "Methods Report for Aircraft Performance Estimation," Northrop Aircraft, Rept. MR-1, 1964.

Corning, G., *Supersonic and Subsonic, CTOL and VTOL, Airplane Design*, published by author, 1960.

Dommasch, D. O., *Airplane Aerodynamics*, Pitman, London, 1961.

Etkin, B., *Dynamics of Flight, Stability and Control*, Wiley, New York, 1959.

Hoerner, S. F., *Fluid–Dynamic Drag*, published by author, 1965.

Nicolai, L. M., *Fundamentals of Aircraft Design*, METS, Inc., 1975.

Perkins, C. D., and Hage, R. E., *Airplane Performance Stability and Control*, Wiley, New York, 1967.

Pope, A., *Aerodynamics of Supersonic Flight*, Pitman, London, 1958.

Prouty, R. W., *Helicopter Aerodynamic* (*Rotor & Wing International*), PJS Publications, Inc., 1985.

Prouty, R.W., *Helicopter Performance, Stability, and Control*, 1995 Edition, Krieger, Malabar, FL, 1995.

Prouty, R. W., *Military Helicopter Design Technology*, Jane's Defence Data, UK, 1989.

Seckel, E., *Stability and Control of Airplanes and Helicopters*, Academic, New York, 1964.

Torenbeek, E., *Synthesis of Subsonic Airplane Design*, Delft Univ., Delft, The Netherlands, 1981.

Wood, K. D., *Aerospace Vehicle Design—Volume I: Aircraft Design*, Johnson Publishing Co., 1966.

SECTION 10

AIR BREATHING PROPULSION DESIGN

The Committee gratefully acknowledges the efforts of the Air Breathing Propulsion Technical Committee, which was responsible for the generation and update of all of the material in this section. Special thanks go to James DeBonis for his help in the preparation of the material, and to the late John Rockensies, who spearheaded this effort, for his careful work and enthusiastic cooperation.

PROPELLER PROPULSION SYMBOLS

$C_P = PW/\rho n^3 D^5$ = power coefficient

$C_S = \sqrt[5]{\rho V^5/PWn^2}$ = speed power coefficient

$C_T = F/\rho n^2 D^4$ = thrust coefficient

$C_Q = Q/\rho n^2 D^5$ = torque coefficient

$J = V/nD$ = advance ratio

$\eta = FV/PW$ = propeller efficiency

$F_a = FV$ = available power, W

$= FV/550$ = available power, hp

where

D = propeller diameter, m (ft)
n = propeller speed, revolutions/s, rps
PW = propeller shaft power, W (lb ft s^{-1} or hp $\times$ 550)
Q = torque, N m (lb ft)
F = thrust, N (lb)
V = aircraft speed, m s^{-1}(ft s^{-1})
ρ = air density at altitude, kg m^{-3} (lb s^2 ft^{-4})

PROPELLER THRUST PER SHAFT HORSEPOWER VS AIRSPEED

As a function of propeller efficiency

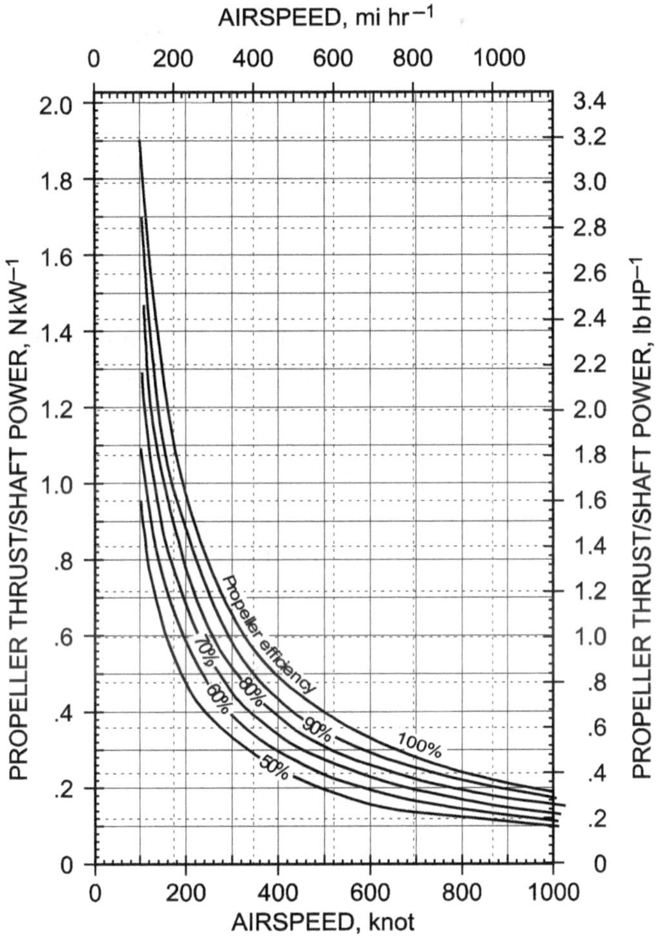

Example

$V = 550$ mph $\eta = 80\%$ $F/PW = 0.55$

JET PROPULSION GLOSSARY

Afterburner	A burner that adds heat to the working fluid after the turbine stages to increase the thrust
Cycle, closed	A thermodynamic cycle in which the working fluid recirculates through the system
Cycle, open	A thermodynamic cycle in which the working fluid passes through the system only once, being thereafter discharged
Gas turbine	An engine consisting of a compressor, burner, or heat exchanger and turbine, using air as the working fluid, producing either shaft horsepower or jet thrust, or both
Intercooler	A heat exchanger that cools the working fluid between stages of compression to decrease the work required for compression
Nacelle	Multifunctional propulsion component consisting of an inlet and fan cowl, engine, exhaust nozzle, thrust reverser, and noise suppressor
Pulse-jet	A jet engine whose thrust impulses are intermittent; usually a simple duct with some type of air-control valves at the front end
Ramjet (Athodyd)	A jet engine consisting of a duct using the dynamic head (due to the forward motion) for air compression and producing thrust by burning fuel in the duct and expanding the result through a nozzle
Regenerator	A heat exchanger that transfers heat from the exhaust gas to the working fluid after compression, before the burner, to increase cycle thermal efficiency
Reheater	A burner (or heat exchanger) that adds heat to the fluid, between turbine stages, to increase the power output
Turbofan	A gas turbine similar to a turboprop except that the shaft work is used to drive a fan or low-pressure compressor in an auxiliary duct, usually annular around the primary duct
Turbojet	A gas turbine whose entire propulsive output is delivered by the jet thrust
Turboprop	A type of gas turbine that converts heat energy into propeller shaft work and some jet thrust

JET PROPULSION ENGINE NOMENCLATURE

A = area, primarily flow area, m^2 (ft^2)
a = sonic velocity, m s^{-1}(ft s^{-1})
BPR = bypass ratio = W_{fan}/W_c
C = coefficient
C_f = gross thrust coefficient (ramjet)
C_{fg} = nozzle gross thrust coefficient
$C_{fg\,peak}$ = nozzle peak gross thrust coefficient
c_p = specific heat at constant pressure, J kg^{-1} K^{-1} (Btu lb^{-1} $°R^{-1}$)
c_υ = specific heat at constant volume, J kg^{-1} K^{-1} (Btu lb^{-1} $°R^{-1}$)
D = drag, N (lb)
D_{cowl} = cowl drag, N (lb)
D_{inter} = interference drag, N (lb)
$D_{spillage}$ = spillage drag, N (lb)
F = thrust, N (lb)
F_g = gross thrust, N (lb)
F_n = net thrust = $F_g - F_r$, N (lb)
F_r = ram drag of engine airflow, N (lb)
f = fuel–air ratio
g = gravitational acceleration, 9.81 m s^{-2} (32.17 ft s^{-2})
h = specific enthalpy, J kg^{-1} (Btu lb^{-1})
I = impulse (s)
J = Joule's constant, 4186 J cal^{-1} (778 ft lb Btu^{-1})
L = length, m (ft)
M = Mach number
N = engine speed (revolutions/min, rpm)
NPF = net propulsive force, N (lb)
P = power, W (Btu s^{-1} or hp)
p = absolute pressure, Pa (lb ft^{-2})
Q = heating value of fuel, J kg^{-1} (Btu lb^{-1})
q = dynamic pressure, Pa (lb ft^{-2})
R = specific gas constant for air, 287.04 J kg^{-1} k^{-1} (53.35 ft lb_f lb^{-1} $°R^{-1}$)
r = radius, m (ft)
s = specific entropy, J kg^{-1} K^{-1} (Btu lb^{-1} $°R^{-1}$)
TSFC = thrust specific fuel consumption, kg $h^{-1}N^{-1}$ (lb_m h^{-1} lb_f^{-1})
T = absolute temperature, K (°R)
V = velocity, m s^{-1} (ft s^{-1})
υ = specific volume, m^3 kg^{-1} (ft^3 lb^{-1})
W = flow rate
w = specific weight, kg m^{-3} (lb ft^{-3})
α = nozzle angularity
γ = ratio of specific heats = c_p/c_v

JET PROPULSION ENGINE NOMENCLATURE, CONTINUED

δ = p/p_0, deflection angle across shock
η = efficiency, usually qualified by subscripts shown in the following list
θ = T/T_0
ρ = density, kg m^{-3} (lb s^2 ft^{-4})
σ = wave angle from incoming flow direction

Subscripts

a = air
act = actual
b = combustion chamber (burner)
c = compressor, core
d = diffuser
e = effective (as in area)
ext = external
f = fuel flow
fan = fan parameters
fan tip = refers to the outer radial dimension of the fan flow path
fc = fan cowl
ff = fan flange
fne = fan nozzle exit
g = gross (when used with thrust, F_g)
hl = hilite plane
i = ideal
int = internal
j = jet
m = mechanical drive
max = maximum
n = net
nac = nacelle
noz = nozzle
p = vehicle velocity (aircraft/ramjet)
r = ram
sp = specific
t = total
th = throat
tur = turbine
υ = velocity
0 = sea level static (standard), or free stream static
0, 1, 2, = location (station number)

Gas Turbine and Turbojet Performance Parameters[a]

Quantity	Symbol	Parameter
Temperature ratio[b]	θ	$\frac{T}{T_0}$
Pressure ratio[c]	δ	$\frac{p}{p_0}$
Thrust	$F(F_g, F_n,$ etc.$)$	$\frac{F}{\delta}$
Fuel flow	W_f	$\frac{W_f}{\delta\sqrt{\theta}}$
Air flow	W_a	$\frac{W_a\sqrt{\theta}}{\delta}$
Fuel/air ratio	W_f / W_a	$\frac{W_f}{W_a\theta}$
rpm	N	$\frac{N}{\sqrt{\theta}}$
Thrust specific fuel consumption (TSFC)[d]	W_f / F	$\frac{W_f}{F\sqrt{\theta}}$
Specific air consumption	W_a / F	$\frac{W_a\sqrt{\theta}}{F}$
Any absolute temperature	T	$\frac{T}{\theta}$
Any absolute pressure	p	$\frac{p}{\delta}$
Any velocity	V	$\frac{V}{\sqrt{\theta}}$

[a]Parameters are corrected to engine-inlet total conditions. Customary usage for T and p is total temperature and total pressure at the front face of the engine, station 2.
[b]T_0 is defined as the sea level standard temperature on the absolute scale of 518.7°R.
[c]p_0 is defined as the sea level standard pressure of 29.92 in Hg or 14.696 psia.
[d]Fuel flow is in pounds/hour.

GAS TURBINE CORRECTION PARAMETERS

Corrected net thrust:

$$F_{n\,(\mathrm{corr})} = \frac{F_{n\,(\mathrm{observed})}}{\delta}$$

Corrected fuel flow:

$$W_{f\,(\mathrm{corr})} = \frac{W_{f\,(\mathrm{observed})}}{\delta\theta^{x}}$$

Corrected thrust specific fuel consumption:

$$\mathrm{TSFC}_{(\mathrm{corr})} = \frac{W_{f\,(\mathrm{corr})}}{F_{n\,(\mathrm{corr})}} = \frac{\mathrm{TSFC}}{\theta^{x}}$$

Corrected air flow:

$$W_{a\,(\mathrm{corr})} = \frac{W_{a\,(\mathrm{observed})}\left(\theta^{x}\right)}{\delta}$$

Corrected exhaust gas temperature:

$$\mathrm{EGT}_{(\mathrm{corr})} = \frac{\mathrm{EGT}_{(\mathrm{observed})}}{\theta^{x}}$$

Corrected rotor speed:

$$N_{(\mathrm{corr})} = \frac{N_{(\mathrm{observed})}}{\theta^{x}}$$

Corrected shaft power:

$$PW_{(\mathrm{corr})} = \frac{PW_{(\mathrm{observed})}\,\theta^{x}}{\delta}$$

The exponent for θ is a function of the engine cycle as developed from theoretical and empirical data. The exponent x is approximately 0.5 for correcting fuel flow and TSFC, air flow, rotor speed, and shaft power, and is approximately 1.0 for correcting temperature.

GAS TURBINE STATION NOTATIONS

SINGLE-SPOOL TURBOJET/TURBOSHAFT

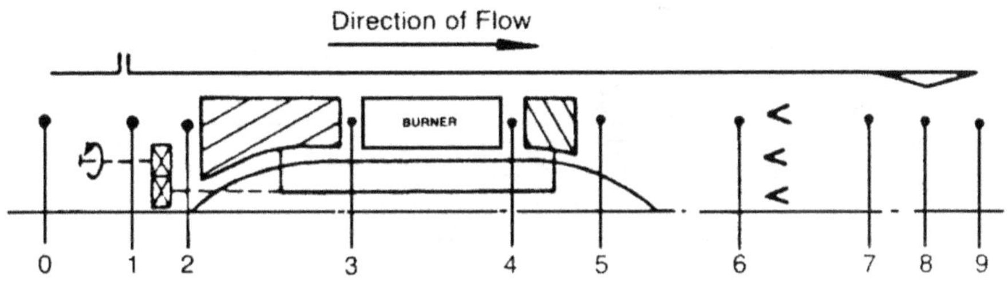

TWIN-SPOOL TURBOFAN

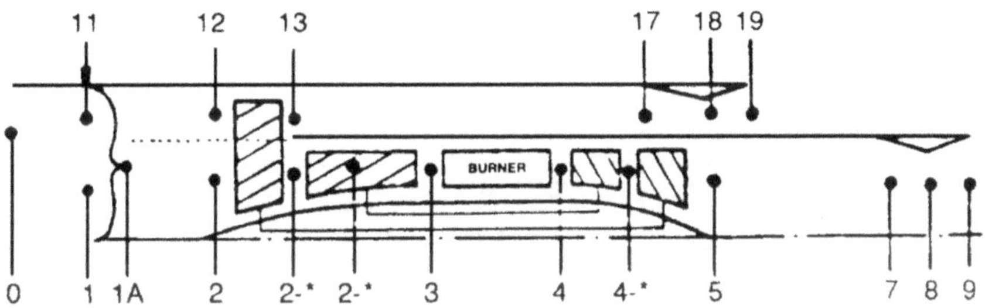

MIXED TWIN-SPOOL TURBOFAN

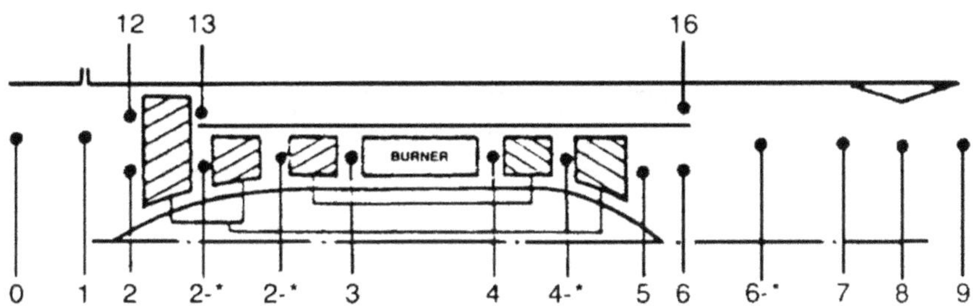

Source: Gas turbine station notations appearing on pages 10-9 and 10-10 are taken from SAE Aerospace Recommended Practice ARP 775A. Society of Automotive Engineers, Warrendale, PA. Reproduced with permission of SAE.

GAS TURBINE STATION NOTATIONS, CONTINUED

TRIPLE-SPOOL TURBOFAN

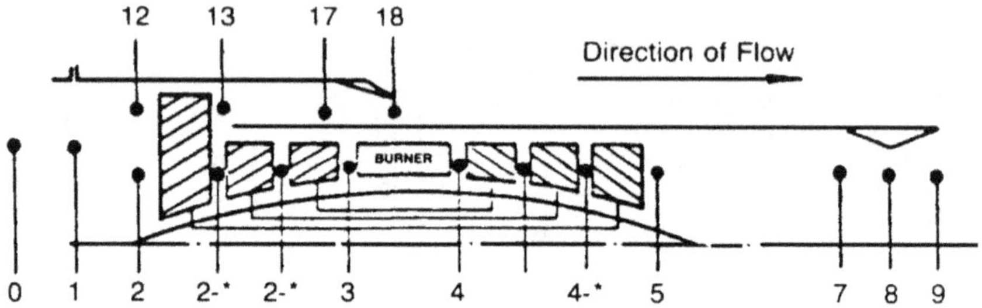

TWIN-SPOOL DUCT HEATER

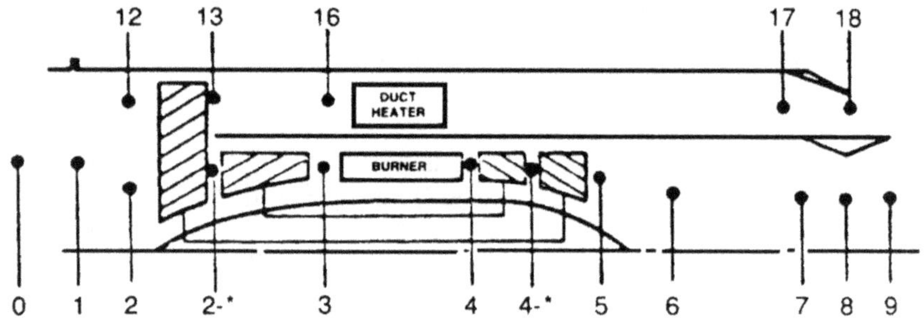

FREE-TURBINE TURBOPROP, TURBOSHAFT

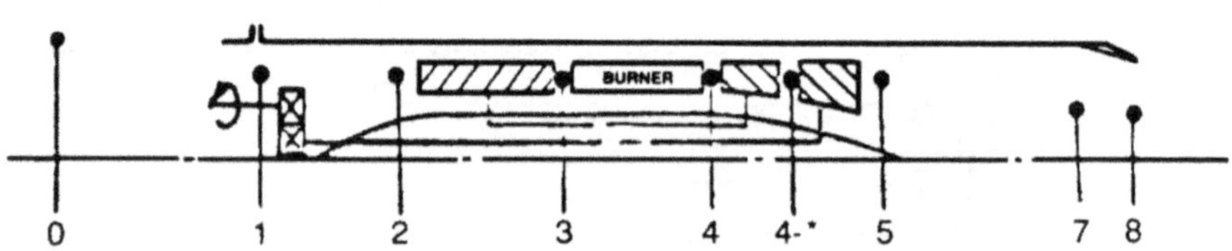

GAS TURBINE FUNDAMENTALS

The following definitions and relationships are commonly used in gas turbine analysis; the number subscripts indicate the station at which the parameter is to be evaluated. Values of $\gamma = 1.4$, $c_p = 0.241$, and $c_v = 0.173$ for air at standard conditions can be used in the equations that follow to yield approximate results. For more exact results, the work of compression and expansion of air can be determined by using tables that reference the properties of air at different temperatures and pressures and take into account variations in γ and c_p.

Compressor power:

$$P_c = W_2 c_{pc} (T_{t3} - T_{t2}) \qquad \text{J s}^{-1} (\text{Btu s}^{-1})$$

Compressor efficiency:

$$\eta_c = \frac{T_{t2}\left[\left(p_{t3}/p_{t2}\right)^{\frac{\gamma-1}{\gamma}} - 1\right]}{T_{t3} - T_{t2}}$$

Turbine power:

$$P_{\text{tur}} = W_4 c_{p\,\text{tur}} (T_{t4} - T_{t5}) \qquad \text{J s}^{-1} (\text{Btu s}^{-1})$$

Turbine efficiency:

$$\eta_{\text{tur}} = \frac{T_{t4} - T_{t5}}{T_{t4}\left[1 - (p_{t5}/p_{t4})^{\frac{\gamma-1}{\gamma}}\right]}$$

Power balance:

$$P_m + W_2 c_{pc} (T_{t3} - T_{t2}) = W_4 c_{p\,\text{tur}} (T_{t4} - T_{t5}) \qquad \text{J s}^{-1} (\text{Btu s}^{-1})$$

where P_m = mechanical drive output power required to drive the fan of a turbofan engine or the power required to drive the propeller shaft of a turboprop engine. Also included in this category are aircraft/engine accessories (i.e., fuel pumps, fuel controls, oil pumps, generators, and hydraulic pumps).

In the analysis of an engine, the requirement of airflow continuity through the engine must be satisfied. Small pressure losses occur in the combustion and exhaust systems and have to be minimized by careful design of these systems. The amount of total pressure loss is characterized by the ratio of total pressure at combustion chamber discharge p_{t4} to the total pressure at the combustion chamber entrance p_{t3}, and is equal to p_{t4}/p_{t3}. Typical values of total pressure loss coefficients are between 0.93 and 0.98. Losses in the exhaust system will be minimized if the tailpipe is made as short as possible, which will reduce wall friction losses.

Sources: The gas turbine fundamentals material appearing on pages 10-11–10-14 is from the *General Electric Aircraft Propulsion Data Book*, AEG.215.4/68 (15 m), pages 3–7; reference data is taken from *Jet Engines: Fundamentals of Theory, Design, and Operations*, pages 128 and 129, by K. Hunecke. Copyright © 1997, Motorbooks International, St. Paul, MN. Reprinted with permission of GE and Motorbooks.

GAS TURBINE FUNDAMENTALS, CONTINUED

Allowance also must be made for the combustion efficiency of the burner. In general, injected fuel does not burn completely and produces less heat than would be possible theoretically. The degree of actual fuel usage is characterized by a combustion efficiency factor η_b giving the amount of heat released by combustion to the heat theoretically available in the fuel. Modern combustion chambers achieve efficiencies between 0.90 and 0.98.

Note that in the following equations subscripts are shown as station 9 indicating convergent–divergent nozzle. For a convergent nozzle the subscript would indicate station 8.

Jet velocity:

$$V_9 = C_v\sqrt{2gJc_pT_{t9}\ [1-(p_9/p_{t9})^{\gamma-1/\gamma}]}$$

where C_v is the coefficient of velocity that is equal to the nozzle efficiency $(\eta_{\text{noz}})^{1/2}$.

Thrust (gross):

$$F_g = \frac{W_9}{g}V_9 + A_9(p_9 - p_0)$$

The pressure-area product here is zero if the nozzle allows complete expansion, that is, if $p_9 = p_0$.

In the inlet system the following equations apply:

$$\frac{T_{t2}}{T_0} = \frac{T_{t0}}{T_0} = 1+\frac{\gamma-1}{2}M^2$$

$$\frac{p_{t2}}{p_0} = \frac{p_{t2}}{p_{t0}}\frac{p_{t0}}{p_0} = \frac{p_{t2}}{p_{t0}}\left[1+\frac{\gamma-1}{2}M^2\right]^{\frac{\gamma}{\gamma-1}}$$

where p_{t2}/p_{t0} is the total pressure recovery of the inlet system.

GAS TURBINE FUNDAMENTALS, CONTINUED

Propulsion of a jet aircraft is accomplished by the principle of reaction, where a gas exhausting at a high velocity through a nozzle generates a force in the opposite direction, which is termed thrust. The amount of thrust depends on the mass of the airflow passing through the engine and the exhaust velocity, the product of which *(W/g × V)* is termed momentum. Changes in momentum per unit time (conservation of momentum theorem) result in a force being generated that is the basis for jet propulsion. Consider an engine installed in a nacelle as indicated in the following figure:

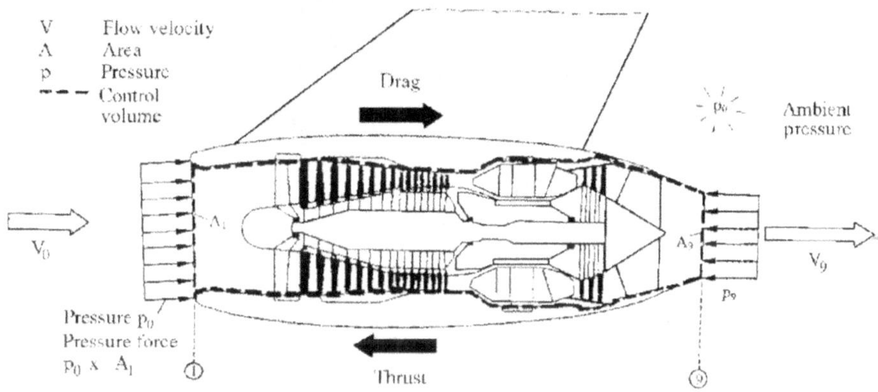

Source: *Jet Engines: Fundamentals of Theory, Design, and Operations*, page 33, by K. Hunecke. Copyright © 1997, Motorbooks International, St. Paul, MN. Reprinted by permission of Motorbooks.

This figure shows a control volume that allows flow to pass from open boundaries from engine station 1–9 but does not transfer momentum effects across solid boundaries that make up the engine casing.

Calculation of the engine thrust can be made knowing the boundary conditions at the inlet and exit of the engine. The conditions internal to the engine are represented by F_{int}, a resultant force where the propulsive thrust can be taken as the integral over those surfaces and components of the engine where thrust is most directly apparent, or chiefly exerted (e.g., inlet, compressor, diffuser, burner, turbine, nozzle, and the surfaces inside and outside of the engine).

Applying the conservation of momentum theorem to the internal flow between stations 1 and 9 gives

$$p_0 A_1 + F_{int} - p_9 A_9 = \frac{W_9}{g} V_9 - \frac{W_1}{g} V_1$$

$$F_{int} = \frac{W_9}{g} V_9 + A_9(p_9 - p_0) - \frac{W_1}{g} V_1 + p_0(A_9 - A_1)$$

GAS TURBINE FUNDAMENTALS, CONTINUED

The net resultant of external pressure forces, including friction acting on the nacelle (D_{ext}), constitutes the aerodynamic drag that must be overcome by the propulsive thrust.

The nacelle propulsive thrust is $F_{nac} = F_{int} - D_{ext}$ since these are the only forces directly acting on the nacelle.

$$F_{nac} = F_{int} - D_{ext} = \frac{W_9}{g} V_9 + A_9 (p_9 - p_0) - \frac{W_1}{g} V_1 + p_0 (A_9 - A_1) - D_{ext}$$

Because certain terms in this expression for nacelle propulsive thrust are dependent on installation and nacelle design details (e.g., inlet area A_1 and D_{ext}), engine net thrust F_n is conventionally defined as

$$F_n = \frac{W_9}{g} V_9 + A_9 (p_9 - p_0) - \frac{W_1}{g} V_1$$

The nacelle drag D_{nac} then is defined such that

$$F_{nac} = F_n - D_{nac}$$

$$D_{nac} = D_{ext} - p_0 (A_9 - A_1)$$

Computation of the drag as just indicated will result in a nacelle propulsive thrust consistent with the conventional definition of engine net thrust.

PROPULSIVE EFFICIENCY

If the performance of the power plant alone is to be considered, its useful output should be taken as $(W/2g)\,V_j^2 - V_p^2$, and its efficiency should be calculated on this basis. However, the power plant cannot be held responsible if the thrust resulting from the increase of kinetic energy is not utilized to the greatest possible extent because the aircraft velocity is low. The efficiency of utilization is measured by the propulsion efficiency defined as

$$\eta_p = \frac{FV_p}{(W/2g)\left(V_j^2 - V_p^2\right)} = \frac{(W/g)\left(V_j - V_p\right)}{(W/2g)\left(V_j^2 - V_p^2\right)}V_p = \frac{2}{1+\left(V_j/V_p\right)}$$

PROPULSIVE EFFICIENCY VS AIRSPEED

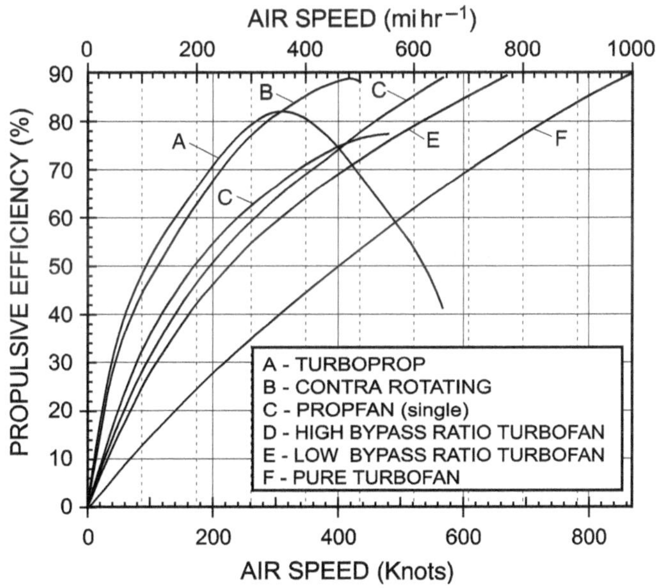

Source: *The Development of Jet and Turbine Aero Engines*, page 15, by William Gunston. Copyright © 1995. Reprinted by permission of William Gunston.

NORMAL SHOCK WAVES

$$M_y^2 = \frac{M_x^2 + \dfrac{2}{\gamma-1}}{\dfrac{2\gamma}{\gamma-1} M_x^2 - 1}$$

$$\frac{p_y}{p_x} = \frac{2\gamma}{\gamma+1} M_x^2 - \frac{\gamma-1}{\gamma+1}$$

$$\frac{T_y}{T_x} = \frac{\left(1+\dfrac{\gamma-1}{2} M_x^2\right)\left(\dfrac{2\gamma}{\gamma-1} M_x^2 - 1\right)}{\dfrac{(\gamma+1)^2}{2(\gamma-1)} M_x^2}$$

$$\frac{\rho_y}{\rho_x} = \frac{p_y}{p_x} \Big/ \frac{T_y}{T_x}$$

$$\frac{V_y}{V_x} = \frac{\rho_x}{\rho_y}$$

$$\frac{p_{ty}}{p_{tx}} = \left[\frac{\dfrac{\gamma+1}{2} M_x^2}{1+\dfrac{\gamma-1}{2} M_x^2}\right]^{\frac{\gamma}{\gamma-1}} \Big/ \left[\frac{2\gamma}{\gamma+1} M_x^2 - \frac{\gamma-1}{\gamma+1}\right]^{\frac{1}{\gamma-1}}$$

Subscripts x and y refer to conditions upstream and downstream of a normal shock, respectively.

OBLIQUE SHOCK WAVES

NOMENCLATURE OF OBLIQUE SHOCK ANALYSIS

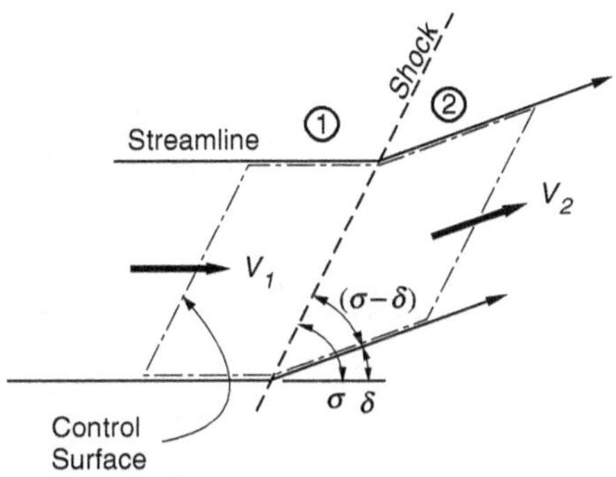

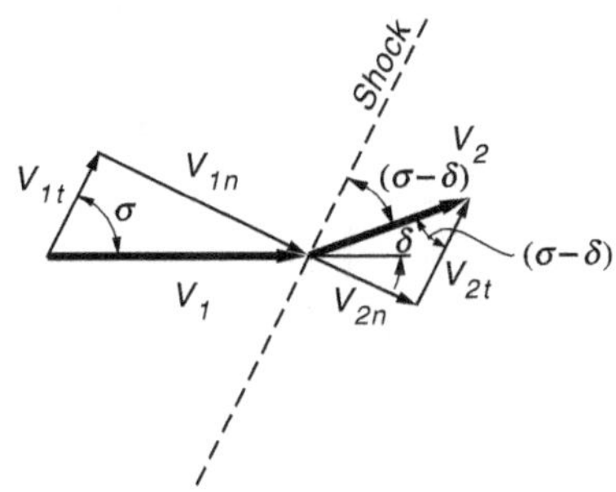

where

V_1 = flow velocity upstream of oblique shock
V_2 = resultant velocity downstream of oblique shock
V_n = component of velocity normal to shock
V_t = component of velocity parallel to shock
δ = deflection angle across shock
σ = wave angle from the incoming flow direction

$$\frac{\rho_2}{\rho_1} = \frac{V_1}{V_2}\frac{\sin\sigma}{\sin(\sigma-\delta)} = \frac{\cos(\sigma-\delta)}{\cos\sigma} = \frac{\sin\sigma}{\sin(\sigma-\delta)} = \frac{\tan\sigma}{\tan(\sigma-\delta)}$$

$$\frac{p_2}{p_1} = 1+\gamma M_1^2(\sin^2\sigma)\left(1-\frac{\rho_1}{\rho_2}\right)$$

Source: The figures on pages 10-17–10-21 (top) are taken from *The Dynamics and Thermodynamics of Compressible Fluid Flow*, Vol. 1, by A. Shapiro, pp. 532, 536, 537, 538.

OBLIQUE SHOCK WAVES, CONTINUED

WAVE ANGLE VS INITIAL MACH NUMBER AND DEFLECTION ANGLE (γ= 1.4)

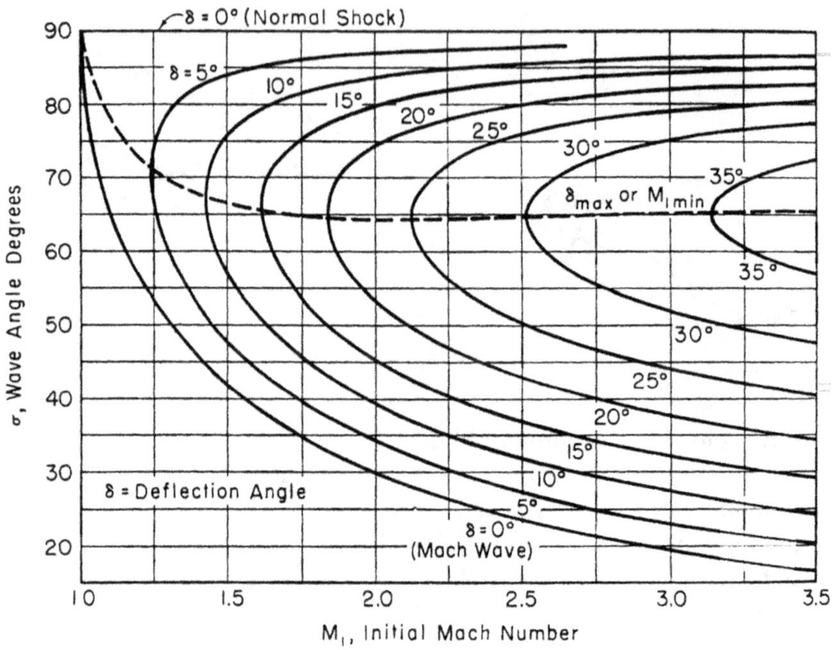

OBLIQUE SHOCK WAVES, CONTINUED

FINAL MACH NUMBER VS INITIAL MACH NUMBER WITH DEFLECTION ANGLE AS PARAMETER ($\gamma = 1.4$)

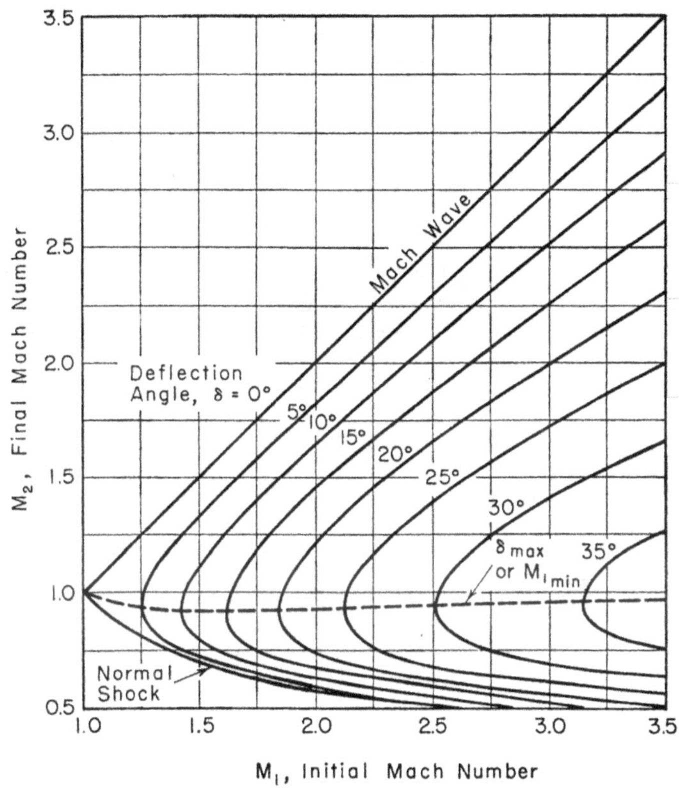

OBLIQUE SHOCK WAVES, CONTINUED

PRESSURE RATIO VS INITIAL MACH NUMBER WITH DEFLECTION ANGLE AS PARAMETER ($\gamma = 1.4$)

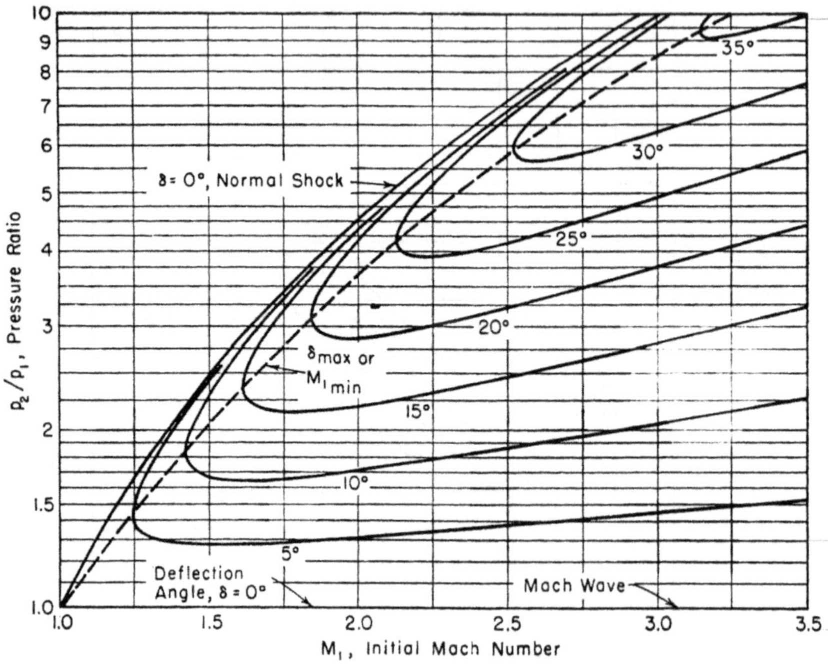

OBLIQUE SHOCK WAVES, CONTINUED

STAGNATION PRESSURE RATIO VS INITIAL MACH NUMBER WITH DEFLECTION ANGLE AS PARAMETER ($\gamma = 1.4$)

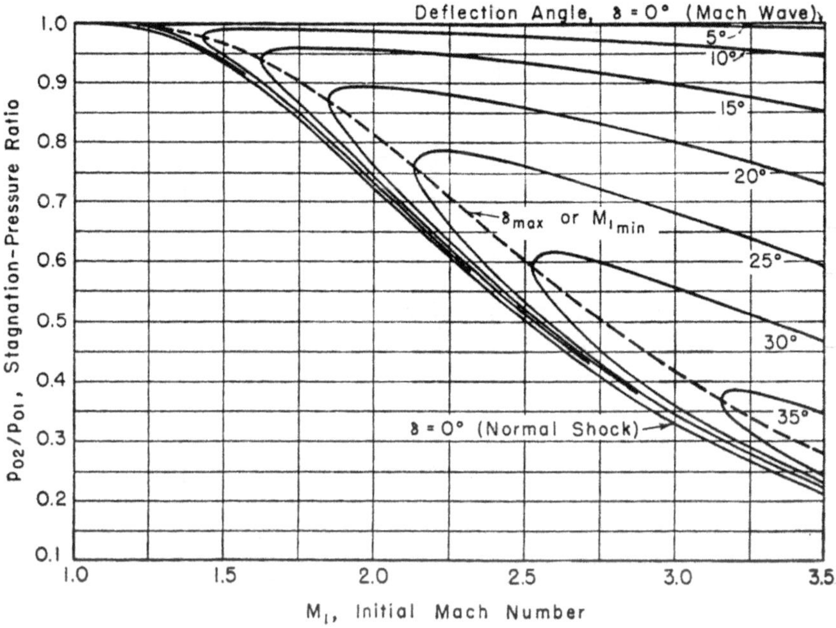

Example

A simplified intake for an aircraft designed to operate at $M = 2.0$ is shown as follows with a corresponding chart that demonstrates the use of the preceding four figures to determine oblique shock parameters.

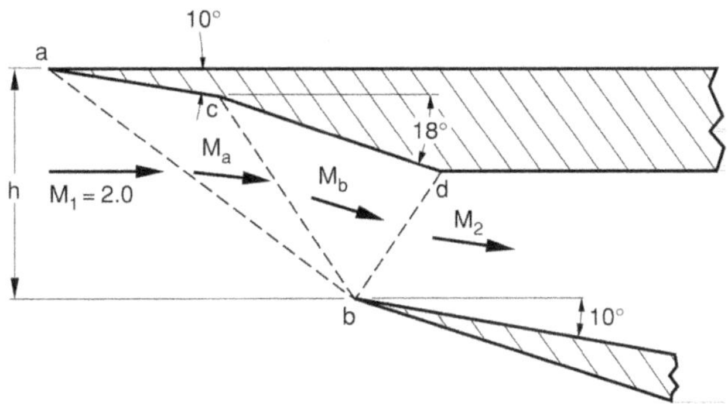

Source: *Jet Propulsion*, page 202, by N. Cumpsty. Copyright © 1997, Cambridge Univ. Press, Cambridge, England, UK. Reprinted by permission of Cambridge Univ. Press.

OBLIQUE SHOCK WAVES, CONTINUED

STAGNATION PRESSURE RATIO VS INITIAL MACH NUMBER WITH DEFLECTION ANGLE AS PARAMETER ($\gamma = 1.4$), CONTINUED

Example, continued

Mach Number (Initial)	Deflection Angle (δ)	Wave Angle (σ)	Wave Line	Mach Number (Final)	Pressure Ratio	Stagnation Pressure Ratio
		see Fig. p. 10-18		see Fig. p. 10-19	see Fig. p. 10-20	see Fig. p. 10-21
$M_1 = 2.0$	10°	39.3°	ab – 39.3° below horizontal	$M_a = 1.64$	1.71	0.96
$M_a = 1.64$	8°	46.6°	cb – 46.6° below plane ac	$M_b = 1.36$	1.49	0.98
$M_b = 1.36$	8°	71.0°	db – 71.0° below plane cd	$M_2 = 0.89$	1.76	0.99

SUBSONIC INLETS AND EXHAUST SYSTEMS

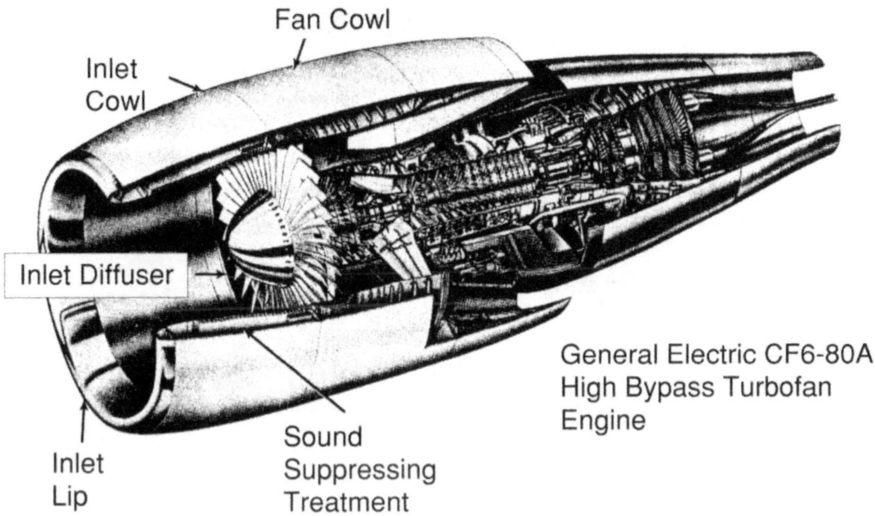

A nacelle and engine cross section is shown to delineate the nacelle component nomenclature. The major components of the nacelle and their functions are as follows: 1) inlet, to provide air to fan with acceptable flow qualities; 2) fan exhaust nozzle, to accelerate flow to produce thrust; 3) core exhaust nozzle, to accelerate flow to produce thrust; and 4) thrust reverser, to decelerate aircraft during landing and rejected takeoff.

Source: The figures on pages 10-23–10-25 are taken from "High Bypass Turbofan Nacelles for Subsonic Transports," UTSI short course in Aeropropulsion, April 2001, by D. A. Dietrich and A. P. Kuchar. Sponsored by AIAA and the Air Breathing Propulsion Technical Committee.

SUBSONIC INLETS AND EXHAUST SYSTEMS, CONTINUED

INLET AND FAN COWL COMPONENT DESIGN PARAMETERS

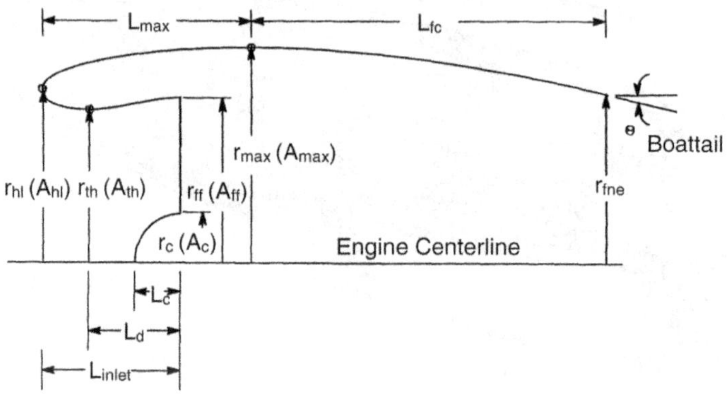

The following are the most significant geometric parameters and their major influence factors used in describing a particular inlet design. These parameters have been determined from the results of an experimental database, flowfield analysis, and wind tunnel testing.

r_{hl}/r_{th} : the low speed angle of attack and crosswind performance
r_{hl}/r_{max} : high speed drag and engine-out cowl drag
$r_{fan\ tip}/r_{th}$: inlet diffuser recovery and influence on distortion

TYPICAL VALUES FOR INLET DESIGN PARAMETERS

Parameter	Typical Values
Maximum M_{th}	0.70–0.77
L_d/r_{ff}	0.56–0.90
r_{hl}/r_{th}	1.11–1.20
r_{hl}/r_{max}	0.75–0.86
L_{max}/r_{max}	0.80–1.30

Typical values are shown for throat Mach number at nominal maximum corrected flow, r_{hl}/r_{th} (internal lip contraction), r_{hl}/r_{max} (forebody diameter ratio), and diffuser area ratio $A_{fan\ tip}/A_{th}$. Specific installation values depend on particular requirements, but the values shown generally reflect the following:

1) Design throat Mach number M_{th} is moderately high to near choking, to reduce nacelle size, assuming moderate-to-high cruise flight speed.

2) There are typical takeoff climb incidence and crosswind requirements, with some tendency toward increasing the ratio r_{hl}/r_{th}, which indicates a thicker lip, proceeding from top to bottom around the periphery.

SUBSONIC INLETS AND EXHAUST SYSTEMS, CONTINUED

TYPICAL VALUES FOR INLET DESIGN PARAMETERS, CONTINUED

3) Forebody r_{hl}/r_{max} values are in the 0.81–0.86 range, depending on installation and design requirements, except for a fan-mounted gearbox, which requires a much thicker bottom cowl, with lower r_{hl}/r_{nax} in the range of 0.75 or less depending on the design of the gearbox.

4) Diffuser area ratio $A_{fan\ tip}/A_{th}$ is approximately 1.2, with the specific value depending on M_{th} selection and engine specific flow defined as $W_{12}/A_{fan\ tip.}$

DEFINITION OF NOZZLE EFFICIENCY

The nozzle efficiency is defined in terms of gross thrust coefficient. The following figure illustrates the engine locations (station numbers) and relevant thermodynamic parameters used in determining the ideal gross thrust. The thrust coefficient is usually estimated analytically and verified by scale model static performance tests. The coefficient and the thermodynamic properties for determining the ideal gross thrust are used to calculate the actual gross thrust of the exhaust nozzle as shown on the following page.

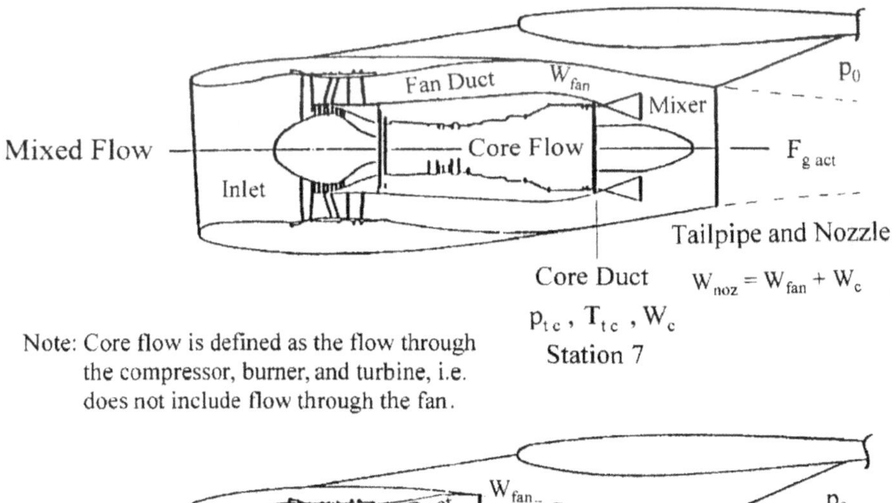

Note: Core flow is defined as the flow through the compressor, burner, and turbine, i.e. does not include flow through the fan.

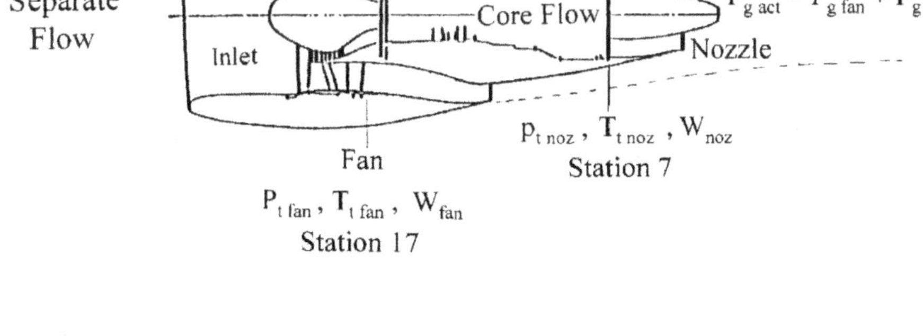

SUBSONIC INLETS AND EXHAUST SYSTEMS, CONTINUED

DEFINITION OF NOZZLE EFFICIENCY, CONTINUED

$$C_{fg} = \frac{F_{g\,\text{act}}}{F_{i\,\text{noz}}} = \text{Nozzle gross thrust coefficient}$$

where

$F_{g\,\text{act}}$ = nozzle actual gross thrust
$F_{i\,\text{noz}}$ = nozzle ideal gross thrust
= $(W_{\text{noz}}/g)\, V_{i\,\text{noz}}$, for mixed flow
= $(W_{\text{fan}}/g)\, V_{i\,\text{fan}} + (W_c/g)\, V_{ic}$, for separate flow

Ideal velocity, where flow is completely expanded to ambient pressure at the gas stream total temperature and total pressure, is represented by the following:

$$V_{i\,\text{noz}} = f(p_{t\,\text{noz}}/p_0, T_{t\,\text{noz}})$$
$$V_{i\,\text{fan}} = f(p_{t\,\text{fan}}/p_0, T_{t\,\text{fan}})$$
$$V_{ic} = f(p_{tc}/p_0, T_{tc})$$

DEFINITION OF ENGINE NET THRUST F_n AND NET PROPULSIVE FORCE NPF

$F_n = F_{g\,\text{act}} - D_r$ = engine net thrust (provided by the engine manufacturer) where $D_r = (W_0/g)\, V_0$ = ram drag and $F_{g\,\text{act}}$ = engine actual gross thrust.

$$\text{NPF} = F_n - D_{\text{cowl}} - D_{\text{spillage}} - D_{\text{inter}}$$

where

D_{cowl} = cowl drag
D_{spillage} = inlet spillage drag
D_{inter} = interference drag

which are determined by the aircraft manufacturer from scale model wind tunnel tests.

MULTI-MISSION INLETS AND EXHAUST SYSTEMS

SUPERSONIC INLET TYPES

Supersonic inlets may each be classified into one of three types—external, mixed, or internal compression—according to whether the supersonic diffusion occurs external to the inlet duct, partly external and partly internal to the duct, or entirely internal to the duct, respectively. The figure compares the three inlet types.

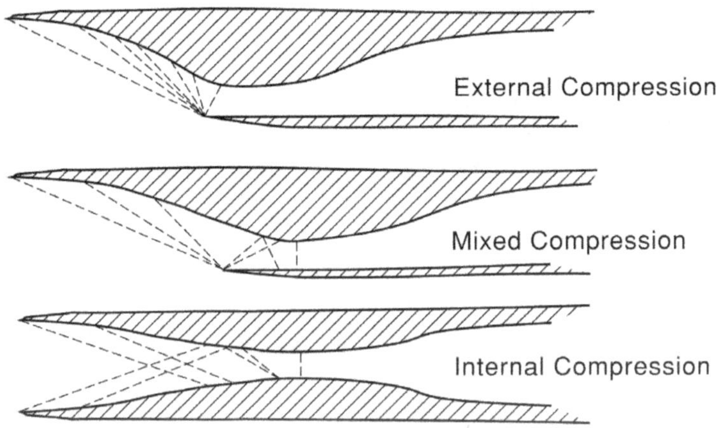

Source: The figures on pages 10-27 and 10-28 are taken from "Inlets and Inlet/Engine Integration," page 257, by J. L. Younghans and D. L. Paul. *Aircraft Propulsion Systems Technology and Design*, AIAA Education Series, edited by Gordon C. Oates.

MULTI-MISSION INLETS AND EXHAUST SYSTEMS, CONTINUED

INLET PRESSURE RECOVERY

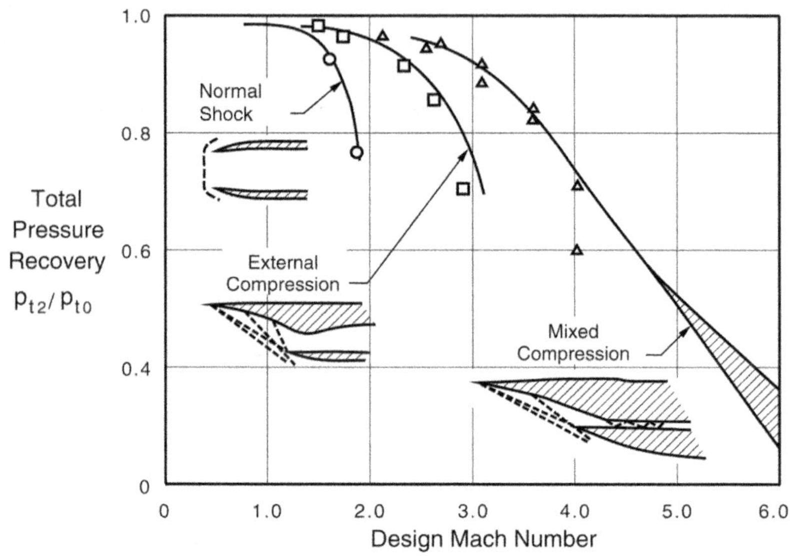

The figure illustrates the achievable pressure recovery trends with design flight Mach number for each of the inlet types. Although the mixed compression inlet provides higher recovery at supersonic conditions, its increased weight, increased control complexity, and increased bleed requirements must be balanced against the improved supersonic recovery. For a multi-mission aircraft, this is seldom a favorable trade.

MULTI-MISSION INLETS AND EXHAUST SYSTEMS, CONTINUED

DEFINITION OF NOZZLE GROSS THRUST COEFFICIENT C_{fg}

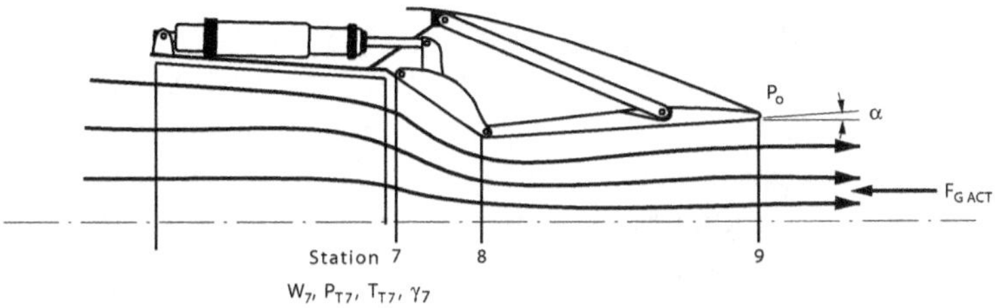

Source: "Inlets and Exhaust Systems for Multi-Mission Applications," UTSI short course in Aeropropulsion, April 2001, by D. A. Dietrich and A. P. Kuchar. Sponsored by AIAA and the Air Breathing Propulsion Technical Committee.

A simplified definition of nozzle gross thrust coefficient and its constituents follows. In practice, the integral of the momentum is difficult to obtain accurately by direct measurement. For convenience, the nonuniformities that occur at the nozzle exit are contained in the peak thrust coefficient $C_{fg\text{ peak}}$, which is defined as the nozzle losses due to angularity α and friction when expanding isentropically to p_0 at station 9; i.e. $p_{9i} = p_0$. The nozzle gross thrust coefficient C_{fg} takes into account off-design operation where at other pressure ratios across station 9 the difference between $C_{fg\text{ peak}}$ and C_{fg} is the expansion loss.

$$C_{fg} = \frac{F_{g\text{ act}}}{F_{i\text{ noz}}} = \frac{C_{fg\text{ peak}}(W_7/g)V_{9i} + (p_{9i} - p_0)A_9}{F_{i\text{ noz}}} = \text{Gross Thrust Coefficient}$$

where

$F_{g\text{ act}}$ = nozzle actual gross thrust
$F_{i\text{ noz}} = (\omega_7/g)V_{i\text{ noz}}$ = nozzle ideal gross thrust
W_7/g = nozzle actual mass flow
$V_{i\text{ noz}} = f(p_{t7}/p_0, T_{t7}, \gamma)$ = fully expanded ideal velocity
$C_{fg\text{ peak}} = f(A_9/A_8, \alpha)$ = peak thrust coefficient
g = gravitational constant
$V_{9i} = f(A_9/A_8, p_{t7}, \gamma)$ = nozzle exit ideal velocity
$p_{9i} = f(A_9/A_8, p_{t7}, \gamma)$ = nozzle exit ideal static pressure
p_0 = ambient pressure
A_9 = nozzle exit area at station 9

Note that C_{fg} and/or $C_{fg\text{ peak}}$ are normally determined by scale model tests in a static thrust stand with adjustments for leakage, Reynolds number, and real gas effects. Other methods for obtaining these factors include historical correlations and computational fluid dynamics analysis.

RAMJET AND SCRAMJET ENGINES

THE RAMJET ENGINE

The ramjet engine operating cycle is basically similar to that of other air breathing engines, consisting of compression, combustion, and expansion. However, it needs a boost or running start before it can operate effectively. Booster speeds of approximately 483 km/h (300 mph) are required for subsonic ramjets; much higher booster speeds are required for supersonic combustion ramjets (scramjets).

Because a ramjet engine obtains compression by virtue of its motion through the air, its operating efficiency is considerably affected by flight speed. Above Mach 2.0 the increased thermal efficiency, coupled with its relative simplicity and attendant lower weight and cost, causes the ramjet engine to be competitive with turbojet engines for certain propulsion applications.

The ramjet engine consists of a diffuser, a combustion chamber, and an exhaust nozzle. Air enters the diffuser, where the fluid is compressed dynamically before mixing with injected fuel. The fuel–air mixture burns in a combustion chamber (burner), where typical exit temperatures reach nearly 4940K (5400°R). Although present materials cannot tolerate temperatures much above 1540K (2000°R), they can be kept much cooler than the main flow by injecting the fuel in a pattern such that there is a cooling layer next to the walls. The hot gases are expelled through the nozzle at high velocity due to the high temperature of the gases and the rise in pressure caused by the deceleration of the airstream from flight velocity to low (subsonic) combustion chamber velocity.

The acceleration of the combustion products passing through the engine provides the reaction force or internal thrust. The net thrust is equal to this internal thrust less the total drag of the external forces and the nacelle–engine combination.

RAMJET AND SCRAMJET ENGINES, CONTINUED

THE RAMJET ENGINE, CONTINUED

The following figure is a schematic of a ramjet engine and illustrates the station numbering:

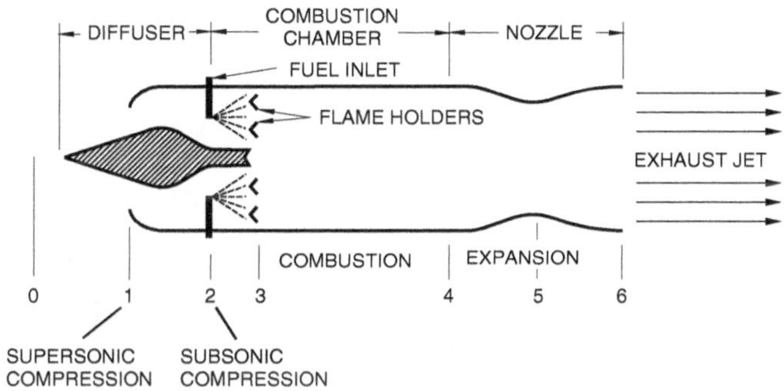

Source: *Mechanics and Thermodynamics of Propulsion*, 2nd Edition, page 149, figure 6-3, by P. Hill and C. Peterson. Copyright © 1992, Addison–Wesley, Reading, MA. Reprinted by permission of Pearson Education, Inc.

The complete ramjet engine system consists of four basic components: 1) inlet diffuser, 2) combustion chamber, 3) nozzle, and 4) control and fuel supply system. The station subscripts are as follows:

0 – Free stream
1 – Diffuser inlet
2 – Combustion chamber inlet or diffuser outlet
3 – Flameholders
4 – Combustion chamber outlet or nozzle inlet
5 – Nozzle throat
6 – Nozzle exit

RAMJET AND SCRAMJET ENGINES, CONTINUED

THE RAMJET ENGINE, CONTINUED

JET ENGINE OPERATIONAL ENVELOPES

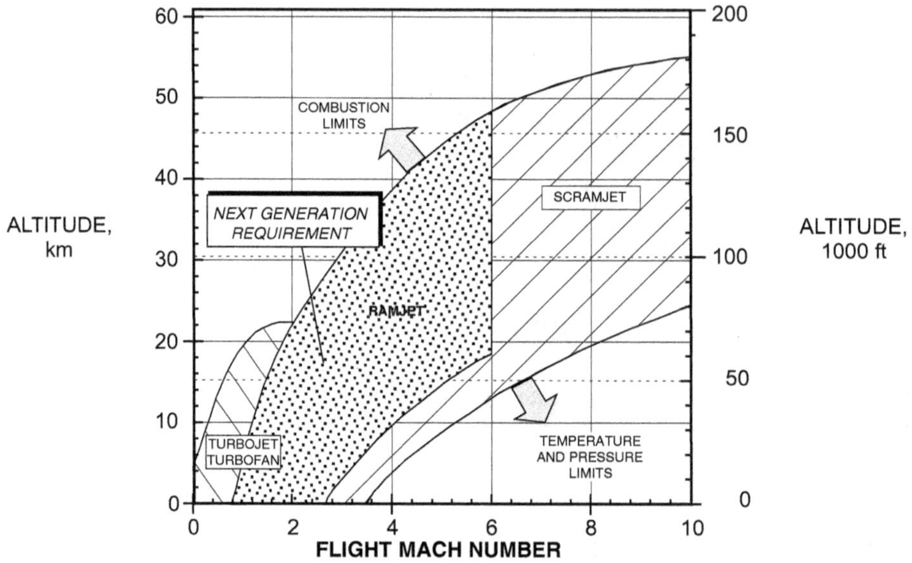

Source: "Advanced Ramjet–Scramjet Technology," UTSI Short Course in Aeropropulsion, Nov. 1981, by A. N. Thomas, Jr. Sponsored by AIAA and the Air Breathing Propulsion Technical Committee.

IDEAL PERFORMANCE OF A RAMJET

The ideal performance of a ramjet is characterized by a cycle shown on a temperature vs entropy diagram (see page 10-36), where compression in the inlet and expansion in the nozzle are isentropic processes (constant entropy), and the heat added in the combustion chamber is a thermodynamically reversible process. Therefore, these processes do not take into account any nonideal conditions where performance in the cycle is degraded because of total pressure losses due to shocks, friction, heat due to friction, and mass diffusion.

The ideal engine thrust is

$$F_n = \frac{W_a}{g}\left[(1+f)\,V_j - V_p\right]$$

Source: The ramjet ideal performance text appearing on pages 10-32 and 10-33 is based on *Mechanics and Thermodynamics of Propulsion,* 2nd Ed., pages 151 and 152, by P. Hill and C. Peterson. Copyright © 1992, Addison–Wesley, Reading, MA. Reprinted by permission of Pearson Education, Inc.

RAMJET AND SCRAMJET ENGINES, CONTINUED

THE RAMJET ENGINE, CONTINUED

IDEAL PERFORMANCE OF A RAMJET, CONTINUED

The specific thrust (thrust per mass of airflow per second) is

$$\frac{F_n}{W_a} = \frac{M(\gamma R T_0)^{\frac{1}{2}}}{(g)^{\frac{1}{2}}}\left[(1+f)(T_{t4}/T_0)^{\frac{1}{2}}\left(1+\frac{\gamma-1}{2}M^2\right)^{-\frac{1}{2}}-1\right]$$

where

F_n/W_a = net thrust/mass airflow/s
T_{t4} = total combustion temperature, K (°R)
T_0 = free stream static temperature, K (°R)

The thrust specific fuel consumption, TSFC, is

$$\text{TSFC} = \frac{W_f}{F_n} = \frac{f \times 3600}{F_n/W_a}$$

where the fuel–air ratio f is defined as

$$f = \frac{(T_{t4}/T_{t0})-1}{(\eta_b Q/C_p T_{t0})-(T_{t4}/T_{t0})}$$

Note that η_b is the combustion efficiency and $\eta_b Q$ is the actual heat release per unit weight of fuel.

RAMJET AND SCRAMJET ENGINES, CONTINUED

THE RAMJET ENGINE, CONTINUED

IDEAL PERFORMANCE OF A RAMJET, CONTINUED

The following performance charts are based on several ideal assumptions: 1) there are no friction or shock losses, 2) there is no dissociation of the combustion products or air, 3) there are no variations in specific heats, 4) there is complete expansion at engine exit to ambient pressure, and 5) hydrocarbon fuel is used.

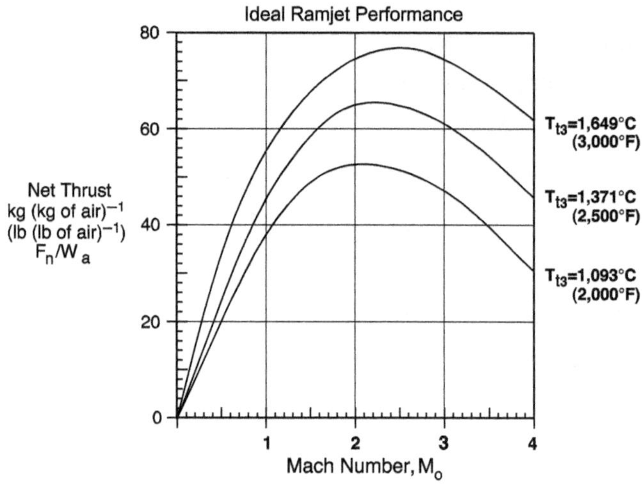

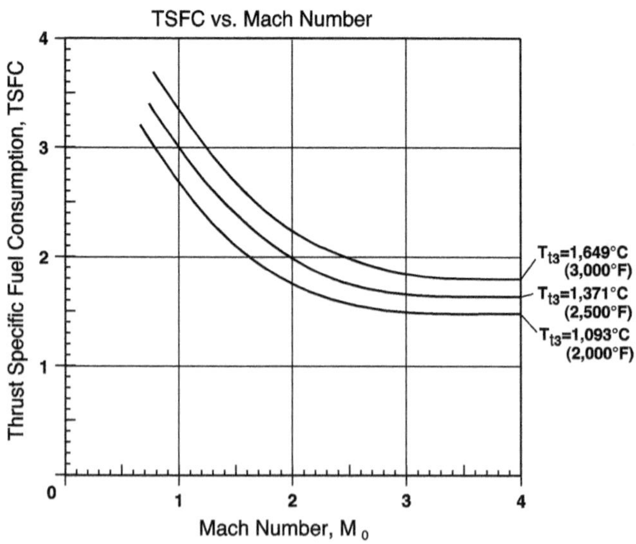

Source: "Ramjet Primer," by D. B. Clark. Wright Aeronautical Division, Curtiss–Wright Corp., Woodbridge, NJ.

RAMJET AND SCRAMJET ENGINES, CONTINUED

SPECIFIC IMPULSE VS MACH NUMBER

Specific impulse I_{sp}, which is defined as F/W_f, also can be expressed in terms of TSFC where I_{sp} = 3600/TSFC.

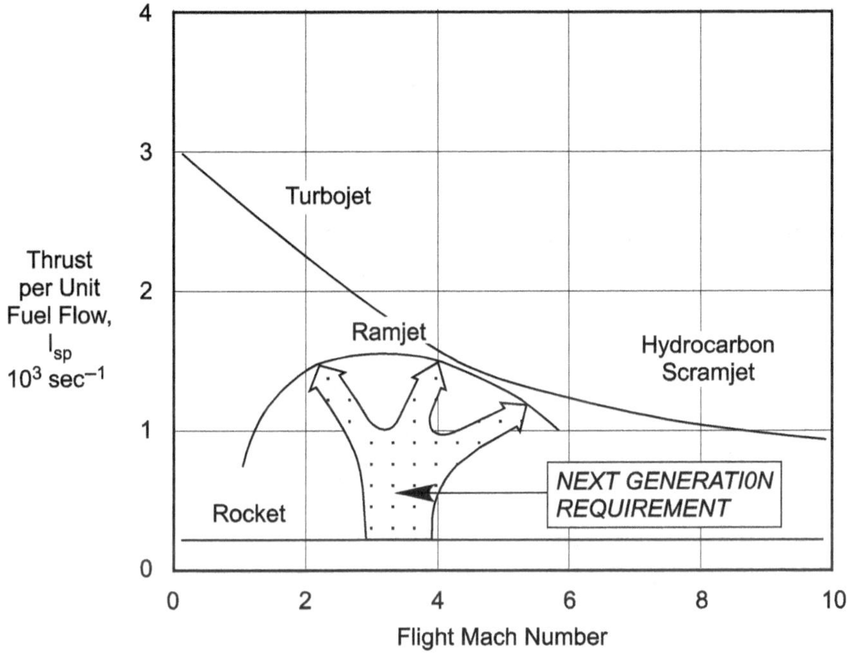

THRUST COEFFICIENT

Aerodynamicists use a nondimensional term called a gross thrust coefficient (C_f) when considering the total integration of a ramjet with a vehicle. This term is defined as the net thrust of the ramjet engine divided by the incompressible impact pressure per unit of frontal area.

$$C_f = \frac{F_n}{\frac{w_0}{2g} V_p^2 A} = \frac{F_n}{\frac{\gamma}{2} M^2 p_0 A}$$

where A = reference area (normally taken on the basis of nominal ramjet diameter).

RAMJET AND SCRAMJET ENGINES, CONTINUED

SPECIFIC IMPULSE VS MACH NUMBER, CONTINUED

THRUST COEFFICIENT, CONTINUED

Note that whereas this term appears to be similar in name to the nozzle gross thrust coefficient (C_{fg}) used in turbojet analysis, it has a different interpretation when used with a ramjet/vehicle installation. The term "gross" in this application indicates that the net thrust F_n of the ramjet engine has been taken into consideration, but the drag D_p of the vehicle has not been factored into the net propulsive power of the integrated installation.

AERODYNAMIC LOSSES INCURRED DURING RAMJET OPERATION

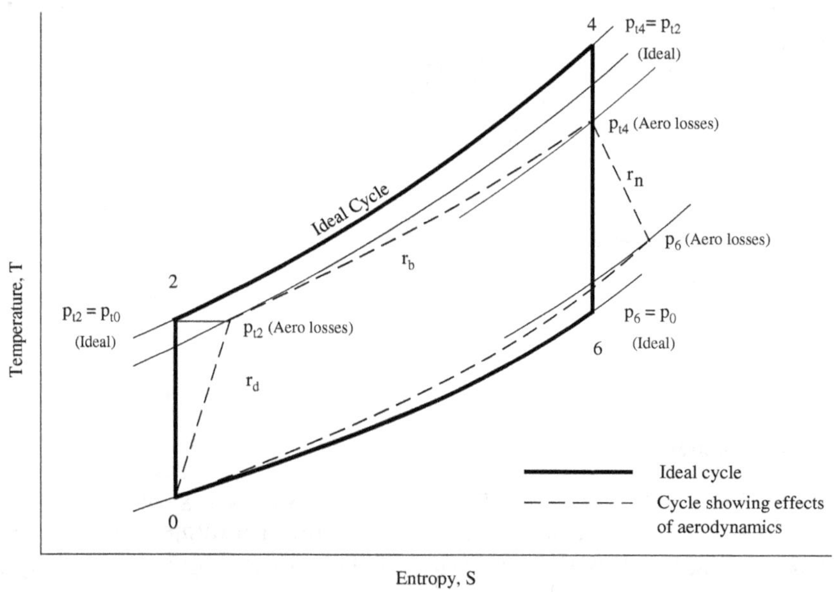

The figure shows the effects of aerodynamic losses on a temperature–entropy (T–S) diagram where the processes are no longer isentropic, thereby resulting in entropy changes.

Source: The temperature–entropy diagram and text on pages 10-36 and 10-37 are based on *Mechanics and Thermodynamics of Propulsion*, 2nd Ed. pages 154–156, by P. Hill and C. Peterson. Copyright © 1992, Addison–Wesley, Reading, MA. Figure reprinted by permission of Pearson Education, Inc.

RAMJET AND SCRAMJET ENGINES, CONTINUED

SPECIFIC IMPULSE VS MACH NUMBER, CONTINUED

AERODYNAMIC LOSSES INCURRED DURING RAMJET OPERATION, CONTINUED

Losses in the ramjet engine are characterized by total pressure losses due to shocks, friction, heat, and mass diffusion, etc. The total pressure loss ratios for inlet (diffuser), combustor (burner), nozzle, and engine are

$$r_d = \frac{p_{t2}}{p_{t0}}$$

$$r_b = \frac{p_{t4}}{p_{t2}}$$

$$r_n = \frac{p_{t6}}{p_{t4}}$$

$$\frac{p_{t6}}{p_{t0}} = r_d r_b r_n$$

The thrust per unit air flow rate (nonideal) is

$$\frac{F_n}{W_a} = \frac{1}{g}\left[(1+f)V_j - V_p\right] + \frac{1}{W_a}(p_6 - p_0)A_6$$

or

$$\frac{F_n}{W_a} = \frac{(1+f)}{(g)^{1/2}}\sqrt{\frac{2\gamma RT_{t4}(m-1)}{(\gamma-1)m}} - \frac{M}{(g)^{1/2}}\sqrt{\gamma RT_0} + \frac{p_6 A_6}{W_a}\left(1 - \frac{p_0}{p_6}\right)$$

where

$$m = \left(1 + \frac{\gamma-1}{2}M^2\right)\left(r_d r_b r_n \frac{p_0}{p_6}\right)^{\frac{(\gamma-1)}{\gamma}}$$

The thrust specific fuel consumption is

$$\text{TSFC} = \frac{f \times 3600}{F_n/W_a}$$

RAMJET AND SCRAMJET ENGINES, CONTINUED

THE SCRAMJET ENGINE

At Mach numbers between 4 and 6, large total pressure losses result from decelerating the flow to subsonic speeds, and increasing chemical dissociation losses occur. The supersonic combustion ramjet (scramjet) avoids these losses by eliminating the flow deceleration and thus becoming more efficient. In the scramjet, the flow enters the combustor after compression in the inlet at supersonic speeds, and combustion is maintained supersonically. The following figures illustrate ramjet/scramjet performance using hydrogen fuel.

RAMJET GROSS THRUST COEFFICIENT VS MACH NUMBER

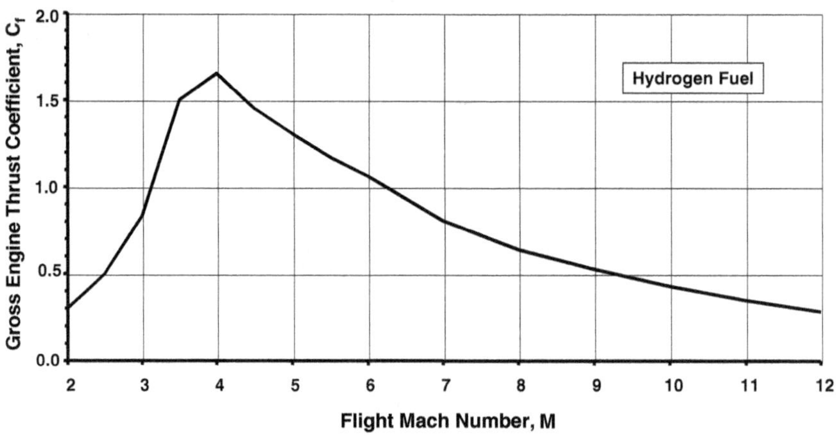

RAMJET GROSS I_{SP} VS MACH NUMBER

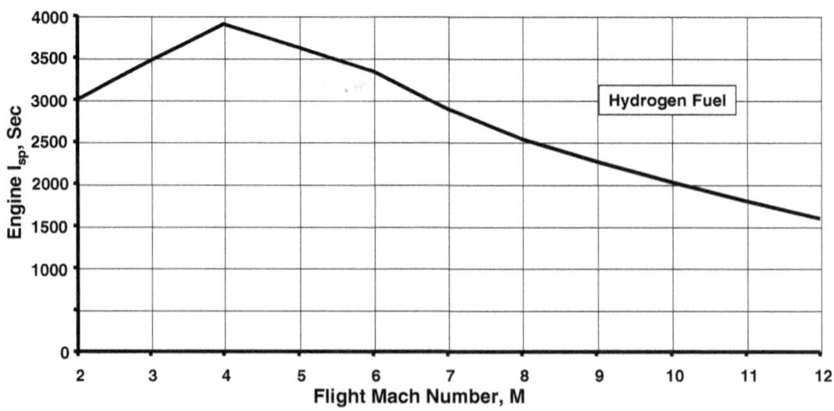

Source: Ramjet charts courtesy of GenCorp Aerojet, Jan. 2000. Reprinted with permission.

SECTION 11

SPACECRAFT AND LAUNCH VEHICLE DESIGN

SPACECRAFT GENERAL INFORMATION

TYPES OF SPACECRAFT MISSIONS AND PAYLOADS

Spacecraft Mission	Payload	Example
Communications		
Full-duplex broadband	Transceiver	Milstar, Intelsat
Message broadcast	Transmitter	DirecTV, GPS
Personal comm	Transceiver	Iridium
Remote Sensing		
Imaging	Images and cameras	LandSat, Space Telescope
Intensity and measurement	Radiometers	SBIRS early warning, Chandra X-Ray Observatory
Topographic mapping	Altimeters	TOPEX/Poseidon
Navigation		
Ranging	Transceiver	TDRS
Nav signal	Clock and transmitter	GPS, GLONASS
Weapons		
Kinetic energy	Warhead	Brilliant Pebbles concept
Directed energy	High-energy weapon	Space-Based Laser concept
In Situ Space		
Crewed	Physical and life sciences	Space Shuttle, MIR
Robotic	Sample collection/return	Mars Sojourner, LDEF
Other		
Microgravity Manufacturing	Physical plant and raw materials	Space Shuttle, MIR
Space power	Solar collector, converter, and transmitter	SPS
Resource utilization	Lunar soil collector and processor	Lunar Base
Tourism	Orbital hotel	Various
Space burial	Remains container	Pegasus XL

Source: *Space Mission Analysis and Design*, 3rd Ed., page 243, table 9-1, by James R. Wertz and Wiley J. Larson. Copyright © 1999 Microcosm, Inc. and W.J. Larson.

TYPICAL SPACECRAFT PROGRAM SEQUENCE

- Mission requirements definition
- Conceptual design
- Preliminary design (PDR)
- Detailed design (CDR)
- Subsystem fabrication
- Integration and test (FRR)
- Launch vehicle integration
- Launch
- Orbital verification
- Operational use
- Disposal

SPACECRAFT GENERAL INFORMATION, CONTINUED

SPACECRAFT BLOCK DIAGRAM

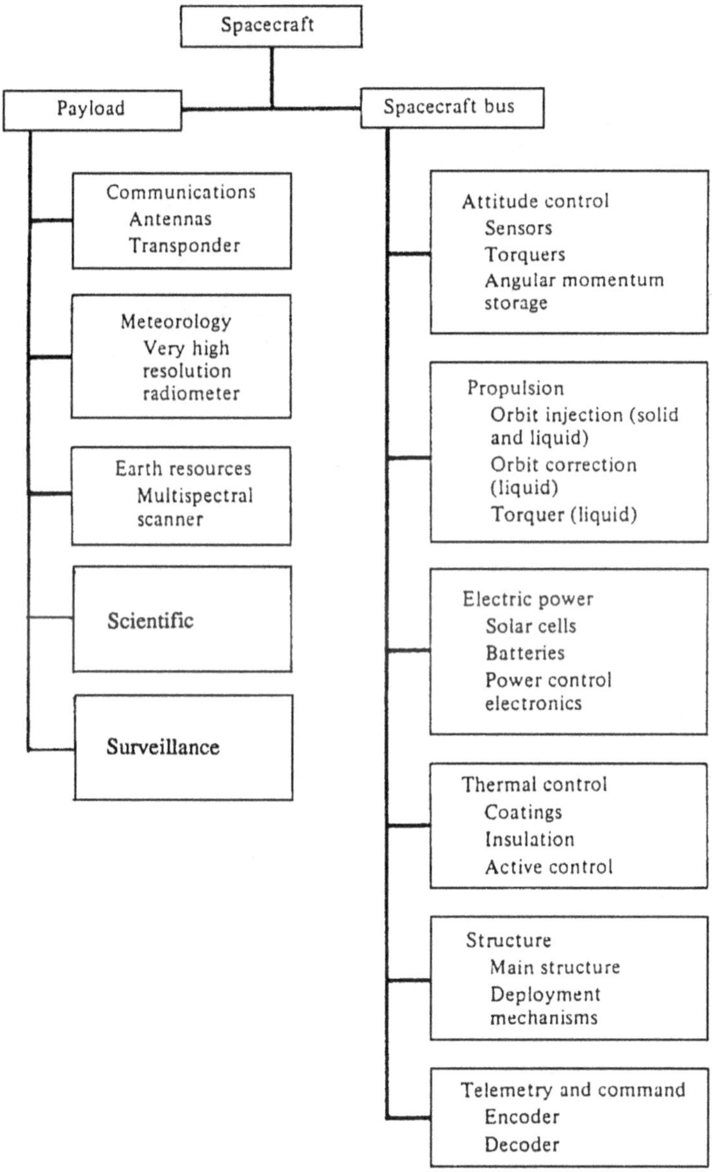

Source: *Design of Geosynchronous, Spacecraft*, page 3, figure 1.2, by B. N. Agrawal.

SPACECRAFT GENERAL INFORMATION, CONTINUED

PRINCIPAL REQUIREMENTS AND CONSTRAINTS FOR SPACECRAFT DESIGN

Requirements and Constraints	Information Needed
Mission:	
Operations Concept	Type, mission approach
Spacecraft Life & Reliability	Mission duration, success criteria
Comm Architecture	Command, control, comm approach
Security	Level, requirements
Programmatic Constraints	Cost and schedule profiles
Payload:	
Physical Parameters	Size, weight, shape, power
Operations Concept	Duty cycle, data rates, fields of view
Pointing	Reference, accuracy, stability
Slewing	Magnitude, frequency
Environment	Max and min temperature, cleanliness
Orbit:	
Defining Parameters	Altitude, inclination, eccentricity
Eclipses	Maximum duration, frequency
Lighting Conditions	Sun angle and viewing conditions
Maneuvers	Size, frequency
Environment:	
Radiation Dosage	Average, peak
Particles & Meteoroids	Size, density
Space Debris	Density, probability of impact
Hostile Environment	Type, level of threat
Launch:	
Launch Strategy	Single, dual; dedicated, shared; use of upper kick stage
Boosted Weight	Launch capabilities
Envelope	Size, shape
Environments	g's, vibration, acoustics, temperature
Interfaces	Electrical and mechanical
Launch Sites	Locations, allowed launch azimuths
Ground-System Interface:	
Degree of Autonomy	Required autonomous operations
Ground Stations	Number, locations, performance
Space Links	Space-to-space link, performance

Adapted from: *Space Mission Analysis and Design*, 3rd Ed., page 305, table 10-3, by James R. Wertz and Wiley J. Larson.

LAUNCH VEHICLES

BRAZIL — CHINA — EUROPE — INDIA — ISRAEL — JAPAN

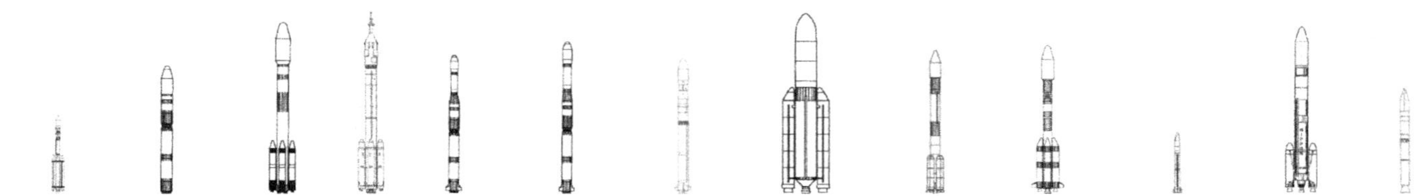

Vehicle	VLS	LM-2C	LM-2E	LM-2F	LM-3	LM-3A	LM-4B	Ariane 5	PSLV	GSLV	Shavit 1	H-IIA 202	M-V
Performance													
LEO Maximum	380 kg (840 lbm)	3900 kg (8600 lbm)	9500 kg (20,950 lbm)	?	?	?	?	18,000 kg (39,700 lbm)	3700 kg (8150 lbm)	5000 kg (11,000 lbm)	225 kg (495 lbm)	9940 kg (21,900 lbm)	1800 kg (4000 lbm)
SSO	80 kg (175 lbm)	?	?	?	?	?	2800 kg (6170 lbm)	12,000 kg (26,400 lbm)	1200 kg (2650 lbm)	2200 kg (4850 lbm)	—	4350 kg (9590 lbm)	?
GTO	—	1400 kg (3100 lbm)[1]	3500 kg (7710 lbm)[1]	—	1500 kg (3300 lbm)	2600 kg (5700 lbm)	—	6800 kg (15,000 lbm)	800 kg (1750 lbm)	2500 kg (5500 lbm)	—	4100 kg (9040 lbm)	1215 kg (2680 lbm)[1]
Cost	$8 million	$20–25 million	$45–55 million	?	$35–40 million	$45–55 million	$25–35 million	$150–180 million	$15–25 million	$35–45 million	$10–15 million	$75 million	$55–60 million
First Flight	1997	1975	1990	2000?	1984	1994	1999	1996	1993	2000	1995	2000	1997
Launch Site(s)	Alcantara	Taiyuan Jiuquan	Xichang	Jiuquan	Xichang	Xichang	Taiyuan	CSG (Kourou)	Sriharikota	Sriharikota	Palmachim	Tanegashima	Kagoshima

Notes:
[1]*Requires available kick motor*
[2]*Assumes new equatorial launch site*

Source: The launch vehicles material appearing on pages 11-5–11-9 is from *International Reference Guide to Space Launch Systems*, 3rd Ed., by S. J. Isakowitz, J. P. Hopkins Jr., and J. B. Hopkins.

LAUNCH VEHICLES, CONTINUED

RUSSIA / UKRAINE

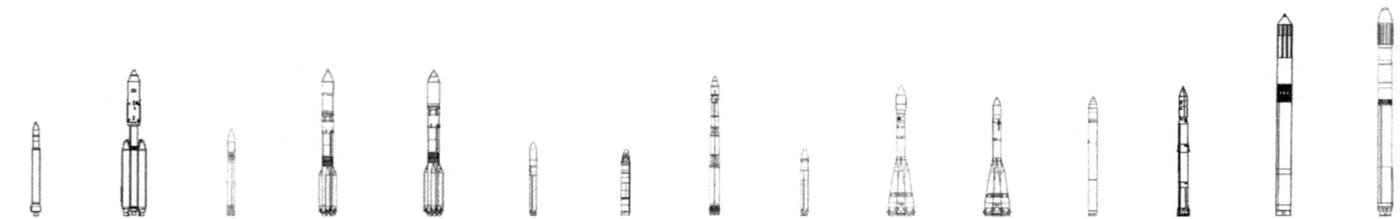

Vehicle	Angara 1.1	Angara 5I	Kosmos 3M	Proton K/Block DM	Proton M/Breeze M	Rockot	Shtil-1	Start-1	Strela	Soyuz U	Molniya	Dnepr-1	Tsiklon 3	Zenit 2	Zenit 3SL
Performance															
LEO Maximum	1600 kg (3500 lbm)	24,500 kg (54,000 lbm)	1500 kg (3300 lbm)	19,760 kg (43,560 lbm)	21,000 kg (46,300 lbm)	1800 kg (3970 lbm)	430 kg (950 lbm)	632 kg (1393 lbm)	1700 kg (3750 lbm)	7000 kg (15,430 lbm)	—	4500 kg (9920 lbm)	4100 kg (9020 lbm)	13,500 kg (29,750 lbm)	?
SSO	?	?	775 kg (1700 lbm)	3620 kg (7980 lbm)	?	1000 kg (2205 lbm)	?	167 kg (368 lbm)	?	2750 kg (6050 lbm) with Ikar	1500 kg (3300 lbm)	140 kg (310 lbm)	—	5000 kg (11,000 lbm)	?
GTO	—	6800 kg (15000 lbm)	—	Up to 4910 kg (10,826 lbm)	5500 kg (12,100 lbm)	—	—	—	—	1350 kg (2975 lbm) with Fregat	—	—	—	—	5000 kg (11,000 lbm)
Cost	?	?	$12 million	$90–98 million	$100–112 million	$13–15 million	$0.1–0.3 million	$9 million	$10.5 million	$30–50 million	$30–40 million	$10–20 million	$20–25 million	$35–50 million	$75–95 million
First Flight	2001	2003	1967	1967	2000	1994	1998	1993	2000	1963	1960	1999	1977	1985	1999
Launch Site(s)	Plesetsk	Plesetsk	Plesetsk	Baikonur	Baikonur	Plesetsk Baikonur	Delta IV Submarine	Svobodny Plesetsk	Svobodny	Baikonur Plesetsk	Baikonur Plesetsk	Baikonur	Plesetsk	Baikonur	Sea Launch Odyssey

Notes:

1 Requires available kick motor

2 Assumes new equatorial launch site

LAUNCH VEHICLES, CONTINUED

UNITED STATES

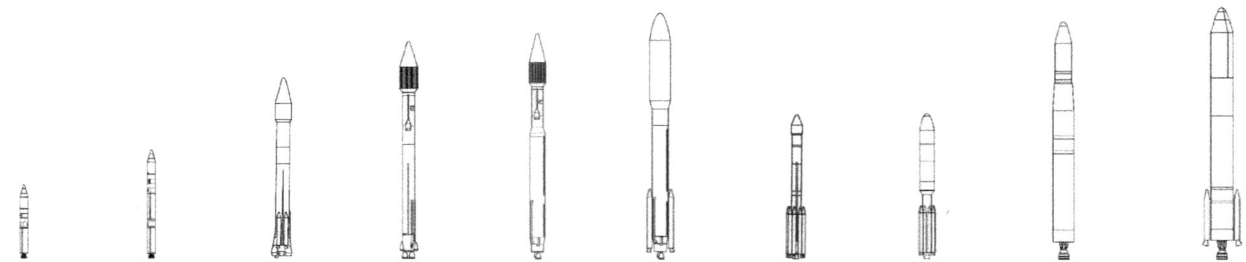

Vehicle	Athena I	Athena II	Atlas IIAS	Atlas IIIB	Atlas V 400	Atlas V 500	Delta II	Delta III	Delta IV Medium	Delta IV Medium Plus
Performance										
LEO Maximum	820 kg (1805 lbm)	2065 kg (4520 lbm)	8618 kg (19,000 lbm)	10,718 kg (23,630 lbm)	12,500 kg (27,550 lbm)	Up to 20,050 kg (44,200 lbm)	Up to 5140 kg (11,330 lbm)	8290 kg (18,280 lbm)	8600 kg (18,600 lbm)	Up to 13,600 kg (30,000 lbm)
SSO	360 kg (790 lbm)	1165 kg (1565 lbm)	?	5885 kg (12974 lbm)	?	?	Up to 3220 kg (7100 lbm)	6100 kg (13450 lbm)	6300 kg (13,900 lbm)	Up to 9600 kg (21,100 lbm)
GTO	—	590 kg (1290 lbm)[1]	3719 kg (8200 lbm)	4477 kg (9870 lbm)	5000 kg (11,000 lbm)	Up to 8200 kg (18,000 lbm)	Up to 1870 kg (4120 lbm)	3810 kg (8400 lbm)	3900 kg (8600 lbm)	Up to 6120 kg (13,500 lbm)
Cost	\$16–17 million	\$22–26 million	\$90–105 million	\$90–105 million	\$75–90 million	\$85–110 million	\$45–60 million	\$85 million	\$75–90 million	\$85–110 million
First Flight	1995	1998	1993	2000	2001	2001	1990	1998	2001	2001
Launch Site(s)	Vandenberg, Cape Canaveral, Kodiak	Vandenberg, Cape Canaveral, Kodiak	Cape Canaveral, Vandenberg	Cape Canaveral	Cape Canaveral, Vandenberg	Cape Canaveral, Vandenberg	Cape Canaveral, Vandenberg	Cape Canaveral	Cape Canaveral, Vandenberg	Cape Canaveral, Vandenberg

Notes:
1 Requires available kick motor
2 Assumes new equatorial launch site

LAUNCH VEHICLES, CONTINUED

UNITED STATES, CONTINUED

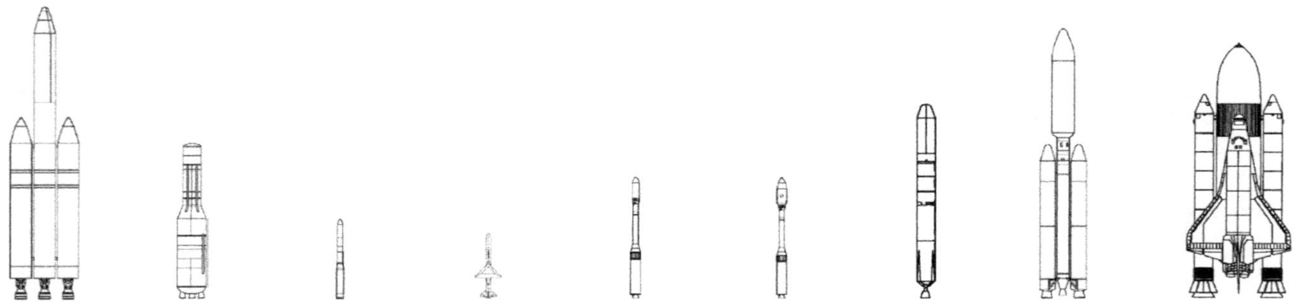

Vehicle	**Delta IV Heavy**	**K-1**	**Minotaur**	**Pegasus XL**	**SSLV Taurus**	**Commercial Taurus**	**Titan II**	**Titan IVB**	**Space Shuttle**
Performance									
LEO Maximum	25,800 kg (56,900 lbm)	4600 kg (10,150 lbm)	639 kg (1408 lbm)	443 kg (977 lbm)	1320 kg (2910 lbm)	1380 kg (3040 lbm)	1900 kg (4200 lbm)	21,680 kg (47,800 lbm)	24,400 kg (53,800 lbm)
SSO	19,200 kg (42,300 lbm)	1250 kg (2750 lbm)	336 kg (740 lbm)	190 kg (420 lbm)	660 kg (1460 lbm)	720 kg (1590 lbm)	(See chapter)	?	—
GTO	12,400 kg (27,400 lbm)	—	—	—	400 kg	448 kg (990 lbm)	—	?	—
Cost	$140–170 million	$17 million	$12.5 million	$12–15 million	$18–20 million	$18–20 million	$30–40 million	$350–450 million	$300 million
First Flight	2003	2000	1999	1994	1994	1998	1986	1997	1981
Launch Site(s)	Cape Canaveral Vandenberg	Woomera Nevada Test Site	Vandenberg Others	Vandenberg Wallops Cape Canaveral Others	Vandenberg Others	Vandenberg Others	Vandenberg	Cape Canaveral Vandenberg	Kennedy Space Center

Notes:

1 Requires available kick motor

2 Assumes new equatorial launch site

LAUNCH VEHICLES, CONTINUED

WORLDWIDE LAUNCH SITES

Launch Site	Map Designation	Country	Latitude (deg min)	Longitude (deg min)
Western Range	Vandenberg AFB, CA	United States	34 36 N	120 36 W
Eastern Range	Cape Canaveral AFS, Cape Kennedy Space Center	United States	28 30 N	80 33 W
Wallops Island	Wallops, VA	United States	37 51 N	75 28 W
Kourou Launch Ctr	Kourou	CNES/ Arianespace	5 32 N	52 46 W
San Marco Launch Platform	San Marco	Italy	2 56 S	40 12 E
Plesetsk	Plesetsk	Russia	62 48 N	40 24 E
Kapustin Yar	Kapustin Yar	Russia	48 24 N	45 48 E
Tyuratam (Baikonur)	Tyuratam	Russia	45 54 N	63 18 E
Thumba Equatorial Station	Thumba	UN/India	8 35 N	76 52 E
Sriharikota	Sriharikota	India	13 47 N	80 15 E
Jiuguan Satellite Launch Ctr	Jiuguan	China	40 42 N	100 00 E
Xichang Satellite Launch Ctr	Xichang (Sichuan)	China	28 12 N	102 00 E
Taiyuan Satellite Launch Ctr	Taiyuan	China	37 46 N	112 30 E
Kagoshima Space Ctr	Kagoshima	Japan/ISAS	31 15 N	131 05 E
Tanegashima Space Ctr	Tanegashima	Japan/NASDA	30 24 N	130 58 E
Woomera Launch Site	Woomera	Australia/U.S.	31 07 S	136 32 E
Israeli Launch Complex	Yavne	Israel	31 31 N	34 27 E

Source: *Space Mission Analysis and Design,* 3rd Ed., page 733, table 18-6, by James R. Wertz and Wiley J. Larson. Copyright ©1999, Microcosm, Inc., and W.J. Larson.

LAUNCH VEHICLES, CONTINUED

SPACE SHUTTLE LAUNCH CONFIGURATION

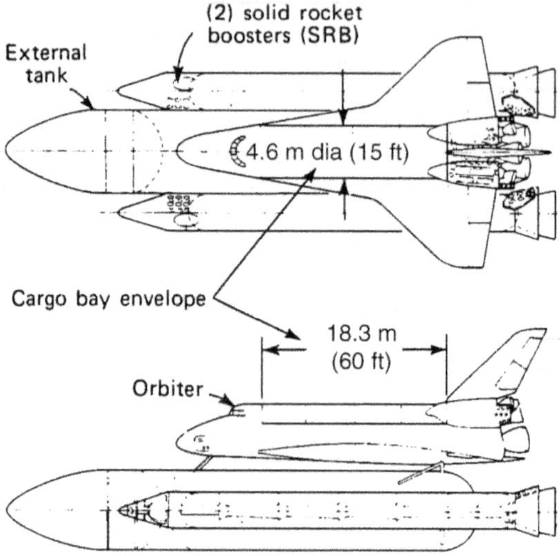

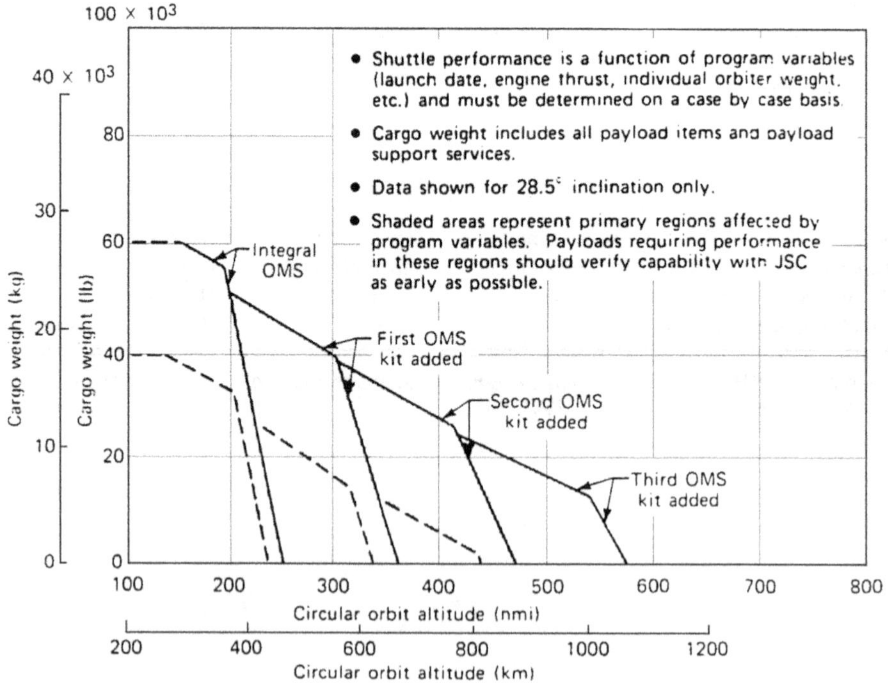

Near-term Space Shuttle cargo weight vs circular orbital altitude–KSC launch

SPACE ENVIRONMENT

SPACE ENVIRONMENT—HARD VACUUM

- Even in low orbit pressure less than that in best laboratory vacuum
- Many materials, especially polymers, may outgas extensively
  - — Change in characteristics
  - — Contamination of cold surfaces by recondensing
- Some plating metals—e.g., cadmium—may migrate to cold areas
- Removal of adsorbed O_2 layer can aggravate galling and allow cold welding between surfaces of similar metals—e.g., stainless steel
- Essentially no corrosion
- No convective heat transfer

ATMOSPHERIC CHARACTERISTICS

ALTITUDE VARIATIONS OF DENSITY, TEMPERATURE AND GRAVITY

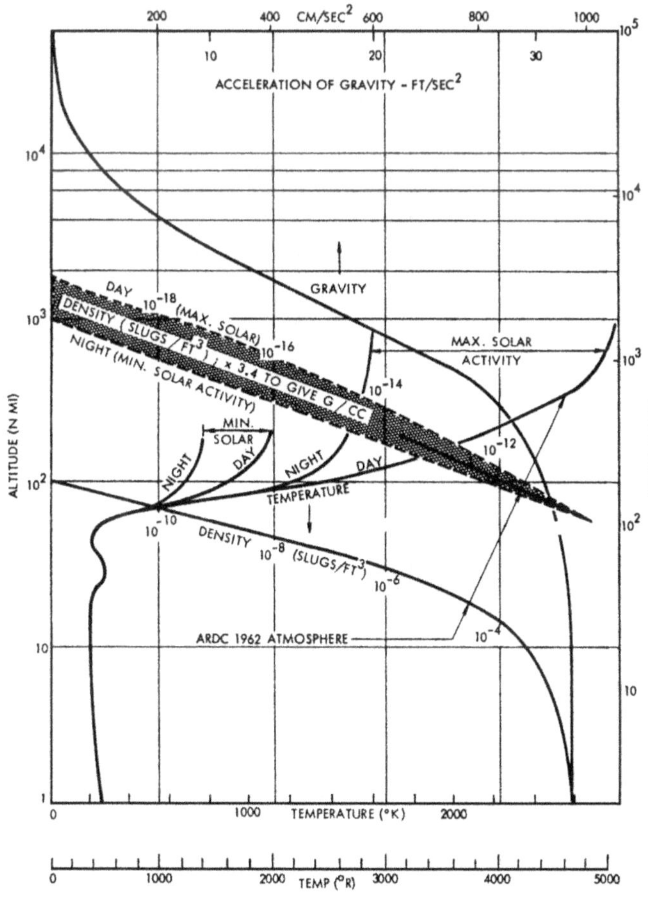

SPACE ENVIRONMENT, CONTINUED

SPACE ENVIRONMENT—RADIATION

- Cosmic Rays
  - — Infrequent, highly energetic
  - — Shielding impractical
  - — Total dose not a problem
  - — Single event upset can be serious
- Solar Flare
  - — Usually not serious for electronics
  - — Human crew requires 2–4 g/cm^2 shielding for average event
  - — Infrequent major event may require 40 g/cm^2 for crew
- Radiation Belts
  - — Charged particles trapped by planetary magnetic field
  - — Earth's Van Allen belts a major problem to both electronics and crew
  - — Jupiter radiation environment very severe

NATURAL RADIATION ENVIRONMENT FOR VARIOUS ORBITS

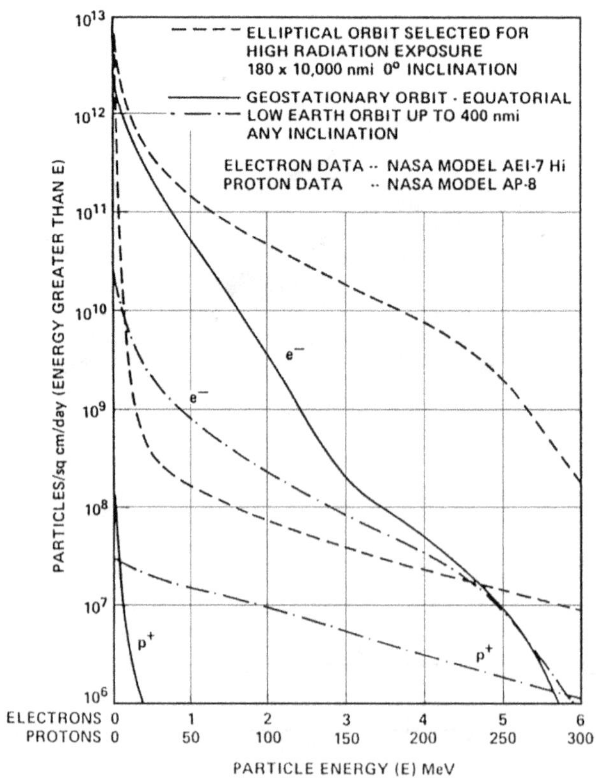

SPACE ENVIRONMENT, CONTINUED

SPACE ENVIRONMENT—RADIATION, CONTINUED

NATURAL RADIATION ENVIRONMENT

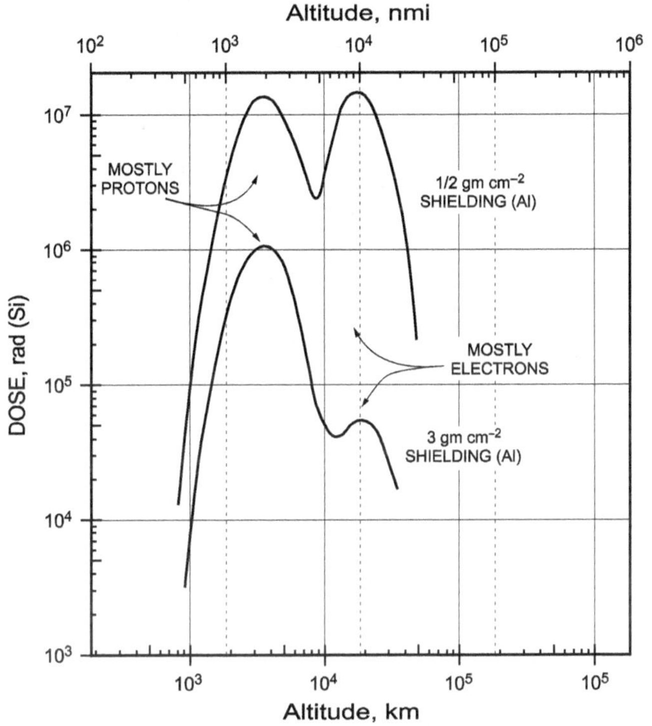

IONIZED PARTICLES/CHARGING

- Atomic oxygen attack up to several hundred kilometers
- Oxygen attack highly detrimental to many polymers—e.g., Kapton
- Spacecraft charging
  - — Very high voltages generated
  - — Affects geosynchronous spacecraft entering eclipse
  - — Discharge can cause electronic state changes or permanent damage

SPACE ENVIRONMENT, CONTINUED

SPACE ENVIRONMENT—OXYGEN ATOM FLUX VARIATION WITH ALTITUDE

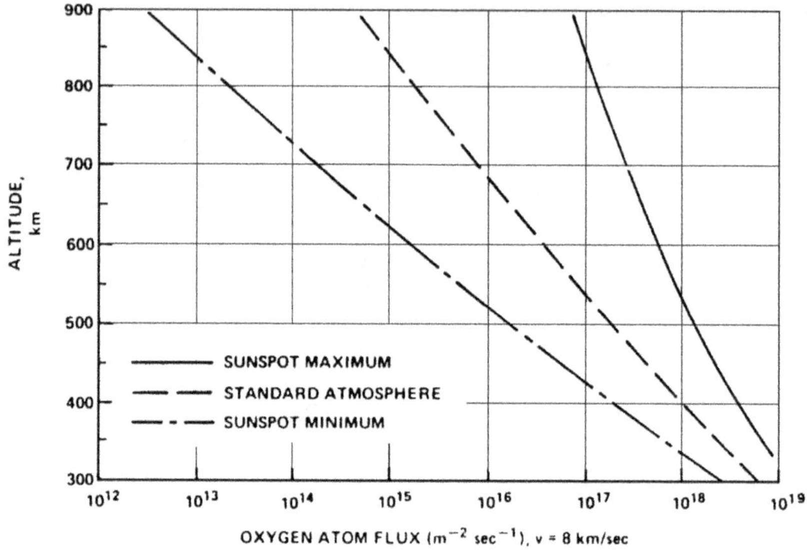

SPACE ENVIRONMENT—METEOROID/DEBRIS

Meteoroid

- Large particles very rare
- Most damage pitting/sandblasting
- Near comet environment can be hazardous
- Planetary bodies increase concentration but also shield

Protection

- "Whipple meteor bumper" very effective for high velocity particles
- Large slow particles very difficult

Debris

- Major, growing problem in low-Earth orbit
- Launch debris, explosions, ASAT tests
- Armoring imposes tremendous mass penalty

SPACE ENVIRONMENT, CONTINUED

SPACE ENVIRONMENT—MICROMETEOROID MODEL

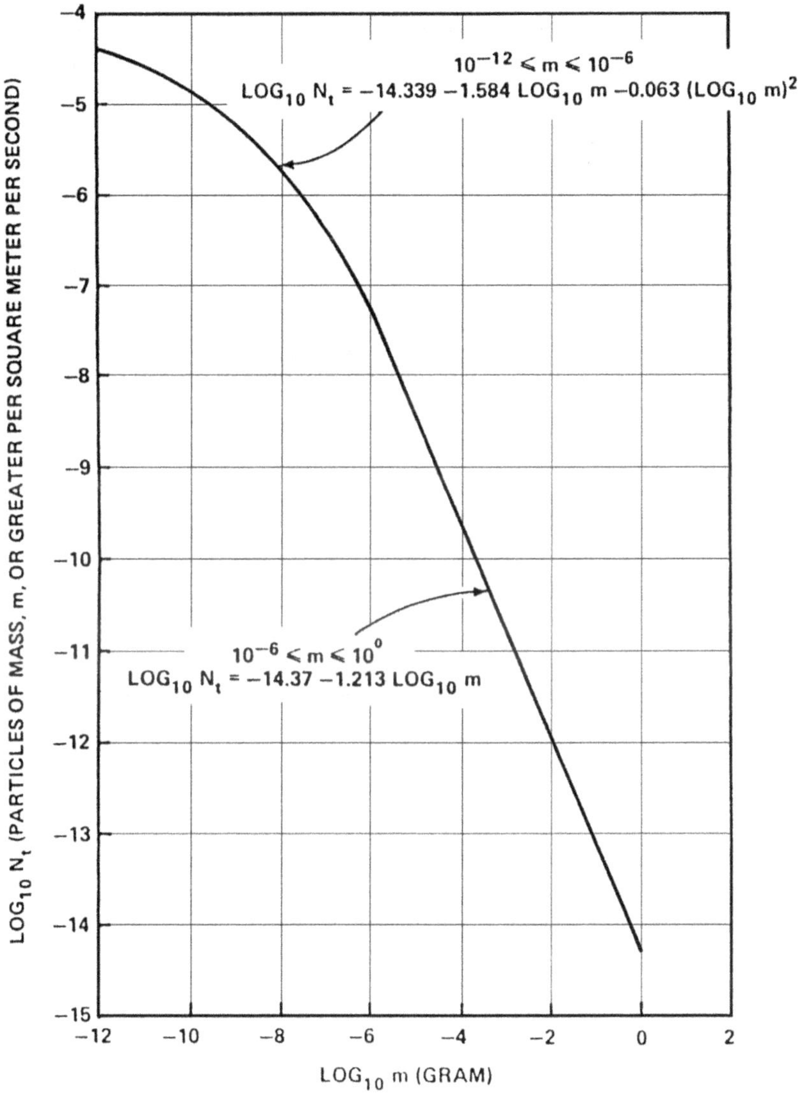

-Average cumulative total meteoroid flux-mass model for 1 A.U.

SPACE ENVIRONMENT, CONTINUED

SPACE ENVIRONMENT METEOROID AND DEBRIS FLUX

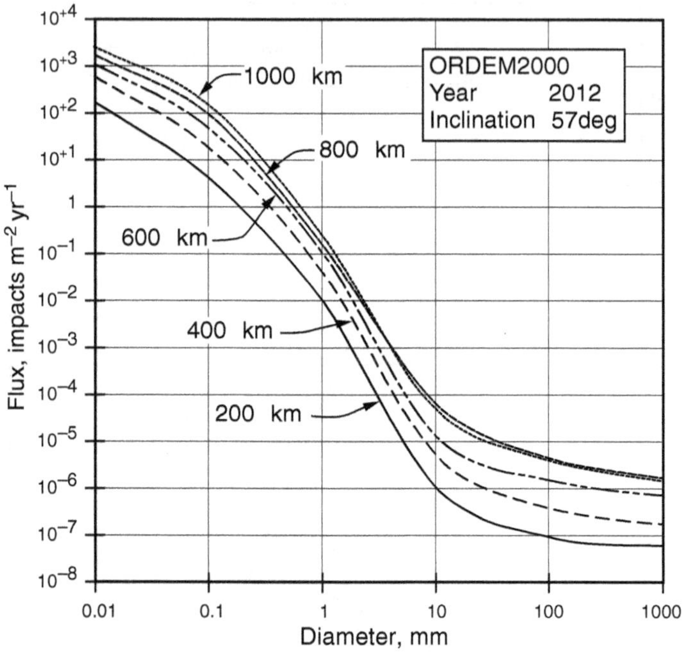

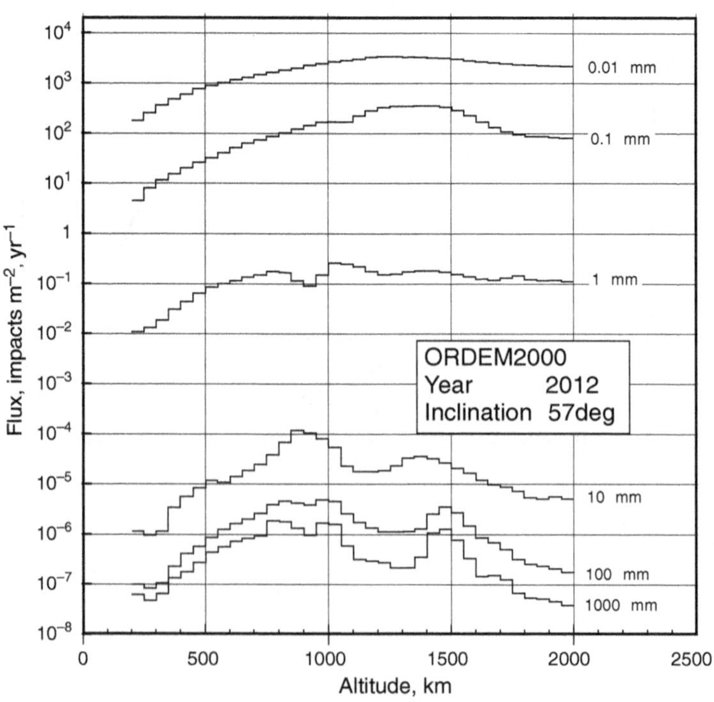

Space Environment, continued

Shielding in Planet Orbit

Body shielding factor ζ

$$\zeta = \frac{1+\cos\theta}{2}$$

$$\text{where } \sin\theta = \frac{R}{R+H}$$

ζ = ratio, shielded to unshielded concentration
R = radius of shielding body
H = altitude above surface

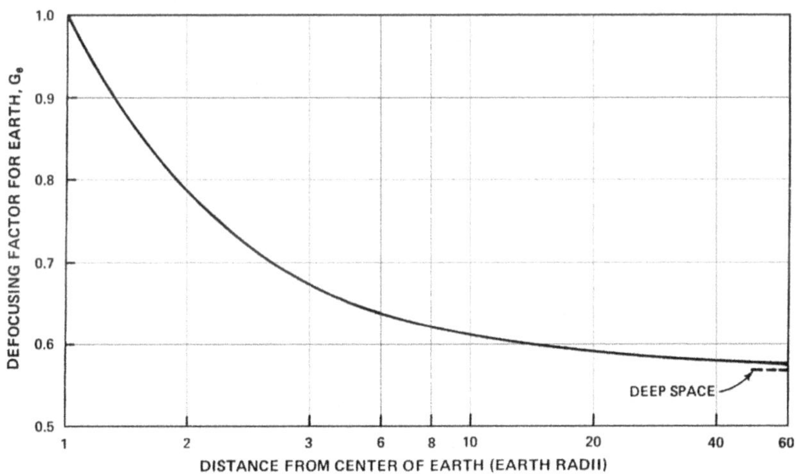

DEFOCUSING FACTOR DUE TO EARTH'S GRAVITY FOR AVERAGE METEOROID VELOCITY OF 20 km/s

SPACE ENVIRONMENT, CONTINUED

ATMOSPHERIC CHARACTERISTICS—THE SOLAR SYSTEM

Object	$\bar{\rho}$[a]	g[a]	P_s[a]	T_s[a]	Major Gases[a]	Minor Gases[a]	Aerosols[a]
Mercury	5.43	3.95×10^2	$\sim 2 \times 10^{-15}$	440	He (~0.98), H (~0.02)[b]		
Venus	5.25	8.88×10^2	90	730 (~230)	CO_2 (0.96), N_2 (~0.035)	H_2O (20–5000) SO_2 (~150), Ar (20–200), Ne (4–20), CO (50)[c], HCl (0.4)[c], HF (0.01)[c]	Sulphuric acid (~35)
Earth	5.52	9.78×10^2	1	288 (~255)	N_2 (0.77), O_2 (0.21), H_2O (~0.01), Ar (0.0093)	CO_2 (315), Ne (18), He (5.2), Kr (1.1), Xe (0.087), CH_4 (1.5), H_2 (0.5), N_2O (0.3), CO (0.12), NH_3 (0.01), NO_2 (0.001), SO_2 (0.0002), H_2S (0.0002), O_3 (~0.4)	Water (~5) Sulphuric acid (~0.01–0.1)[d] Sulphate, sea salt Dust, organic (~0.1)[d]
Mars	3.96	3.73×10^2	0.007	218 (~212)	CO_2 (0.95), N_2 (0.027) Ar (0.016)	O_2 (1300), CO (700), H_2O (~300), Ne (2.5), Kr (0.3), Xe (0.08), O_3 (~0.1)	Water ice (~1)[e] Dust (~0.1–10)[e] CO_2 ice(?)[e]
Moon	3.34	1.62×10^2	$\sim 2 \times 10^{-14}$	274	Ne (~0.4), Ar (~0.4), He (~0.2)	—	—

(continued)

[a]Reading from left to right, the variables are the object's mean density (g cm^{-3}); acceleration of gravity (cm s^{-2}); surface pressure (bar); surface temperature (K), the numbers in parentheses are values of effective temperature; major gas species, the numbers in parentheses are volume mixing ratios; minor gas species, the numbers in parentheses are fractional abundance by number in units of ppm; aerosol species, the numbers in parentheses are typical values of the aerosols' optical depth in the visible. [b]These mixing ratios refer to typical values at the surface. [c]These mixing ratios pertain to the region above the cloud tops. [d]The sulphuric acid aerosol resides in the lower stratosphere, while the sulphate, etc., aerosols are found in the troposphere, especially in the bottom boundary layer. [e]The ice clouds are found preferentially above the winter polar regions. Dust particles are present over the entire globe. [f]These mixing ratios pertain to the stratosphere.

Source: *The Origin and Evolution of Planetary Atmospheres* by A. Henderson-Sellers.

SPACE ENVIRONMENT, CONTINUED

ATMOSPHERIC CHARACTERISTICS—THE SOLAR SYSTEM, CONTINUED

Object	$\bar{\rho}$[a]	g[a]	P_s[a]	T_s[a]	Major Gases[a]	Minor Gases[a]	Aerosols[a]
Jupiter	1.34	2.32×10^3	$\gg 100$[f]	(129)	H_2 (~0.89), He (~0.11)	HD (20), CH_4 (~2000), NH_3 (~200), H_2O (1?), C_2H_6 (~5)[f], CO (0.002), GeH_4 (0.0007), HCN (0.1), C_2H_2 (~0.02)[f], PH_3 (0.4)	Stratospheric "smog" (~0.1) Ammonia ice (~1) Ammonium hydrosulphide (~1) Water (~10)
Saturn	0.68	8.77×10^2	$\gg 100$[f]	(97)	H_2 (~0.89), He (~0.11)	CH_4 (3000), NH_3 (~200) C_2H_6 (~2)[f]	Same aerosol layers as for Jupiter
Uranus	1.55	9.46×10^2	$\gg 100$[f]	(58)	H_2 (~0.89)	CH_4 He (~0.11)	Same aerosol layers as for Jupiter, but thinner smog layer, plus possibly methane ice
Neptune	2.23	1.37×10^3	$\gg 100$[f]	(56)	H_2 (~0.89)	CH_4 He (~0.11)	Same aerosol layers as for Juipiter plus possibly methane ice
Titan	~1.4	$\sim 1.25 \times 10^2$	$2 \times 10^{-12} \rightarrow \sim 1$	~85	CH_4 (0.1–1)	C_2H_6 (~2)	Stratospheric "smog" (~10)
Io	3.52	1.79×10^2	$\sim 1 \times 10^{-10}$	~110	SO_2 (~1)	——	——

[a]Reading from left to right, the variables are the object's mean density (g cm^{-3}); acceleration of gravity (cm s^{-2}); surface pressure (bar); surface temperature (K), the numbers in parentheses are values of effective temperature; major gas species, the numbers in parentheses are volume mixing ratios; minor gas species, the numbers in parentheses are fractional abundance by number in units of ppm; aerosol species, the numbers in parentheses are typical values of the aerosols' optical depth in the visible. [b]These mixing ratios refer to typical values at the surface. [c]These mixing ratios pertain to the region above the cloud tops. [d]The sulphuric acid aerosol resides in the lower stratosphere, while the sulphate, etc., aerosols are found in the troposphere, especially in the bottom boundary layer. [e]The ice clouds are found preferentially above the winter polar regions. Dust particles are present over the entire globe. [f]These mixing ratios pertain to the stratosphere.

Source: *The Origin and Evolution of Planetary Atmospheres* by A. Henderson-Sellers. Copyright © 1983, IOP Publishing Limited, Bristol, England. Reprinted by permission of IOP Publishing Limited.

SPACE ENVIRONMENT, CONTINUED

ATMOSPHERIC CHARACTERISTICS—THE SOLAR SYSTEM, CONTINUED

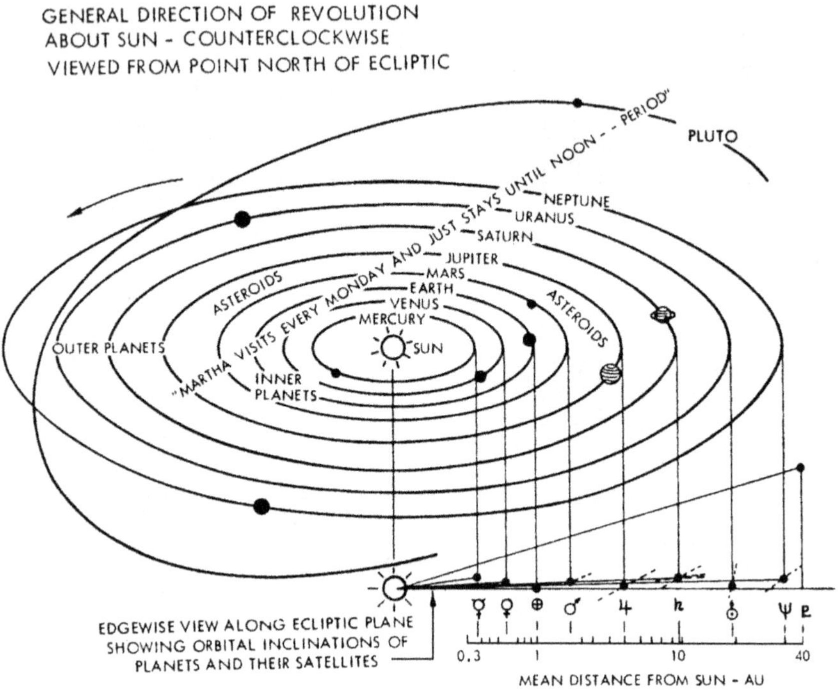

SPACE ENVIRONMENT, CONTINUED

GRAVITATION

Newton's law of universal gravitation:

$$F = G\,Mm/r^2$$

where

F = universal gravitational force of attraction between two bodies
M, m = masses of the two bodies
G = universal gravitational constant = 6.664×10^{-8} dyne-cm^2/g^2
= 6.904×10^{-19} kg m s^2
r = distance between the two bodies

EARTH DATA

Mean equatorial radius, r_e	= 5378.14 km (3963.1 statute mile)
Polar radius	= 6378.1 km (3450.0 statute mile)
Mass, M_e	= 5.98×10^{24} kg (1.32×10^{25} lbm)
Average density	= 5522 kg/m^3 (344.727 lbm/ft^3)
Mean orbital velocity	= 29.79 km/s (18.51 st mile/s)
Mean surface velocity at equator	= 0.465 km/s (0.289 st mile/s)
1′ of arc on surface at equator	= 1 n mile
	= 1.852 km

Defining a gravitational parameter μ as

$$\mu = M_e G = 398{,}600.4 \text{ km}^3/\text{s}^2$$

allows the following expression for spacecraft velocity V in a circular Earth orbit:

$$V = \sqrt{\mu/(r_e + \text{altitude})}$$

ORBITAL MECHANICS

The following figure depicts the nomenclature associated with elliptical orbits:

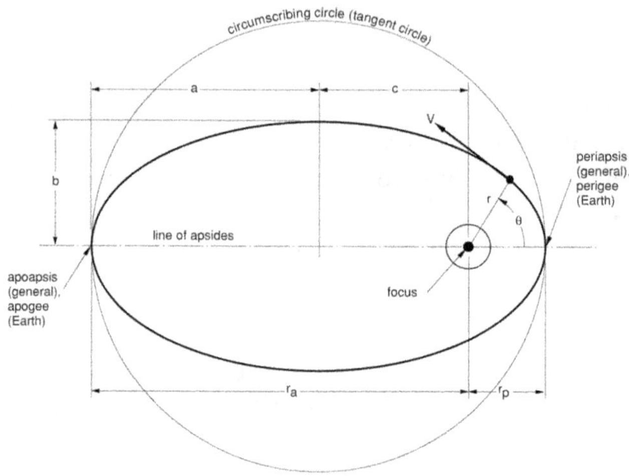

e = eccentricity = $c/a = (r_a - r_p)/(r_a + r_p)$

$V = \sqrt{(2\mu/r) - (\mu/a)}$

n = mean motion = $\sqrt{\mu/a^3}$ (angular velocity the spacecraft would have if traveling on a tangent circle instead of the ellipse)

HOHMANN ORBIT TRANSFER

Two-impulse, minimum energy transfer between coplanar circular orbits as shown in the following figure:

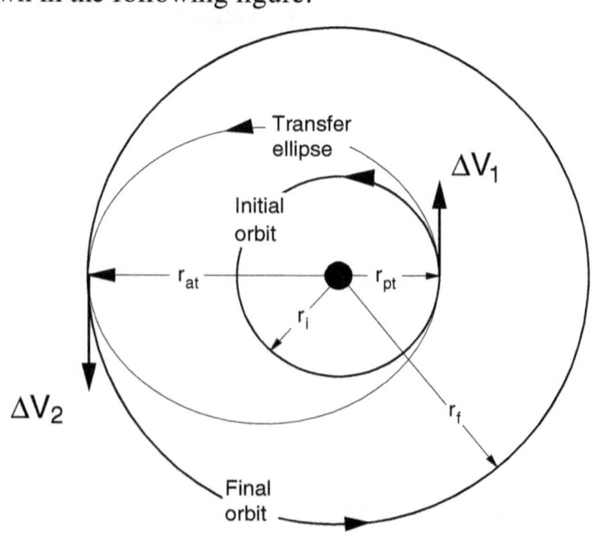

ORBITAL MECHANICS, CONTINUED

VELOCITY CHANGE ΔV REQUIRED FOR ORBIT ALTITUDE AND INCLINATION CHANGES

The chart below may be used to determine the change in velocity ΔV required to change the altitude or inclination of a low-Earth circular orbit. ΔV then can be used to calculate the mass of propellant required to execute the change. Note that the relationships are approximately linear for altitudes between 200 and 1000 km:

$$|\Delta V_a| = \sqrt{\mu/r_f} - \sqrt{\mu/r_i}\text{, km/s}$$

$$|\Delta V_i| = 2V_i \sin(\Delta i/2)\text{ km/s}$$

where

$|\Delta V_a|$ = ΔV required for altitude change, m/s
$|\Delta V_i|$ = ΔV required for inclination change, m/s
a = altitude, km
R_e = Earth mean equatorial radius, km
R_i = initial radius, km $= R_e + a_{\text{initial}}$
R_f = final radius, km $= R_e + a_{\text{final}}$
V_i = orbital velocity at beginning of maneuver, km/s
Δi = desired change in inclination, deg

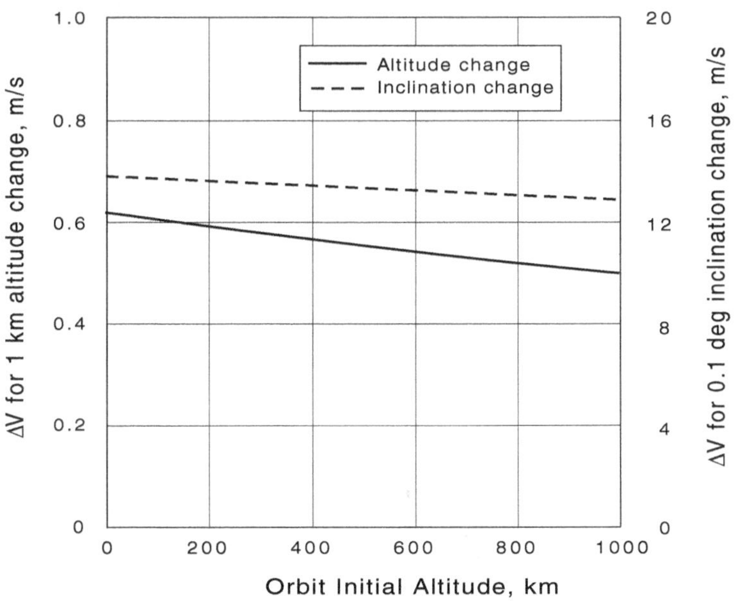

ORBITAL MECHANICS, CONTINUED

ELLIPTICAL ORBIT PARAMETERS

$$\tan\frac{\theta}{2} = \sqrt{\frac{1+e}{1-e}}\tan\frac{E}{2}$$

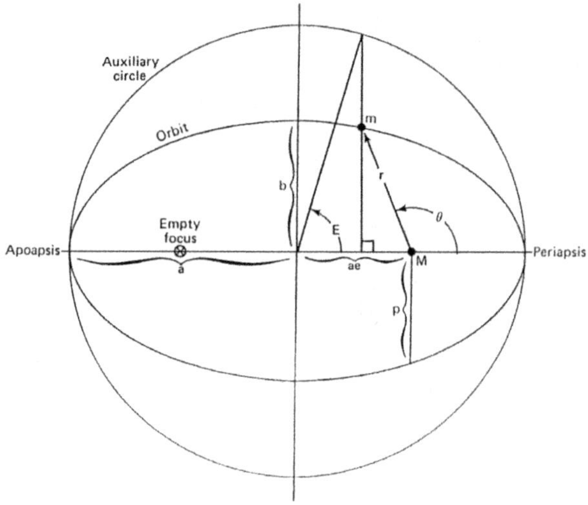

E = eccentric anomaly

Circle	Ellipse	Parabola	Hyperbola
$a = b > 0$	$a > b > 0$	$a = \infty$	$a < 0$
$e = 0$	$e^2 = 1 - \frac{b^2}{a^2}$	$e = 1$	$e^2 = 1 + \frac{b^2}{a^2}$

$$r = \frac{p}{1+e\cos\theta} = \frac{r_p(1+e)}{1+e\cos\theta} = \frac{a(1-e^2)}{1+e\cos\theta}$$

$$p = a(1 - e^2) = h^2/\mu$$

$$r_p = a(1 - e) = \text{periapsis radius}$$

$$r_a = a(1 + e) = \text{apoapsis radius}$$

$$a = \frac{r_a + r_p}{2} \qquad V^2_{\text{CIRC}} = \frac{\mu}{r}$$

$$e = \frac{r_a - r_p}{r_a + r_p} \qquad V^2_{\text{ESC}} = \frac{2\mu}{r}$$

Source: The orbital mechanics material appearing on pages 11-24–11-26 is from *Space Vehicle Design* by M. D. Griffin and J. R. French.

ORBITAL MECHANICS, CONTINUED

ELLIPTICAL ORBIT PARAMETERS, CONTINUED

$$V^2 = \mu\left(\frac{2}{r} - \frac{1}{a}\right)$$

$$E_t = \frac{V^2}{2} - \frac{\mu}{r} = -\frac{\mu}{2a}$$

$$e^2 = 1 + 2E_t \frac{h^2}{\mu^2}$$

$$E = -e \sin E - n(t - t_p) = 0 \quad \text{(Kepler's equation)}$$

where

t_p = time of periapsis
n = mean motion = $\sqrt{\mu/a^3}$
$M \triangleq n(t - t_p)$ = mean anomaly
E = eccentric motion

for $E = 2\pi, \gamma = t - t_p = 2\pi\sqrt{a^3/\mu}$

for small e, $\theta \cong M + 2e \sin M + (5e^2/4) \sin 2M$

ORBITAL ELEMENTS

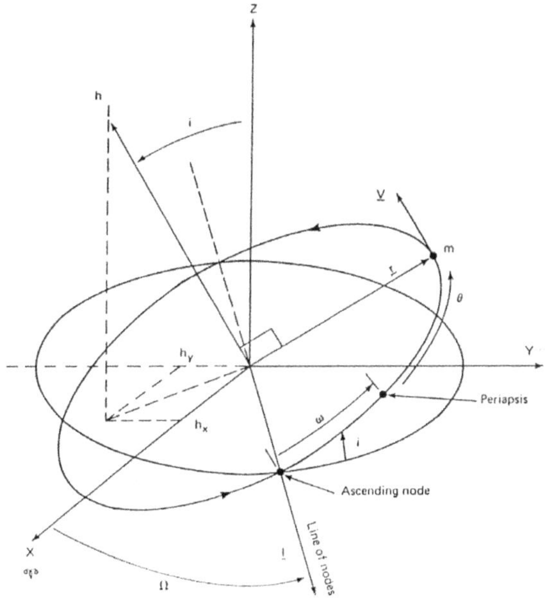

ORBITAL MECHANICS, CONTINUED

HYPERBOLIC ORBIT GEOMETRY

$$\tan\frac{\theta}{2} = \sqrt{\frac{e+1}{e-1}}\tanh\frac{F}{2}$$

V_∞ = hyperbolic approach and departure velocity relative to target body M

θ_a = true anomaly of asymptotes

β = B-plane miss distance; offset in plane ⊥ to arrival asymptote

ψ = turning angle of passage

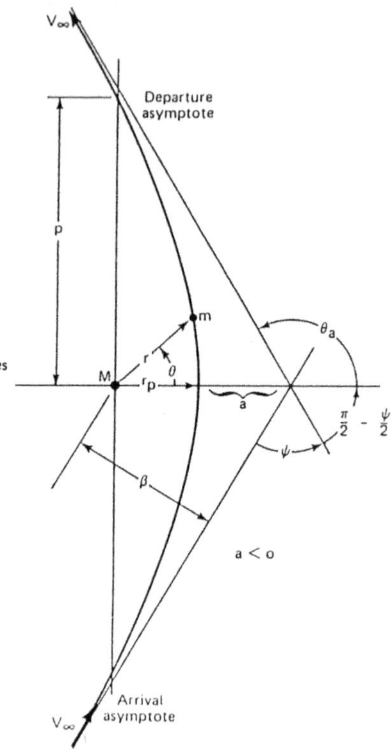

$$(e \sinh F) - F - n(t - t_p) = 0$$

$$n = \sqrt{\mu/(-a)^3}$$

Key point: Hyperbolic orbit must be oriented in external frame to account for arrival/departure conditions.

At "infinity," the asymptote conditions are as follows:

$$\theta_a = \cos^{-1}(-1/e)$$

$$V_\infty^2 = 2E_t = (-\mu/a)$$

$$h = \beta V_\infty$$

Key point: Hyperbolic passage alters $V_{-\infty}$, but not V_∞.

$$\psi = 2\sin^{-1}(1/e) = 2\theta_a - \pi = \text{turning angle}$$

$$\Delta V = 2V_\infty \sin\psi/2 = 2V_\infty/e$$

ORBITAL MECHANICS, CONTINUED

HYPERBOLIC ORBIT GEOMETRY, CONTINUED

This is the basis for "gravity assist" maneuvers.

$$e = 1 + \left(r_p V_\infty^2/\mu\right) = \sqrt{\left[1 + \left(\beta V_\infty^2/\mu\right)^2\right]}$$

$$\beta = r_p \sqrt{\left[1 + \left(2\mu/r_p V_\infty^2\right)\right]}$$

$$\left(r_p/\beta\right) = \left(-\mu/BV_\infty^2\right) = \sqrt{\left\{1 + \left(\mu/BV_\infty^2\right)^2\right\}}$$

SATELLITE LIFETIME VS ALTITUDE

$$F = ma$$

$$F_d = M_{SC}\frac{\Delta V}{\Delta t} = \frac{\sigma C_d V^2 A_f}{2}$$

$$\text{Satellite life} = \Delta t = \frac{M_{SC}\,\Delta V}{C_d A_f \rho V^2}\ \text{seconds}$$

where

F_d = atmospheric drag force, N
M_{SC} = spacecraft mass, kg
ΔV = change in velocity, km/s
Δt = time increment (satellite lifetime), s
ρ = atmospheric density, kg/m^3
C_d = drag coefficient, assumed to be 2.0
V = orbital velocity, km/s
A_f = projected area of spacecraft normal to velocity vector, m^2
C_b = ballistic coefficient = $BC = M_{SC}/C_d A_f$, kg/m^2

The following chart shows the approximate lifetime of a satellite, as determined by the atmospheric drag. End of life is defined as loss of 50 km of altitude; no additional propulsion is assumed.

ORBITAL MECHANICS, CONTINUED

SATELLITE LIFETIME VS ALTITUDE, CONTINUED

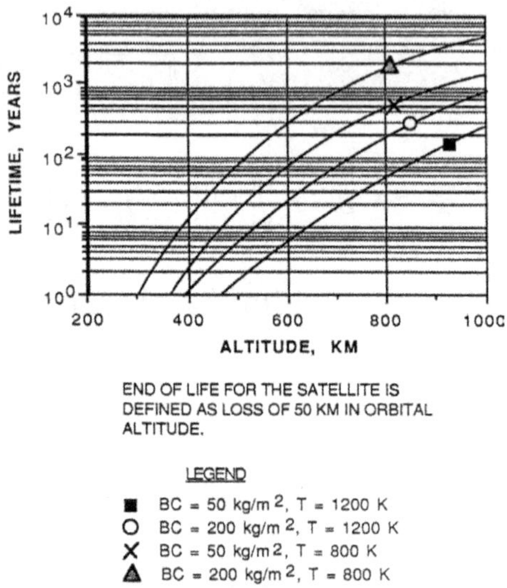

NODAL REGRESSION AND PERIGEE ROTATION

The equatorial bulge of the Earth causes a component of the gravitational force to be out of the orbit plane, causing the orbit plane to precess gyroscopically; the resulting orbit rotation is called regression of nodes. The bulge also causes rotation of apsides, or perigee rotation, for elliptical orbits. Regression of nodes, $\mathrm{d}\Omega/\mathrm{d}t$, and perigee rotation, $\mathrm{d}\omega/\mathrm{d}t$, are approximated by

$$\frac{\mathrm{d}\Omega}{\mathrm{d}t} \cong \frac{-3nJ_2R_0^2\cos(i)}{2a^2(1-e^2)^2}$$

$$\frac{\mathrm{d}\omega}{\mathrm{d}t} \cong \frac{3nJ_2R_0^2\left(4-5\sin^2(i)\right)}{4a^2(1-e^2)^2}$$

where

$\mathrm{d}\Omega/\mathrm{d}t$ = rate of change of the longitude of the ascending node, rad/s
$\mathrm{d}\omega/\mathrm{d}t$ = rate of rotation of the line of apsides (or perigee)
R_0 = mean equatorial radius of the central body
a = semimajor axis of the orbit
i = inclination of the orbit
e = eccentricity of the orbit
J_2 = zonal coefficient = 0.00108263 for the Earth
n = mean motion = $\sqrt{\mu/a^3}$

The following two charts show the approximate nodal regression for circular Earth orbits and the approximate perigee rotation for elliptical Earth orbits, respectively.

ORBITAL MECHANICS, CONTINUED

NODAL REGRESSION AND PERIGEE ROTATION, CONTINUED

NODAL REGRESSION

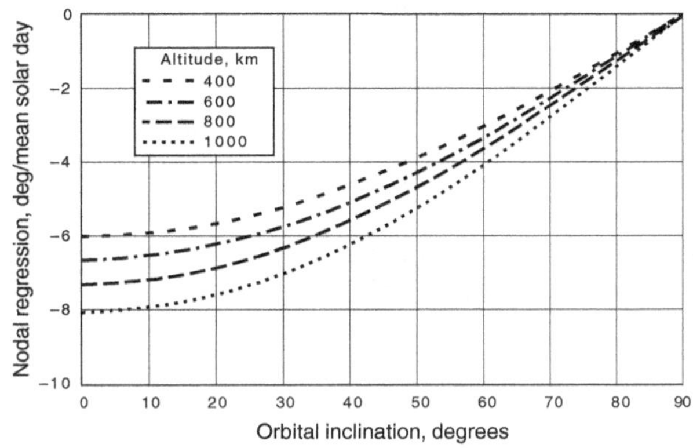

Nodes move westward for direct orbits (0 deg $\leq i \leq$ 90 deg) and eastward for retrograde orbits (90 deg $\leq i \leq$ 180 deg).

ROTATION OF PERIGEE

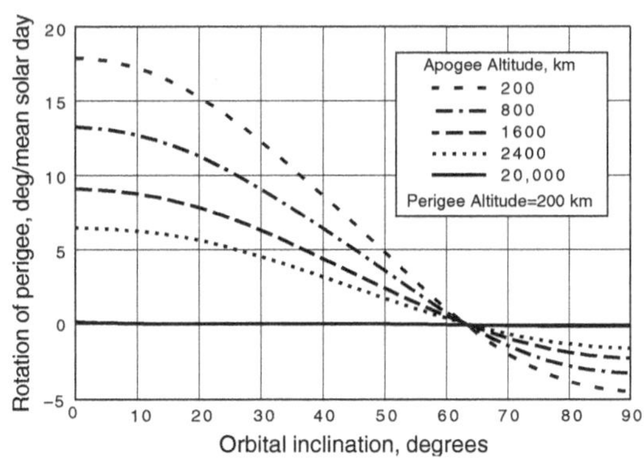

ORBITAL MECHANICS, CONTINUED

INCLINATION FOR SUN-SYNCHRONOUS ORBITS

The oblateness of Earth can cause fairly large secular changes in the location of the longitude of the ascending node. Solving for the inclination i in the previous equation for dΩ/dt results in

$$i \cong \pi - a\cos\left[\frac{2a^2 \mathrm{d}\Omega/\mathrm{d}t}{3nJ_2R_0^2} e^2(1+e^2)\right]$$

With dΩ/dt set to the rotational rate of the Earth about the sun, about 1 deg/day, the following chart results, which shows the approximate inclination of sun-synchronous, circular, or elliptical orbits:

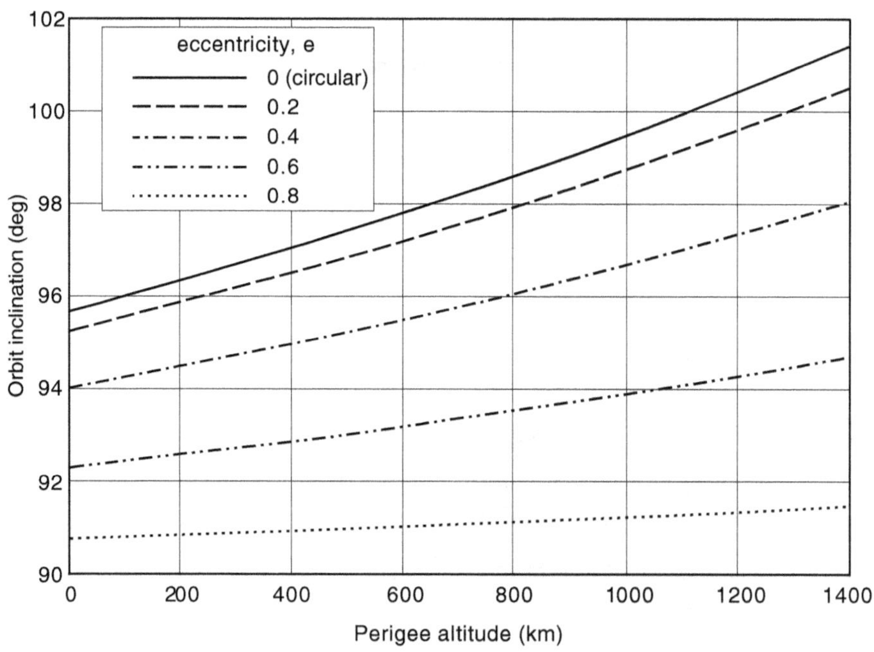

ORBITAL MECHANICS, CONTINUED

ECLIPSE TIME VS SUN ANGLE

The following chart shows the approximate duration of satellite eclipse, if any, for various circular, low-Earth orbit altitudes, as a function of the sun/orbit angle β:

$$P = 2\pi\sqrt{R^3/\mu}$$

$$T_e = \frac{P}{\pi}\cos^{-1}\left[\sqrt{1 - R_e^2/R^2}\,/\cos(\beta)\right]$$

where

T_e = eclipse duration, s
P = orbital period, s
R_e = radius of the Earth, km
R = distance from the center of the Earth to the spacecraft, km
β = sun/orbit angle, deg

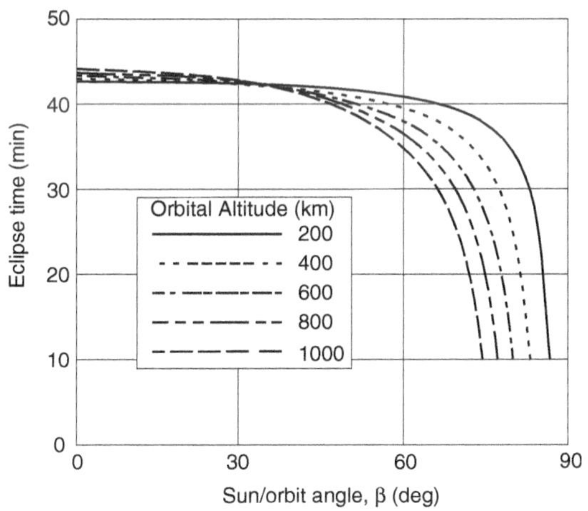

PROPULSION SYSTEMS

TYPES OF PROPULSION SYSTEMS

COLD GAS

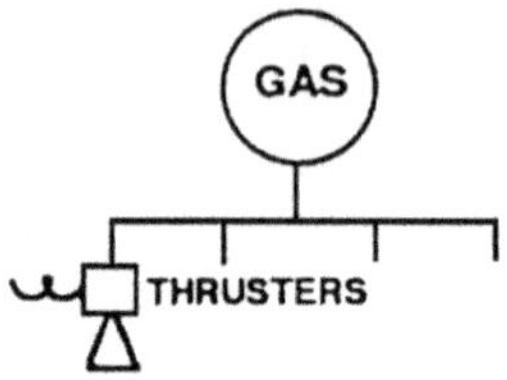

Characteristics:

$I_{sp} \cong 50$ s

$$I_{sp} \cong K\sqrt{\frac{T}{M}}$$

$F < 4.45$ N (1 lb)

MONOPROPELLANTS

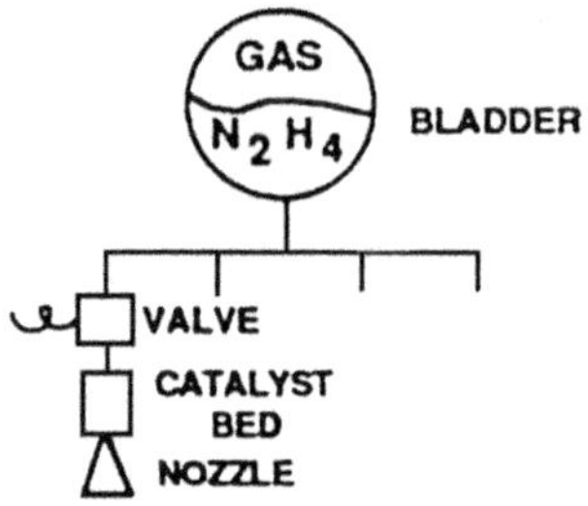

Characteristics:

$I_{sp} \cong 225$ s

$1 < F < 2669$ N (600 lb)

BIPROPELLANTS

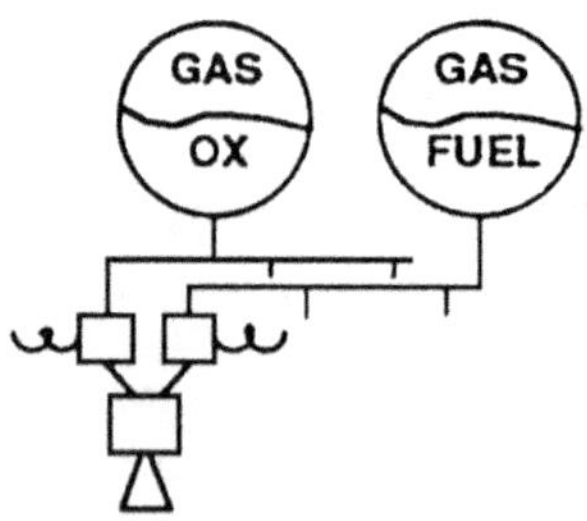

Characteristics:

$I_{sp} \cong 310–460$ s

Source: The propulsion systems material appearing on pages 11-32–11-48 is from *Rocket Propulsion Elements, An Introduction to the Engineering of Rockets*, 6th Ed., by G. P. Sutton.

PROPULSION SYSTEMS, CONTINUED

TYPES OF PROPULSION SYSTEMS, CONTINUED

SOLIDS

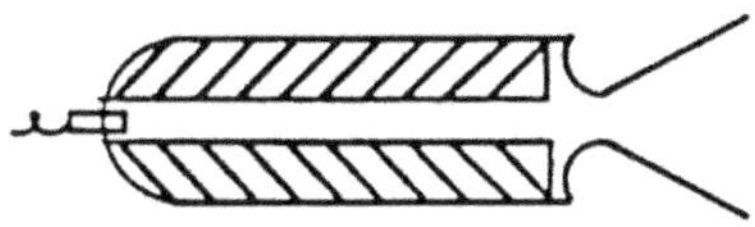

Characteristics:
$I_{sp} \cong 250\text{–}290$ s

DUAL MODE PROPULSION SYSTEM

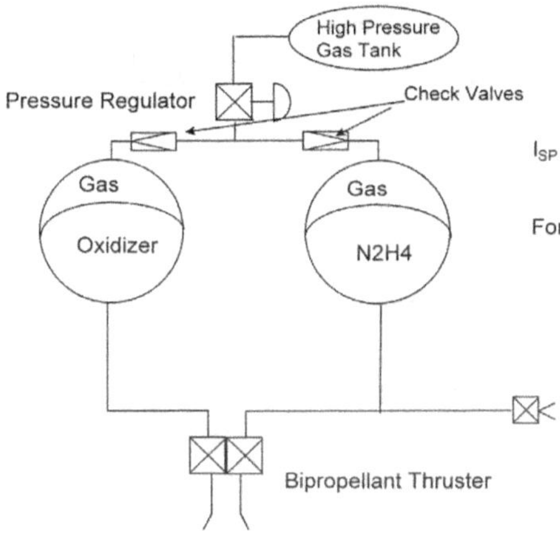

I_{SP} - Monopropellant: 180–500 s
Bipropellant: 300–330 s

Force - Monopropellant:.22–222 N (.05–50 lb)
Bipropellant: 22.2–489 N (5–110 lb)

RESISTOJET (ELECTROTHERMAL HYDRAZINE THRUSTER) SYSTEM

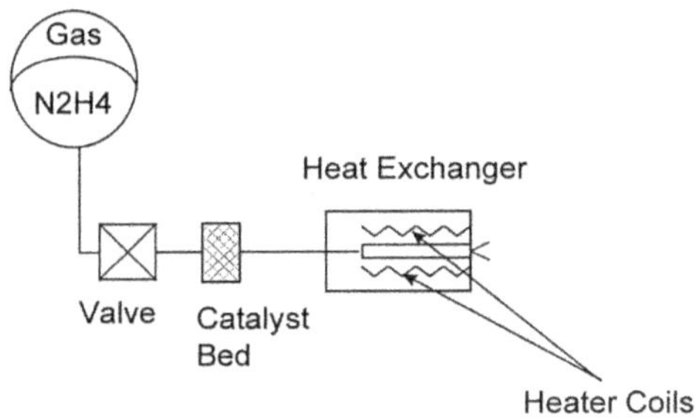

Characteristics:
$I_{sp} = 280\text{–}300$ s
$F < .44$ N (.1 lb)

PROPULSION SYSTEMS, CONTINUED

SUMMARY OF KEY EQUATIONS FOR IDEAL ROCKETS

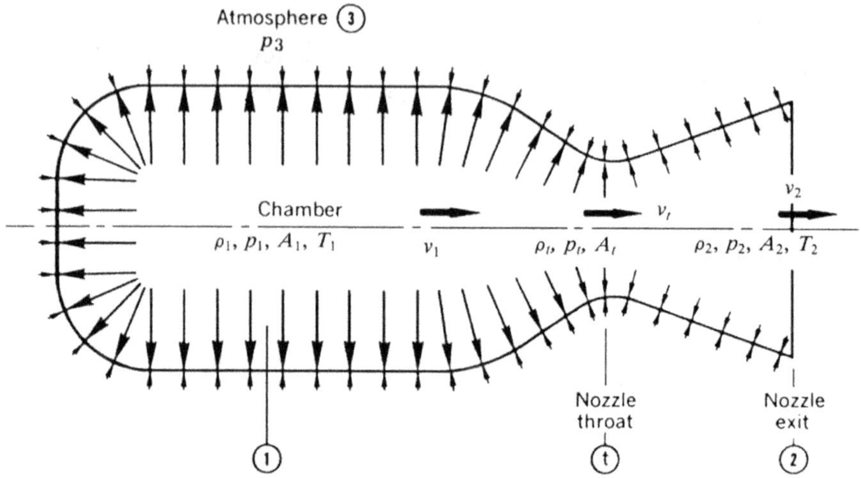

Note: Four subscripts are used above (1, 2, 3, *t*). They are shown in circles and refer to the specific locations. Thus, p_1 is the chamber pressure, p_2 is the nozzle exit pressure, p_3 is the external fluid of atmospheric pressure, and p_t is the nozzle throat pressure. T_1 is the combustion chamber absolute temperature.

Parameter	Equations
Average exhaust velocity, v_2 (m/s or ft/s) (assume that $v_1 = 0$)	$v_2 = c - (p_2 - p_3)A_2/\dot{m}$ When $p_2 = p_3$, $v_2 = c$ $v_2 = \sqrt{[2k/(k-1)]RT_1\left[1-(p_2/p_1)^{(k-1)/k}\right]}$ $= \sqrt{2(h_1 - h_2)}$
Effective exhaust velocity, c (m/s or ft/s)	$c = c^* C_F = F/\dot{m} = I_{sp} g_0$ $c = v_2 + (p_2 - p_3)A_2/\dot{m}$
Thrust, F (N or lbf)	$F = c\dot{m} = cm_p/t_b$ $F = C_F p_1 A_t$ $F = \dot{m}v_2 + (p_2 - p_3)A_2$ $F = \dot{m} I_{sp} g_0$
Characteristic exhaust velocity, c^* (m/s or ft/s)	$c^* = c/C_F = p_1 A_t/\dot{m}$ $c^* = \dfrac{\sqrt{kRT_1}}{k\sqrt{[2/(k+1)]^{(k+1)(k-1)}}}$ $c^* = I_{sp} g_0/C_F$

(continued)

SUMMARY OF KEY EQUATIONS FOR IDEAL ROCKETS, CONTINUED

Parameter	Equations
Trust coefficient, C_F (dimensionless)	$C_F = c/c^* = F/(p_1 A_t)$
	$C_F = \sqrt{\frac{2k^2}{k-1}\left(\frac{2}{k+1}\right)^{(k+1)/k-1)}\left[1-\left(\frac{p_2}{p_1}\right)^{(k-1)/k}\right]} + \frac{p_2 - p_3}{p_1}\frac{A_2}{A_t}$
Mass flow rate, $\dot{m}$ (kg/s or lb/s)	$\dot{m} = \frac{A_t v_t}{V_t} = p_1 A_1/c^*$
	$= A_t p_1 \frac{k\sqrt{[2/(k+1)]^{(k+1)/(k-1)}}}{\sqrt{kRT_1}}$
Total impulse, I_t [N-s or lbf-s]	$I_t = \int_0^{t_b} F dt$
	$= Ft_b\,(F \text{ constant over } t_b)$
Specific impulse, I_{sp} [N-s/(kg × 9.8066 m/s^2) or lbf-s/lbm or s]	$I_{sp} = c/g_0 = c^* C_F/g_0$
	$I_{sp} = F/\dot{m}g_0 = F/\dot{w}$
	$I_{sp} = v_2/g_0 + (p_2 - p_3)A_2/\dot{m}g_0)$
	$I_{sp} = I_t(m_p g_0) = I_t/w_p$
Propellant mass fraction (dimensionless)	$\zeta = m_p/m_0 = \frac{m_0 - m_f}{m_0}$
	$\zeta = 1 - \text{MR}$
Mass ratio of vehicle or stage, MR (dimensionless)	$\text{MR} = \frac{m_f}{m_0} = \frac{m_0 - m_p}{m_0}$
	$= m_f/(m_f + m_p)$
	$m_0 = m_f + m_p$
Vehicle velocity increase in gravity free vacuum, Δv (m/s or ft/s) (assume that $v_0 = 0$)	$\Delta v = -c\ \ln \text{MR} = +c\ \ln \frac{m_0}{m_f}$
	$= c\ \ln m_0/(m_0 - m_p)$
	$= c\ \ln(m_p + m_f)/m_f$
	$= I_{sp}\, g_0 \ln \text{MR}$
Propellant mass flow rate, $\dot{m}$ (kg/s or lb/s)	$\dot{m} = Av/V = A_1 v_1/V_1$
	$= A_t v_t/V_t = A_2 V_2/V_2$
	$\dot{m} = F/c = p_1 A_t/c^*$
	$\dot{m} = p_1 A_t k\sqrt{\frac{[2/k+1)]^{(k+1)/(k-1)}}{\sqrt{kRT_1}}}$
	$\dot{m} = m_p/t_b$

(continued)

PROPULSION SYSTEMS, CONTINUED

SUMMARY OF KEY EQUATIONS FOR IDEAL ROCKETS, CONTINUED

Parameter	Equations
Mach number, M (dimensionless)	$M = v/a$ $= v/\sqrt{kRT}$ At throat $v = a$ and $M = 1.0$
Nozzle area ratio, ε	$\varepsilon = A_2/A_t$ $\varepsilon = \frac{1}{M_2}\left[\frac{1+\frac{k-1}{2}M_2^2}{1+\frac{k-1}{2}}\right]^{(k+1)/(k-1)}$
Isentropic flow relationships for stagnation and free-stream conditions	$T_0/T = (p_0/p)^{(k-1)/k} = (V/V_0)^{k-1}$ $T_x/T_y = (p_x/p_y)^{(k-1)/k} = (V_x/V_y)^{k-1}$

where

R = $R'/\mathfrak{M}$
R' = universal gas constant
= 8314 J/kg mol-K (1544 ft-lb/mol-°R)
$\mathfrak{M}$ = molecular weight of reaction gases, kg (lbm)
k = ratio of specific heats = c_p/c_v
c_p = specific heat at constant pressure
c_v = specific heat at constant volume
T = temperature, K (°R)
ρ = density, kg/m^3 (lbm/ft^3)
V = specific volume, m^3/kg (ft^3/lbm)
p = pressure, N/m^2 (lb/ft^2)
h = enthalpy = $c_p\,T$
F = thrust, N (lbf)
g_0 = acceleration due to gravity on Earth = 9.81 m/s (32.2 ft/s^2)
t_b = total burn time
m_p = total propellant mass
w_p = total propellant weight
A = cross-sectional area
m_0 = mass at start of burn
m_f = mass at end of burn
v = flow velocity
a = speed of sound
T_0, k_0, V_0 = stagnation conditions

PROPULSION SYSTEMS, CONTINUED

SUMMARY OF KEY EQUATIONS FOR SOLID ROCKET MOTORS

Mass flow rate, kg/s (lbm/s):

$$\dot{m} = A_b r \rho_b$$

$$= \frac{\mathrm{d}}{\mathrm{d}t}(\rho_1 V_1) + A_t p_1 \sqrt{\frac{k}{RT_1}\left(\frac{2}{k+1}\right)^{(k+1)/(k-1)}}$$

Burn rate r, in./s:

$$r = a p_1^n$$

Propellant mass m_p, kg (lbm):

$$m_p = \int \dot{m}\ \mathrm{d}t = \rho_b \int A_b r\ \mathrm{d}t$$

Assuming:

$$\frac{\mathrm{d}}{\mathrm{d}t}(\rho_t V_c) \ll A_t p_1 \sqrt{\frac{k}{RT_1}\left(\frac{2}{k+1}\right)^{(k+1)/(k-1)}}$$

(Rate of change of gas mass in motor cavity is small relative to mass flow through the nozzle.)

Then,

$$\frac{A_b}{A_t} = K \begin{bmatrix} \text{Ratio of burning area} \\ \text{to nozzle throat area} \end{bmatrix}$$

$$= \frac{p_1^{(1-n)} \sqrt{k\left[2/(k+1)\right]^{(k+1)/(k-1)}}}{\rho_b a \sqrt{RT_1}}$$

It follows:

$$p_1 = \left[K \rho_b a c^*\right]^{1/(1-n)}$$

where

A_b = burn area of propellant grain
ρ_b = solid propellant density prior to motor start
V_1 = combustion chamber gas volume
a = burn rate coefficient
n = burn rate exponent

PROPULSION SYSTEMS, CONTINUED

TYPICAL PERFORMANCE OF PROPULSION TYPES

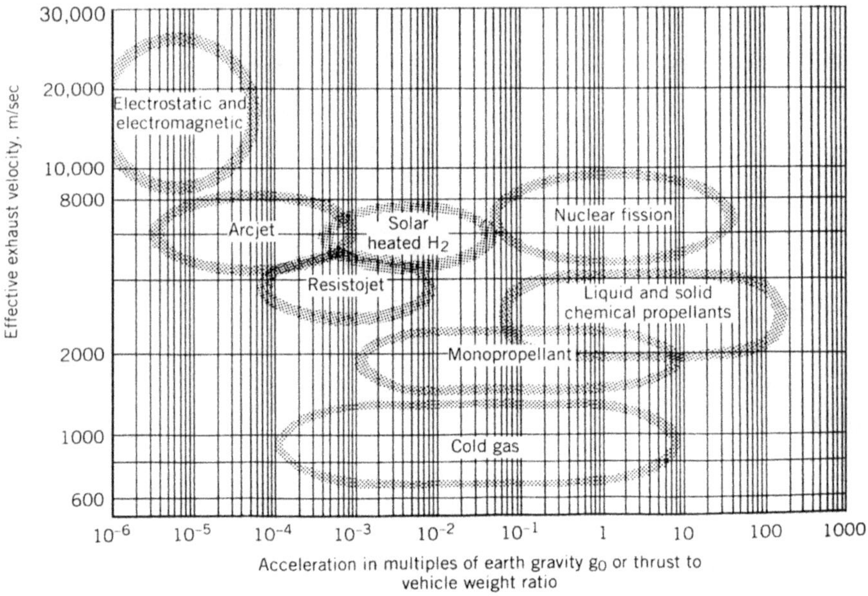

PROPULSION SYSTEMS, CONTINUED

RANGES OF TYPICAL PERFORMANCE PARAMETERS FOR VARIOUS ROCKET PROPULSION SYSTEMS

Engine Type	Specific Impulse,[a] s	Maximum Temperature, °C	Thrust-to-Weight Ratio[b]	Propulsion Duration	Specific Power,[c] kW/kg	Typical Working Fluid	Status of Technology
Chemical—solid or liquid bipropellant	200–410	2500–4100	10^{-2}–100	Seconds to a few minutes	10^{-1}–10^{3}	Liquid or solid propellants	Flight proven
Liquid monopropellant	180–223	600–800	10^{-1}–10^{-2}	Seconds to minutes	0.02–200	N_2H_4	Flight proven
Nuclear fission	500–860	2700	10^{-2}–30	Same	10^{-1}–10^{3}	H_2	Development was stopped
Resistojet	150–300	2700	10^{-2}–10^{-4}	Days	10^{-3}–10^{-1}	H_2, N_2H_4	Flight tested
Arc heating—electrothermal	280–1200	5500	10^{-4}–10^{-2}	Days	10^{-3}–1	N_2H_4, H_2, NH_3	Flight tested
Electromagnetic	1200–6000	—	10^{-6}–10^{-4}	Weeks	10^{-3}–1	H_2	Several have flown
Ion—electrostatic	1200–5000	—	10^{-6}–10^{-4}	Months	10^{-3}–1	Xe	Several have flown
Solar heating	400–700	1300	10^{-3}–10^{-2}	Days	10^{-2}–1	H_2	Not yet flight tested

[a]At $p_1 = 100$ psia and optimum gas expansion at sea level ($p_2 = p_3 = 14.7$ psia).
[b]Ratio of thrust force to full propulsion system sea level weight (with propellants, but without payload).
[c]Kinetic energy per unit exhaust mass flow.

PROPULSION SYSTEMS, CONTINUED

PROPERTIES OF GASEOUS PROPELLANTS USED FOR AUXILIARY PROPULSION

Propellant	Molecular Mass	Density,[a] kg/m^3	Theoretical Specific Impulse,[b] s
Hydrogen	2.0	26.55	284
Helium	4.0	53.1	179
Methane	16.0	211.5	114
Nitrogen	28.0	370.5	76
Air	28.9	382.5	74
Argon	39.9	529.5	57
Krypton	83.8	1111.5	50

[a]At 34.464 Mpa (5000 psia) and 20° C.
[b]In vacuum with nozzle area ratio of 50:1 and initial temperature of 20° C.

Engine Type	η_{int}	I_{sp}	υ_2, m/s	$\dot{m}$, kg/s	Power Input, kW
Chemical rocket	0.50	300	2352	0.0425	117
Nuclear fission	0.50	800	6860	0.0145	682
Arc-electrothermal	0.50	1200	10,780	0.0093	1351
Ion electrostatic	0.90	5000	49,000	0.0020	2668

$$\eta_{int} = \frac{\text{power of the jet (output)}}{\text{power input}}$$

$$= \frac{\frac{1}{2}\dot{m}C^2}{P} = \frac{FI_{sp}g_0}{2P}$$

PROPULSION SYSTEMS, CONTINUED

THEORETICAL PERFORMANCE OF LIQUID ROCKET PROPELLANT COMBINATIONS

Oxidizer	Fuel	Mixture Ratio By Mass	Mixture Ratio By Volume	Average Specific Gravity	Chamber Temp, K	c^*, m/s	$\mathfrak{M}$, kg/mol	I_{sp}, s	k
Oxygen	75% Ethyl	1.30	0.98	1.00	3177	1641	23.4	267	1.22
	alcohol	1.43	1.08	1.01	3230	1670	24.1	279	
	Hydrazine	0.74	0.66	1.06	3285	1871	18.3	301	1.25
		0.90	0.80	1.07	3404	1892	19.3	313	
	Hydrogen	3.40	0.21	0.26	2959	2428	8.9	387	1.26
		4.02	0.25	0.28	2999	2432	10.0	390	
	RP-1	2.24	1.59	1.01	3571	1774	21.9	286	1.24
		2.56	1.82	1.02	3677	1800	23.3	300	
	UDMH	1.39	0.96	0.96	3542	1835	19.8	295	1.25
		1.65	1.14	0.98	3594	1864	21.3	310	

(continued)

Note: Combustion chamber pressure—1000 psia (6895 kN/m^2). Nozzle exit pressure—14.7 psia (1 atm). Optimum nozzle expansion ratio. Adiabatic combustion and isentropic expansion of ideal gas. Compositions expressed in mass percent. The density at the boiling point was used for those oxidizers or fuels that boil below 20°C at 1 atm pressure. For every propellant combination, there are two sets of values listed; the upper line refers to frozen equilibrium, the lower line to shifting equilibrium.

PROPULSION SYSTEMS, CONTINUED

THEORETICAL PERFORMANCE OF LIQUID ROCKET PROPELLANT COMBINATIONS, CONTINUED

Oxidizer	Fuel	Mixture Ratio By Mass	Mixture Ratio By Volume	Average Specific Gravity	Chamber Temp, K	c^*, m/s	$\mathfrak{M}$, kg/mol	I_{sp}, s	k
Fluorine	Hydrazine	1.83	1.22	1.29	4553	2128	18.5	334	1.33
		2.30	1.54	1.31	4713	2208	19.4	363	
	Hydrogen	4.54	0.21	0.33	3080	2534	8.9	398	1.33
		7.60	0.35	0.45	3900	2549	11.8	410	
Nitrogen tetroxide	Hydrazine	1.08	0.75	1.20	3258	1765	19.5	283	1.26
		1.34	0.93	1.22	3152	1782	20.9	292	
	50% UDMH° 50% hydrazine	1.62	1.01	1.18	3242	1652	21.0	278	1.24
		2.00	1.24	1.21	3372	1711	22.6	288	
Red fuming nitric acid	RP-1	4.1	2.12	1.35	3175	1594	24.6	258	1.22
		4.8	2.48	1.33	3230	1609	25.8	268	
	50% UDMH- 50% hydrazine	1.73	1.00	1.23	2997	1682	20.6	272	1.22
		2.20	1.26	1.27	3172	1701	22.4	279	

Note: Combustion chamber pressure—1000 psia (6895 kN/m^2). Nozzle exit pressure—14.7 psia (1 atm). Optimum nozzle expansion ratio. Adiabatic combustion and isentropic expansion of ideal gas. Compositions expressed in mass percent. The density at the boiling point was used for those oxidizers or fuels that boil below 20°C at 1 atm pressure. For every propellant combination, there are two sets of values listed; the upper line refers to frozen equilibrium, the lower line to shifting equilibrium.

PROPULSION SYSTEMS, CONTINUED

TYPICAL TURBOPUMP FEED SYSTEM CYCLES FOR LIQUID-PROPELLANT ROCKET ENGINES

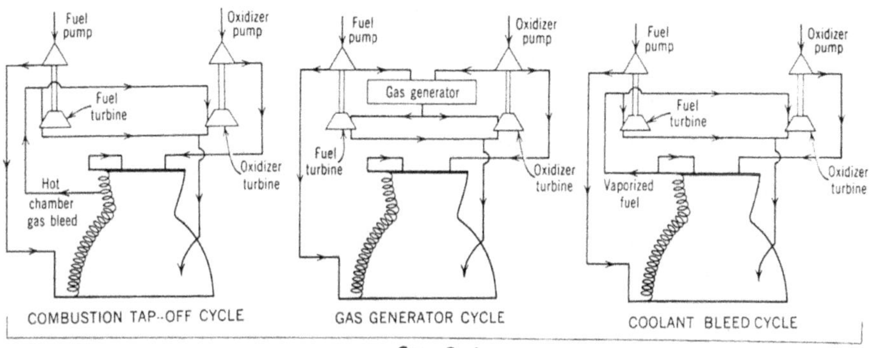

Open Cycles

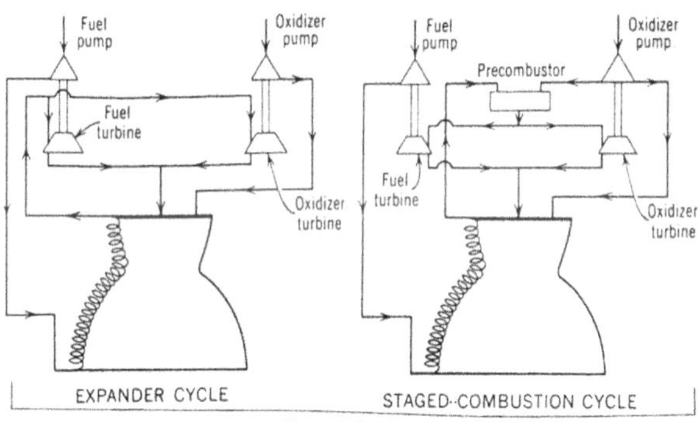

Closed Cycles

PROPULSION SYSTEMS, CONTINUED

BURNING RATE VS CHAMBER PRESSURE FOR SOME TYPICAL PROPELLANTS AT SEVERAL PROPELLANT TEMPERATURES

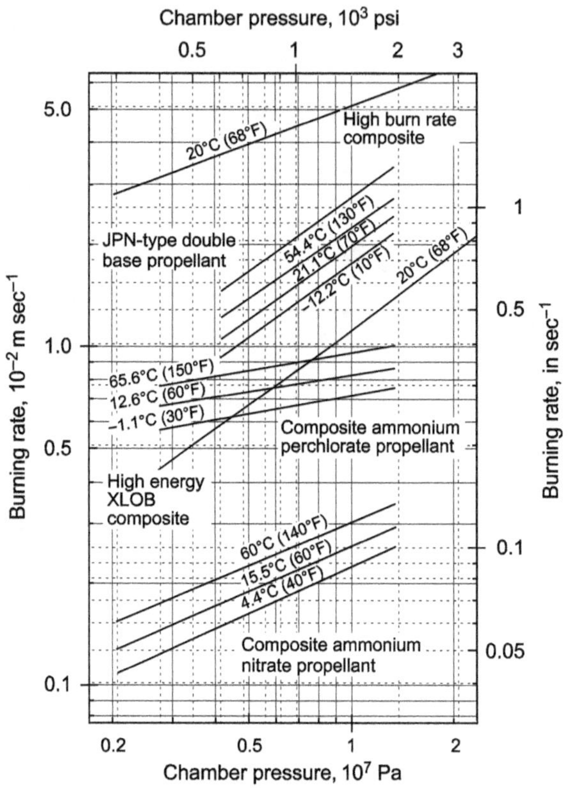

Burn rate temperature sensitivity at constant pressure:

$$\sigma_p = 1/r\left(\frac{\delta r}{\delta T}\right)_p$$

Burn rate temperature sensitivity at constant K:

$$\pi_K = 1/p_c\left(\frac{\delta r}{\delta T}\right)_K$$

Generally,

$$0.08\% < \sigma_p < 0.80\%$$

$$0.12\% < \pi_K < 0.50\%$$

PROPULSION SYSTEMS, CONTINUED

TYPICAL GRAIN DESIGN CONFIGURATIONS

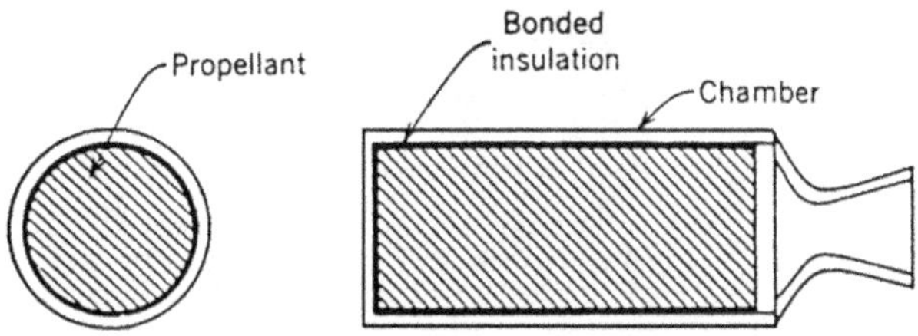

End-burner (case bonded), neutral burn

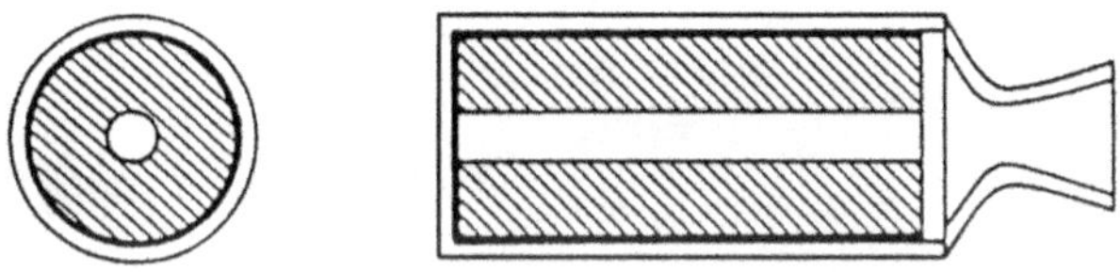

Internal burning tube (case bonded and end restricted), progressive

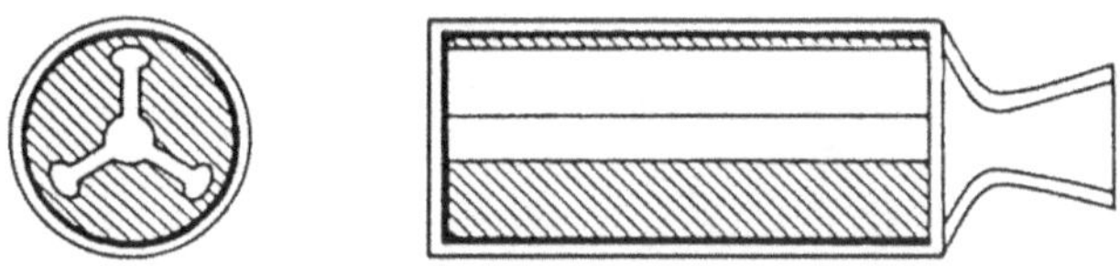

Dogbone (case bonded), neutral burn

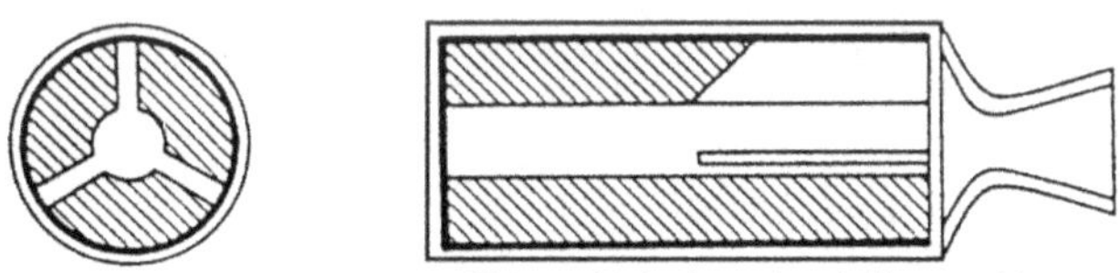

Slots and tube (case bonded), neutral burn

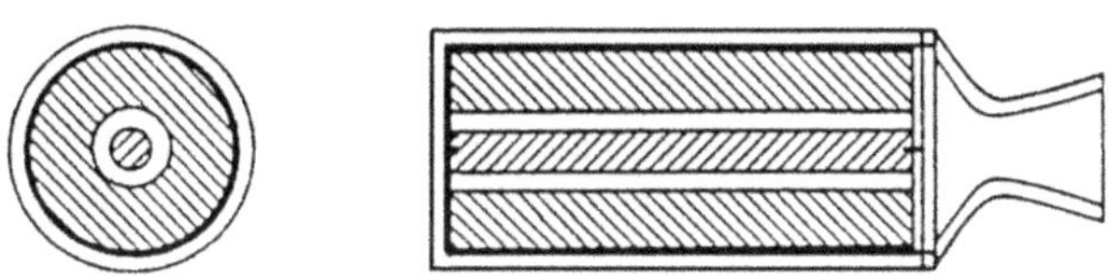

Rod and tube (case bonded), neutral burn

PROPULSION SYSTEMS, CONTINUED

TYPICAL GRAIN DESIGN CONFIGURATIONS, CONTINUED

Star (neutral)

Wagon Wheel (neutral)

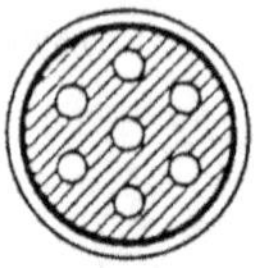

Multiperforated (progressive-regressive)

Dendrite (case bonded)

CLASSIFICATION OF PROPELLANT GRAINS ACCORDING TO PRESSURE-TIME CHARACTERISTICS

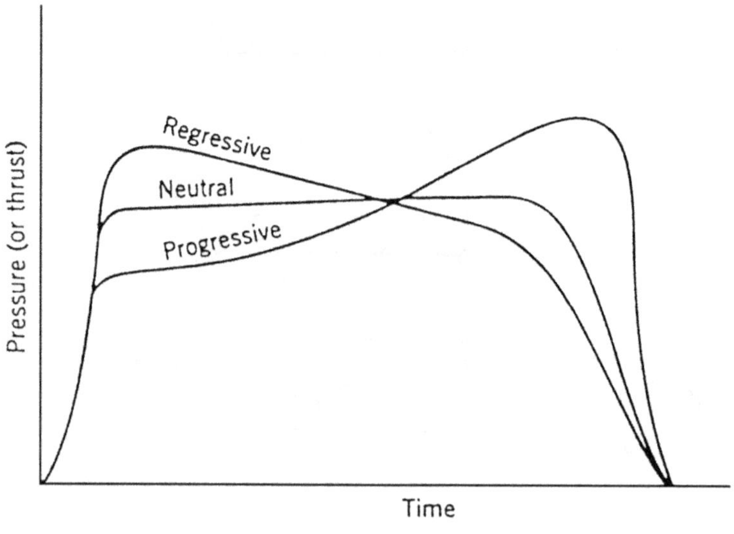

PROPULSION SYSTEMS, CONTINUED

THEORETICAL PERFORMANCE OF TYPICAL SOLID ROCKET PROPELLANT COMBINATIONS

Oxidizer	Fuel	ρ,[a] g/cm^3	T_1, K	c^*,[b] m/s	$\mathfrak{M}_c$, kg mol	I_{sp},[b] s	k
Ammonium nitrate	11% binder and 7% additives	1.51	1282	1209	20.1	192	1.26
Ammonium perchlorate 78–66%	18% organic polymer binder and 4–20% aluminum	1.69	2816	1590	25.0	262	1.21
Ammonium perchlorate 84–68%	12% polymer binder and 4–20% aluminum	1.74	3371	1577	29.3	266	1.17

[a] ρ, average specific gravity of propellant.
[b] Conditions for I_{sp} and c^*: combustion chamber pressure, 6.9 mPa (absolute); nozzle exit pressure, atmospheric; optimum nozzle expansion ratio; frozen equilibrium.

PROPULSION SYSTEMS, CONTINUED

CHARACTERISTICS OF SOME OPERATIONAL SOLID PROPELLANTS

Propellant Type[a]	I_{sp} Range,[b] s	Flame Temp, °C	Density $\times 10^{-3}$ g mm^{-3}	Metal Content, wt%	Burning Rate,[c] mm s^{-1}	Pressure Exponent, n	Hazard Classification	Stress, MPa / Strain, % −51°C	Stress, MPa / Strain, % +66°C	Processing Method
DB	220–230	2260	1.605	0	11.4	0.30	1.1 or 1.3	31.716/2	3.378/60	Extruded
DB/AP/AI	260–265	3593	1.799	20–21	19.8	0.40	1.3	18.961/5	0.827/50	Extruded
DB/AP-HMX/AI	265–270	3704	1.799	20	14.0	0.49	1.1	16.375/3	0.345/33	Solvent cast
PVC/AP/AI	260–265	3093	1.771	21	11.4	0.35	1.3	2.544/150	0.262/220	Cast or extruded
PS/AP/AI	240–250	2760	1.716	3	7.90	0.33	1.3	2.206/11	0.683/42	Cast
PU/AP/AI	260–265	2982–3316	1.771	16–20	6.90	0.15	1.3	8.067/6	0.517/33	Cast
PBAN/AP/AI	260–263	3204	1.771	16	14.0	0.33	1.3	3.585/16 (at–23°C)	0.490/28	Cast
CTPB/AP/AI	260–265	3093–3204	1.771	15–17	11.4	0.40	1.3	2.241/26	0.607/75	Cast
HTPB/AP/AI	260–265	3093–3204	1.854	4–17	10.2	0.40	1.3	6.274/50	0.621/33	Cast
PBAA/AP/AI	260–265	2982/3316	1.771	14	8.10	0.35	1.3	3.447/13	0.283/31	Cast
AN/Polymer	180–190	1260	1.467	0	7.60	0.60	1.3	1.379/5	n/a	Cast

[a]AI, aluminum; AN, ammonium nitrate; AP, ammonium perchlorate; CTPB, carboxy-terminated polybutadiene; DB, double base; HMX, cyclotetramethylene tetranitramine; HTPB, hydroxy-terminated polybutadiene; PBAA, polybutadiene-acrylic acid polymer; PBAN, polybutadiene-acrylic acid acrylonitrile terpolymer; PS, polysulfide; PU, polyurethane; PVC, polyvinyl chloride.

[b]At 6.895 MPa expanding to 0.101 MPa (1 atmosphere).

[c]At 6.895 MPa.

PROPULSION SYSTEMS, CONTINUED

CHARACTERISTICS OF SOME OPERATIONAL SOLID PROPELLANTS

Propellant Type[a]	I_{sp} Range,[b] s	Flame Temp, °F	Density, lb in.$^{-3}$	Metal Content, wt%	Burning Rate,[c] in. s^{-1}	Pressure Exponent, n	Hazard Classification[d]	Stress, psi / Strain, % –60°F	Stress, psi / Strain, % +150°F	Processing Method
DB	220–230	4100	0.058	0	0.45	0.30	1.1 or 1.3	4600/2	490/60	Extruded
DB/AP/Al	260–265	6500	0.065	20–21	0.78	0.40	1.3	2750/5	120/50	Extruded
DB/AP-HMX/Al	265–270	6700	0.065	20	0.55	0.49	1.1	2375/3	50/33	Solvent cast
PVC/AP/Al	260–265	5600	0.064	21	0.45	0.35	1.3	369/150	38/220	Cast or extruded
PS/AP/Al	240–250	5000	0.062	3	0.31	0.33	1.3	320/11	99/42	Cast
PU/AP/Al	260–265	5400–6000	0.064	16–20	0.27	0.15	1.3	1170/6	75/33	Cast
PBAN/AP/Al	260–263	5800	0.064	16	0.55	0.33	1.3	520/16 (at–10°F)	71/28	Cast
CTPB/AP/Al	260–265	5600–5800	0.064	15–17	0.45	0.40	1.3	325/26	88/75	Cast
HTPB/AP/Al	260–265	5600–5800	0.067	4–17	0.40	0.40	1.3	910/50	90/33	Cast
PBAA/AP/Al	260–265	5400/6000	0.064	14	0.32	0.35	1.3	500/13	41/31	Cast
AN/Polymer	180–190	2300	0.053	0	0.3	0.60	1.3	200/5	n/a	Cast

[a]Al, aluminum; AN, ammonium nitrate; AP, ammonium perchlorate; CTPB, carboxy-terminated polybutadiene; DB, double base; HMX, cyclotetramethylene tetranitramine; HTPB, hydroxy-terminated polybutadiene; PBAA, polybutadiene–acrylic acid polymer; PBAN, polybutadiene–acrylic acid acrylonitrile terpolymer; PS, polysulfide; PU, polyurethane; PVC, polyvinyl chloride.
[b]At 1000 psia expanding to 14.7 psia.
[c]At 1000 psia.

ATTITUDE CONTROL

TYPES OF ATTITUDE CONTROL SYSTEMS

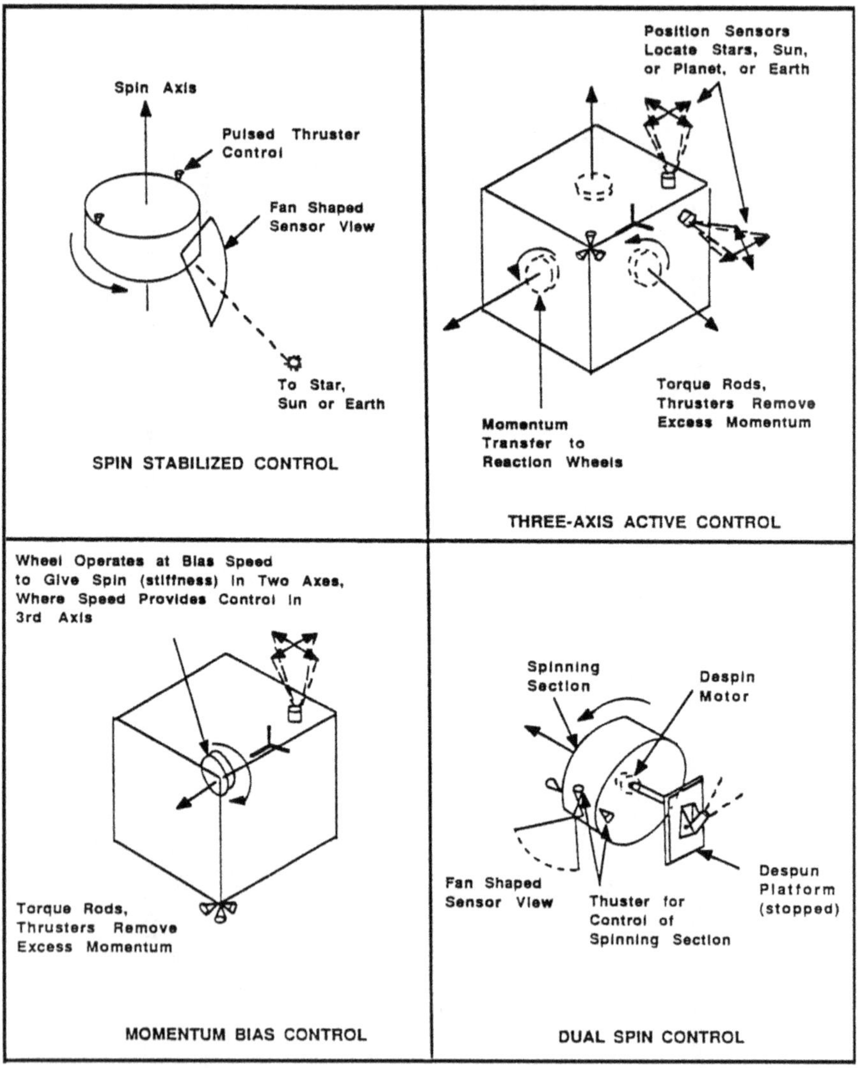

ATTITUDE CONTROL, CONTINUED

TYPES OF ATTITUDE CONTROL SYSTEMS, CONTINUED

Method	Typical Accuracy	Remarks
Spin stabilized	0.1 deg	Passive, simple; single axis inertial; low cost
Gravity gradient	1–3 deg	Passive, simple; central body oriented; low cost
Jets	0.1 deg	Consumables; fast; high cost
Magnetic	1 deg	Near Earth; slow; low weight, low cost
Reaction wheels	0.01 deg	Internal torque, requires other momentum control; high power, cost

COORDINATE FRAMES

The attitude of a satellite defines the relation between a reference coordinate frame and a satellite body-fixed coordinate frame.

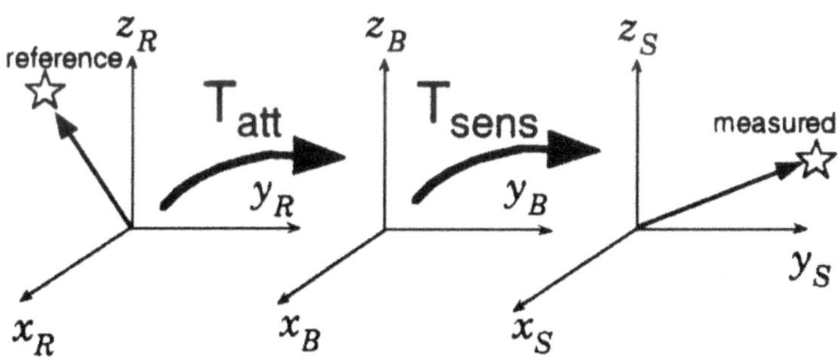

Measurements of reference vectors (e.g., sun, stars, local vertical) made in the sensor frame are used to compute the attitude matrix T_{att}. T_{att} is usually parameterized in Euler angles for analysis or Euler symmetric parameters (quaternions) for numerical computations.

The choice of an Euler angle sequence for the transformation T_{att} is dependent on the application, such as inertially stabilized, spin stabilized, or vertically stabilized. Proper choice of the Euler angle sequence will simplify the formulation. A typical Euler angle sequence (θ, ϕ, ψ) is shown in the following figure.

Source: The attitude control material appearing on pages 11-50–11-54 is from *Space Vehicle Design*, pages 282, 283, 288, 289, by M. D. Griffin and J. R. French.

ATTITUDE CONTROL, CONTINUED

COORDINATE FRAMES, CONTINUED

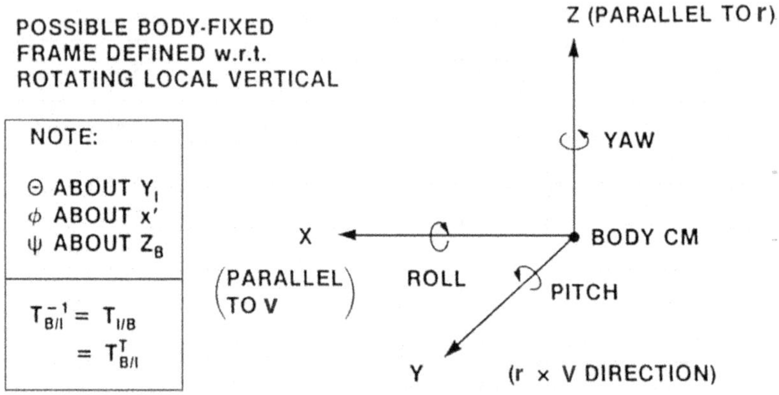

(PITCH, ROLL, YAW) = (θ, ϕ, ψ) ⟶ EULER ANGLES

Transformation from body to "inertial" frame:

$$
T_{B/I} = \begin{pmatrix} C\psi & S\psi & 0 \\ -S\psi & C\psi & 0 \\ 0 & 0 & 1 \\ & (\text{Yaw}) & \end{pmatrix} \begin{pmatrix} 1 & 0 & 0 \\ 0 & C\phi & S\phi \\ 0 & -S\phi & C\phi \\ & (\text{Roll}) & \end{pmatrix} \begin{pmatrix} C\theta & 0 & -S\theta \\ 0 & 1 & 0 \\ S\theta & 0 & C\theta \\ & (\text{Pitch}) & \end{pmatrix}
$$

$C\phi$ denotes $\cos(\phi)$ and $S\theta$ denotes $\sin(\theta)$. Note that the matrix multiplications of the transformation are noncommutative.

ANGULAR MOMENTUM

$$
\boldsymbol{H}_{\text{total}} = \sum_{i=1}^{n} \mathbf{r}_i \times m_i \dot{\mathbf{r}}_i \quad \begin{pmatrix} \text{collection of} \\ \text{point masses} \\ \text{at } \mathbf{r}_i \end{pmatrix}
$$

The net torque from all forces is

$$
\boldsymbol{T} = \sum_{i=1}^{n} \rho_i \times \mathbf{F}_i = \sum_{i=1}^{n} \rho_i \times m_i \ddot{\mathbf{r}}_i
$$

$$
\boldsymbol{T} = \frac{\mathrm{d}\boldsymbol{H}}{\mathrm{d}t} = (\dot{\boldsymbol{H}})_{\text{body frame}} + \omega \times \boldsymbol{H}
$$

$$
\dot{\boldsymbol{H}} = \boldsymbol{T} - \omega \times (\boldsymbol{I}\omega)
$$

where $\dot{\boldsymbol{r}}$ is the time derivative of vector $\boldsymbol{r}$, and $\dot{\boldsymbol{H}}$ is the time derivative of the angular momentum $\boldsymbol{H}$.

ATTITUDE CONTROL, CONTINUED

COORDINATE FRAMES, CONTINUED

EULER'S EQUATIONS

In a body-fixed principal axis frame, with the origin at the center of mass,

$$\dot{H}_x = I_x\dot{\omega}_x = T_x + (I_y - I_z)\omega_y\omega_z$$
$$\dot{H}_y = I_y\dot{\omega}_y = T_y + (I_z - I_x)\omega_z\omega_x$$
$$\dot{H}_z = I_z\dot{\omega}_z = T_z + (I_x - I_y)\omega_x\omega_y$$

No general solution exists. Particular solutions exist for simple torques. Computer simulation is usually required.

An important special case is the torque-free motion of a (nearly) symmetric body spinning primarily about its symmetry axis, such as a spin-stabilized satellite.

Thus,

$$\omega_x, \omega_y \ll \omega_z \triangleq \Omega$$
$$I_x \cong I_y$$

And

$$\dot{\omega}_x \cong -K_x\Omega\omega_y$$
$$\dot{\omega}_y \cong K_y\Omega\omega_x$$
$$\dot{\omega}_z \cong O$$
$$\Rightarrow \omega_x = \omega_y = A\cos\omega_n t$$

where

$$K_x = \frac{I_z - I_y}{I_x}$$
$$K_y = \frac{I_z - I_x}{I_y}$$
$$\omega_n = \sqrt{K_x K_y}\Omega$$

Also,

$$I_z > I_x = I_y \qquad \text{or} \qquad I_z < I_x = I_y$$

yields ω_n real and sinusoidal motion at the nutation frequency ω_n.

ATTITUDE CONTROL, CONTINUED

COORDINATE FRAMES, CONTINUED

EULER'S EQUATIONS, CONTINUED

Two cases exist:

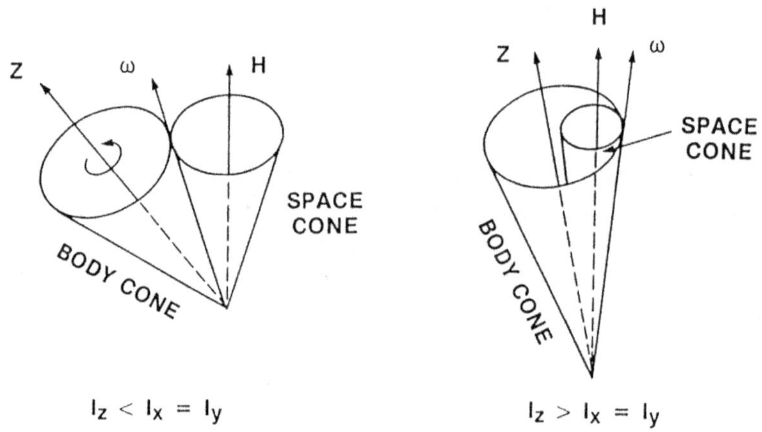

The preceding figure shows the relationship between the nominal spin axis Z, the inertial angular velocity vector ω, and the angular momentum vector $\boldsymbol{H}$. In the torque-free case, the angular momentum $\boldsymbol{H}$ is fixed in space, and the angular velocity vector ω rotates about $\boldsymbol{H}$. For a disk-shaped body ($I_x = I_y < I_z$), the precession rate is faster than the spin rate. For a rod-shaped body ($I_x = I_y > I_z$), the precession rate is faster than the spin rate.

$$\Omega_P = \omega_z \left[1 - \frac{I_T - I_S}{I_T} \right] = \omega_z \left[\frac{I_S}{I_T} \right]$$

where

ω_z = spin rate
Ω_P = precession rate

SPIN AXIS STABILITY

$$E_{\text{total}} = \underbrace{\frac{1}{2}\left(\sum_{i=1}^{n} m_i \right) R^2}_{E_{\text{trans}}} + \underbrace{\frac{1}{2} \sum_{i=1}^{n} m_i \dot{\rho}_i^2}_{E_{\text{rot}}}$$

For a rigid body, center of mass coordinates, with $\vec{\omega}$ resolved in body axis frame,

$$E_{\text{rot}} = \frac{1}{2} \omega \cdot \boldsymbol{H} = \frac{1}{2} \omega^T I \omega$$

The results above are valid only for a rigid body. When any flexibility exists, energy dissipation will occur, and ω will lose its significance.

ATTITUDE CONTROL, CONTINUED

SPIN AXIS STABILITY, CONTINUED

$$\boldsymbol{H} = I\omega \Rightarrow \text{constant}$$

$$E_{\text{rot}} = \frac{1}{2}\omega^T I\omega \Rightarrow \text{decreasing}$$

Spin goes to maximum I, minimum ω

Conclusion: Stable spin is possible only about the axis of maximum inertia. A classic example is Explorer 1.

Reference measurement methods are given in the following table:

Reference	Typical Accuracy	Remarks
Sun	1 min	Simple, reliable, low cost; not always visible
Earth	0.1 deg	Orbit dependent; usually requires scan; relatively expensive
Magnetic field	1 deg	Economical; orbit dependent, low altitude only; low accuracy
Stars	0.001 deg	Heavy, complex, expensive, accurate
Inertial space	0.01 deg/h	Rate only; good short-term ref; high weight, power, cost

POWER SYSTEMS

POWER SUPPLY OPERATING REGIMES

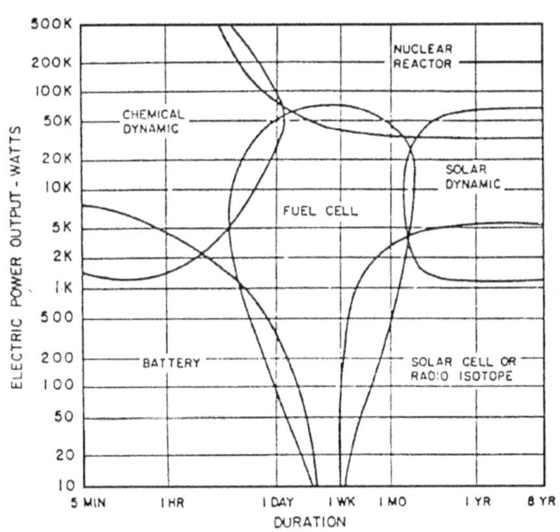

POWER SYSTEMS, CONTINUED

CHARACTERISTICS OF ISOTOPES FOR POWER PRODUCTION

Isotope	Half-Life	Compound	Compound Power, W/g	Dose Rate,[a] mR/h Bare	Dose Rate,[a] mR/h 3-cm of Uranium
Beta emitters					
Cobalt 60	5.27 yr	Metal	3.0[b]	3×10^8	3×10^6
Strontium 90	28 yr	$SrTiO_3$	0.2	6×10^6	1×10^4
Promethium 147[c]	2.67 yr	Pm_2O_3	0.3	1×10^5	1.0
Thulium 170	127 days	Tm_2O_3	1.75	4×10^6	50
Alpha emitters					
Polonium 210	138 days	Metal matrix	17.6	760	1.8
Plutonium 238	86 yr	PuO_2	0.35	5	0.03
Curium 242	163 days	Cm_2O_3 in metal matrix	15.5	280	2
Curium 244	18.4 yr	Cm_2O_3	2.5	600	32

[a]At 1 m from 5-thermal-kW source. [b]200 Ci/g metal. [c]Aged, 1 half-life.

ISOTOPE ELECTRICAL POWER VS TIME

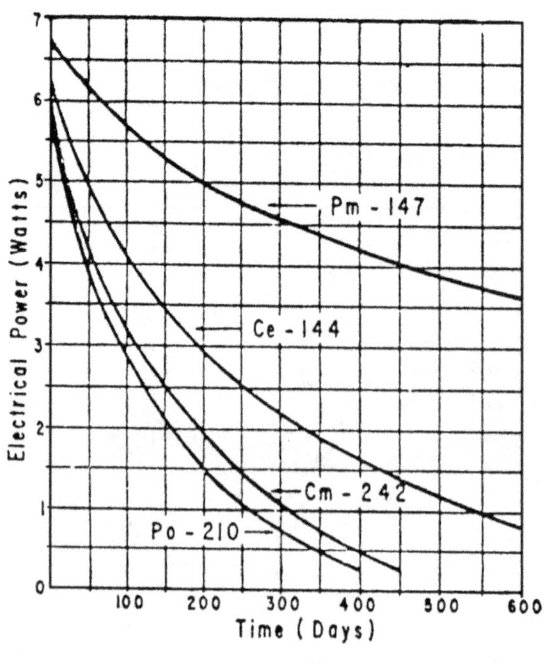

POWER SYSTEMS, CONTINUED

NI-CD BATTERY LIFE

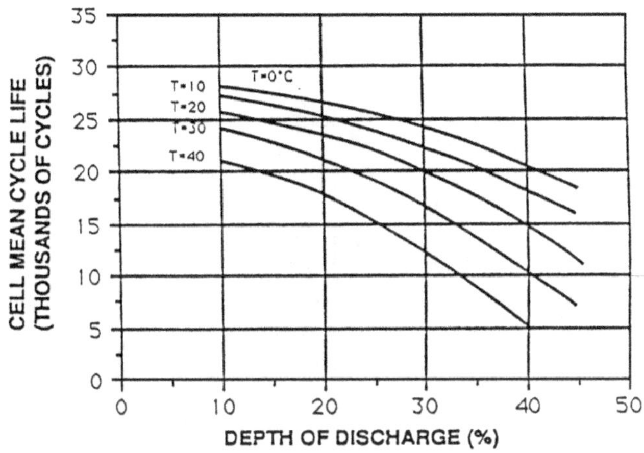

Source: "Spacecraft Systems Design Handbook" by W.J. Larson. Copyright ©1988, W. J. Larson. Reproduced with permission of the author.

CHARACTERISTICS OF DIFFERENT STORAGE CELLS

Types of Cells	Electrolyte	Nominal Voltage/ Cell, V	Energy Density, Wh/kg	Temp, °C	Cycle Life at Different Depth of Discharge Levels 25%	50%	75%	Whether Space Qualified
Ni-Cd	Diluted potassium hydroxide (KOH) solution	1.25	25–30	−10–40	21,000	3000	800	Yes
Ni-H_2	KOH solution	1.30	50–80	−10–40	>15,000	>10,000	>4000	Yes[a]
Ag-Cd	KOH solution	1.10	60–70	0–40	3500	750	100	Yes
Ag-Zn	KOH solution	1.50	120–130	10–40	2000	400	75	Yes
Ag-H_2	KOH solution	1.15	80–100	10–40	>18,000	—	—	No
Pb-acid	Diluted sulfuric acid	2.10	30–35	10–40	1000	700	250	—

[a]Ni-H_2 cells are employed onboard the Navigational Technology Satellite (NTS-2) and other geosynchronous satellites. However, these cells have not been used on any low-Earth orbit satellites.

POWER SYSTEMS, CONTINUED

EXPECTED RADIATION DOSE

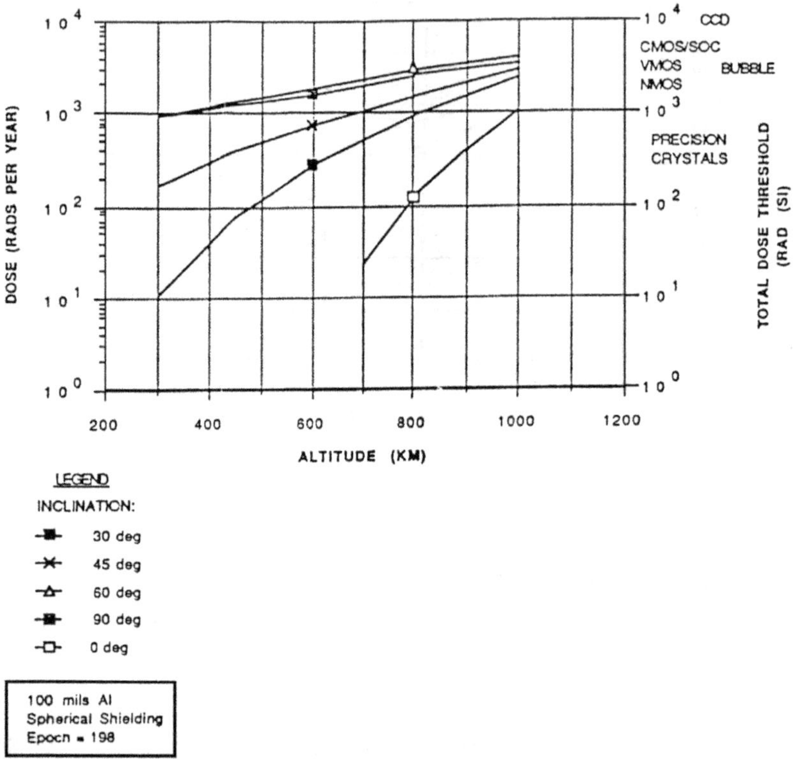

Source: "Spacecraft Systems Design Handbook" by W. J. Larson. Copyright ©1988, W. J. Larson. Reproduced with permission of the author.

CHARACTERISTICS OF A TYPICAL SOLAR CELL FOR VARIOUS TEMPERATURES

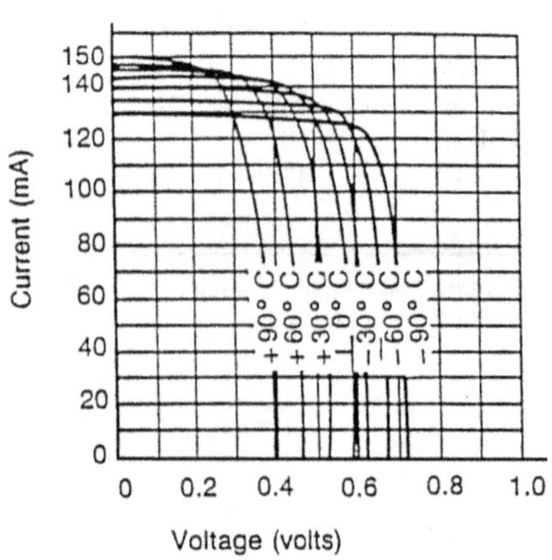

Power Systems, Continued

Characteristics of a Typical Solar Cell Subjected to Successive Doses of 1 MeV Electrons

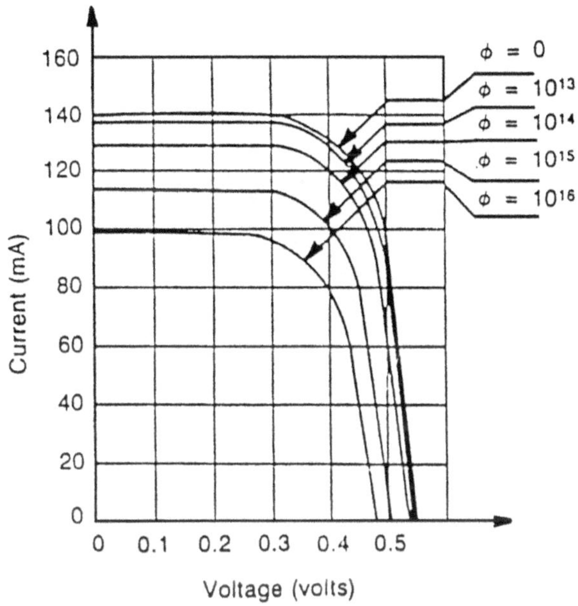

Source: "Spacecraft Systems Design Handbook" by W. J. Larson. Copyright © 1988, W. J. Larson. Reproduced with permission of the author.

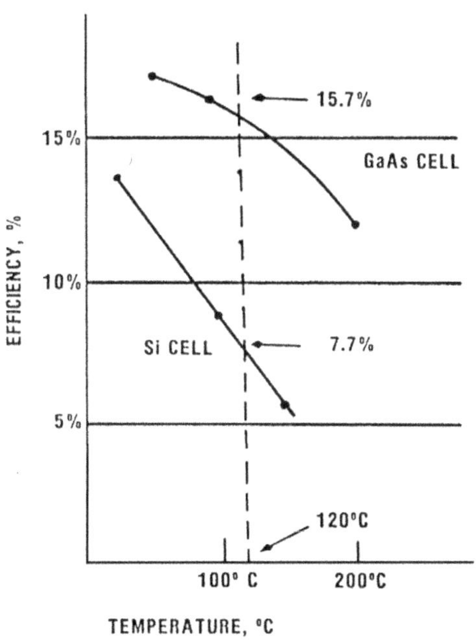

POWER SYSTEMS, CONTINUED

NADIR POINTED SPACECRAFT WITH A FLAT, ROOFTOP SOLAR ARRAY

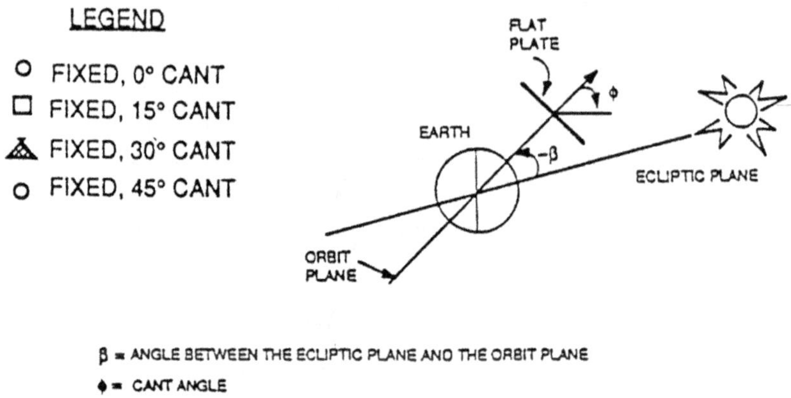

ECLIPSE TIME VS BETA

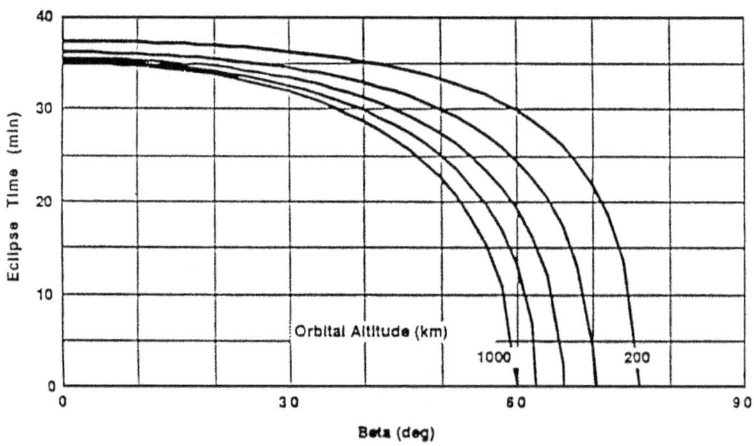

Source: "Spacecraft Systems Design Handbook" by W. J. Larson. Copyright © 1988, W. J. Larson. Reproduced with permission of the author.

$$T_e = (P/180) * \cos^{-1}\left[1 - \left(R_e^2/R^2\right)^{0.5}/\cos(\beta)\right]$$

$$P = 2 * \pi * \left[R^3/398{,}600.8\right]^{0.5}/60$$

where

T_e = eclipse time, min
P = orbital period, min
R_e = radius of Earth, 6378.135 km
R = distance from the center of Earth to the s/c, km
β = sun/orbit angle, deg

POWER SYSTEMS, CONTINUED

ALTITUDE EFFECTS ON ECLIPSES

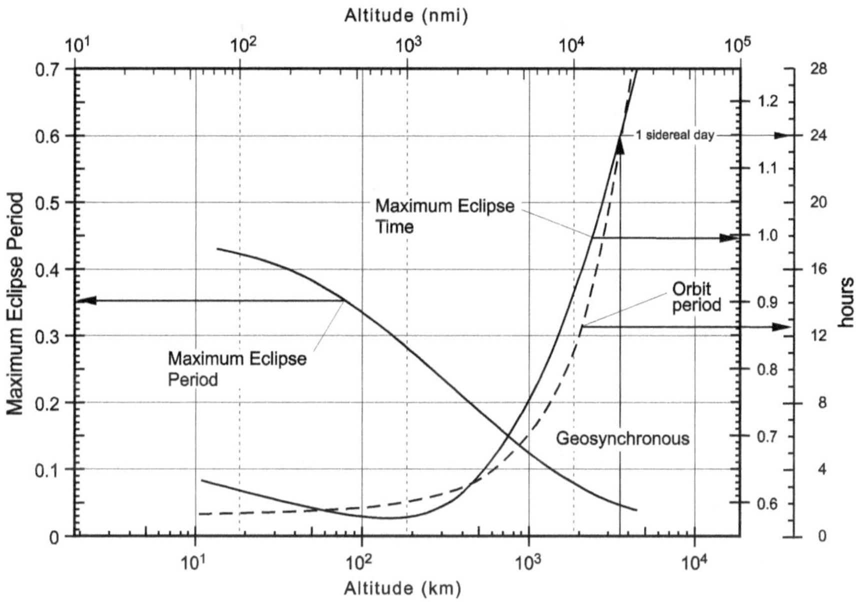

SOLAR ARRAY AVERAGE POWER VS BETA ANGLE

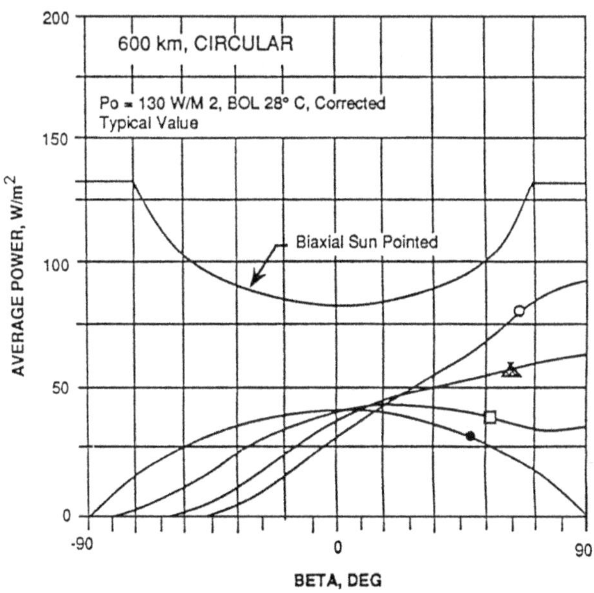

THERMAL CONTROL

FUNDAMENTALS OF HEAT TRANSFER

CONDUCTION

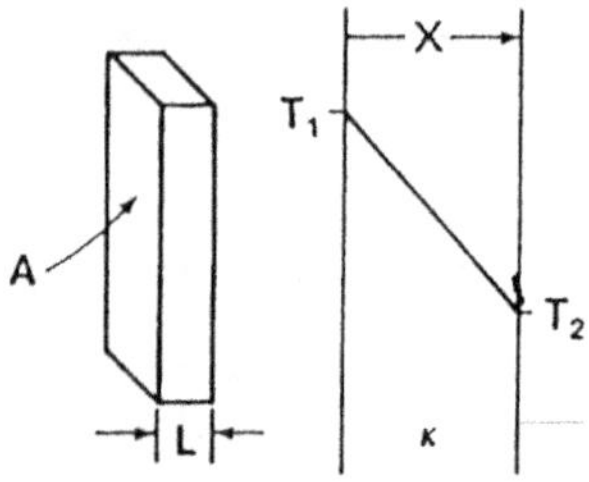

where

$\dot{Q} = \text{power} = \dfrac{\text{energy}}{\text{time}} = W$

κ = thermal conductivity (W/m K)
A = surface area, m^2
T = temperature, K

$$\dot{Q} = -\kappa A \frac{dT}{dx}$$ 1-D conduction equation (Fourier's Law)

$$\dot{q} \triangleq \frac{\dot{Q}}{A} = -\kappa \frac{dT}{dx}$$ 1-D heat flux equation

$$\underline{\dot{q}} = -\kappa \nabla \underline{T}$$ 3-D conduction equation

$$\nabla^2 T = \frac{\rho C}{\kappa}\frac{\partial T}{\partial t} - \frac{\dot{g}_{\text{int}}}{\kappa}$$ from 1st law thermodynamics

where

ρ = density, kg/m^3
C = specific heat, J/kg K
$\dot{g}_{int}$ = internal power source, W/m^3

CASES

No source ⇒ diffusion equation
Source but steady state ⇒ Poisson equation
Steady, no source ⇒ Laplace equation

Source: The thermal control material appearing on pages 11-61–11-64 is from *Space Vehicle Design,* pages 375, 378, 380, 381, by M. D. Griffin and J. R. French.

THERMAL CONTROL, CONTINUED

FUNDAMENTALS OF HEAT TRANSFER, CONTINUED

CONVECTION

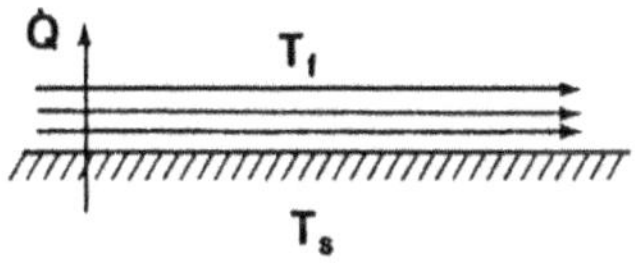

where

h = convection or film coefficient, W/m^2 K
$\dot{Q}$ = power, W
A = surface area, m^2
ΔT = temperature difference for process, K

$\dot{Q} = hA\Delta T$ 1-D convection equation

$T_a > T_f$,

$\Delta T = T_s - T_f$ forced convection condition

RADIATION

$$\dot{Q} = \varepsilon A \sigma T^4 \qquad \text{Stefan–Boltzmann law}$$

where

σ = Stefan–Boltzmann constant 5.6696×10^{-8} W/m^2 K^4
A = surface area, m^2
ε = emissivity (blackbody $\triangleq 1$)

$$E \triangleq \frac{\dot{Q}}{A} = \text{hemispherical total emissive power, W/m}^2$$

$$E_{\lambda b} = \frac{2\pi hc^2}{\lambda 5\left(e^{hc/\lambda kT} - 1\right)} = \frac{\text{Planck's law (vacuum)}}{\text{hemispherical spectral emissive power}}$$

where

$h = 6.626 \times 10^{-34}$ J s
$k = 1.381 \times 10^{-23}$ J/K
$c = 2.9979 \times 10^{8}$ m/s

THERMAL CONTROL, CONTINUED

FUNDAMENTALS OF HEAT TRANSFER, CONTINUED

RADIATION, CONTINUED

Wien's displacement law—peak intensity of $E_{\lambda b}$:

$$(\lambda T)_{\text{MAXIMUM}} = \frac{1}{4.965114}\frac{hc}{k} = 2.8978 \times 10^{-3} \text{ m K}$$

Blackbody radiation fundamentals:

- Neither reflects nor transmits incident energy; perfect absorber at all wavelengths, angles
- Equivalent blackbody temperature sun ≅ 5780 K; Earth ≅ 290 K

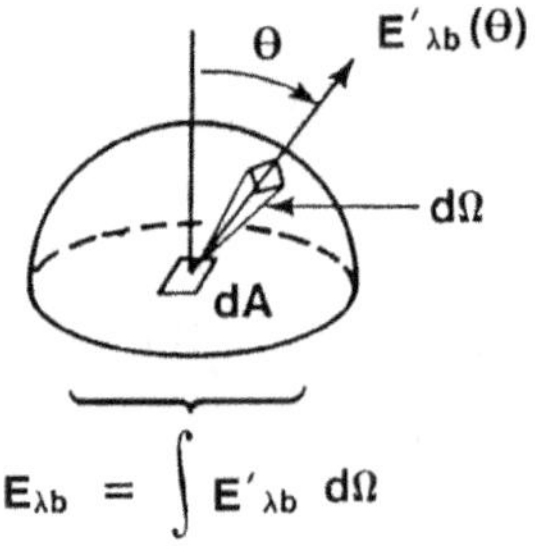

Lambertian surface:

= Spectral directional radiant intensity, W/m^2 μm sr

$i'_{\lambda b}(\lambda) = \frac{1}{\pi} E_{\lambda b}(\lambda)$:

= Energy per unit time, per unit solid angle, per unit projected area $dA_{\perp}$, per unit wavelength
= Directional spectral emissive power

$E'_{\lambda b} = i'_{\lambda b} \cos\theta$:

= Energy per unit time, per unit wavelength, per unit solid angle, per unit area dA

THERMAL CONTROL, CONTINUED

SURFACE PROPERTIES FOR PASSIVE THERMAL CONTROL[a]

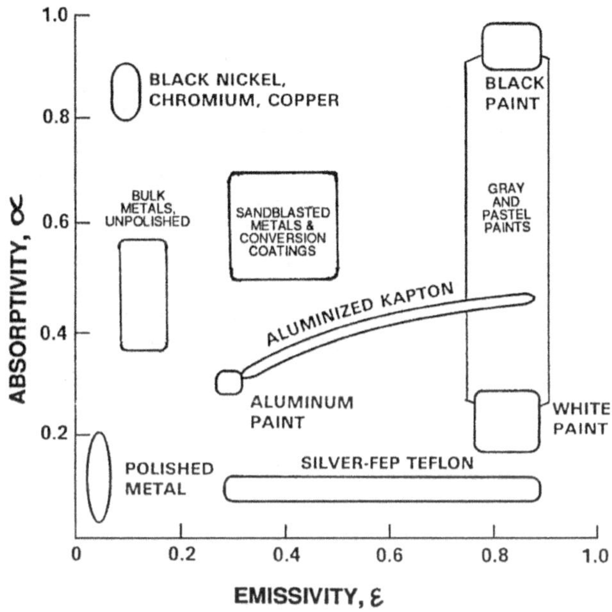

[a]Absorptivity values are at BOL (beginning of life).

RADIATION PROPERTIES OF VARIOUS SURFACES

Surface Description	Solar Absorptivity Alpha	Infrared Emissivity Epsilon
Al foil coated w/10 Si	0.52	0.12
Si solar cell 1 mm thick	0.94	0.32
Si solar cell 3 mil glass	0.93	0.84
Stainless steel, type 410	0.76	0.13
Flat black epoxy paint	0.95	0.80
White epoxy paint	0.25	0.88
2 mil Kapton w/vacuum dep Al	0.49	0.71
first surface mirror		
vacuum deposited gold	0.3	0.03
vacuum deposited Al	0.14	0.05
second surface mirror		
6 mil Telflon, vacuum dep Ag	0.09	0.75
6 mil Telflon, vacuum dep Al	0.14	0.75

Note: Absorptivity values are nominal (at BOL).

THERMAL CONTROL, CONTINUED

EQUILIBRIUM TEMP VS ABSORPTIVITY TO EMITTANCE RATIO $\beta = (\alpha/\varepsilon)^{a}$

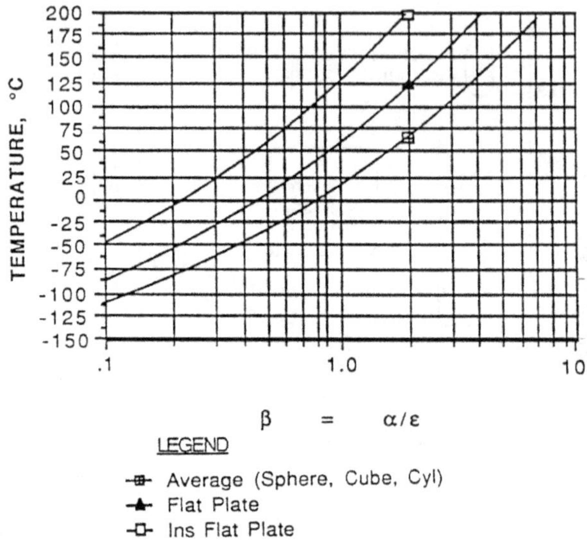

[a]Does not account for shadowing, albedo, Earthshine, or internal heat dissipation. It is appropriate for MLI covers, solar arrays dissipation, and passive structure.

Source: "Spacecraft Systems Design Handbook" by W. J. Larson. Copyright © 1988, W. J. Larson. Reproduced with permission of the author.

EQUATIONS

$$T_{ES} = \left[\left(\beta * S * A_P / A_T * A_T \right) / \left(A_R * \sigma \right) \right]^{0.25}$$

$$T_{ESA} = \left[\left(\left(\alpha - F_P * e \right) * A_F * S * A_P \right) / \left(\varepsilon_F * A_F + \varepsilon_B * A_B \right) \right]^{0.25}$$

where

β = absorbance/emissivity, α/ε

S = solar constant = 1356 W/m^2

A_P = projected area, m^2

A_T = total area, m^2

A_R = radiator area, m^2

A_B = back of solar array area, m^2

A_F = frontal area of solar array, m^2

σ = Boltzmann constant = 5.97×10^{-8} W/m^2

α = absorbance

F_P = packing factor for solar array

e = solar cell efficiency

ε_F = emissivity of front of solar array

ε_B = emissivity of back of solar array

THERMAL CONTROL, CONTINUED

EQUILIBRIUM TEMP VS ABSORPTIVITY TO EMITTANCE RATIO, $(\beta = \alpha/\varepsilon)$, CONTINUED

ASSUMPTIONS

$A_F = A_B$
$F_P = 0.95$
$A_T = A_R$
Unblocked view to space
Isothermal spacecraft
No MLI

INPUTS REQUIRED

β or temperature, °C

SPACECRAFT SYSTEMS DESIGN AND ENGINEERING THERMAL CONTROL

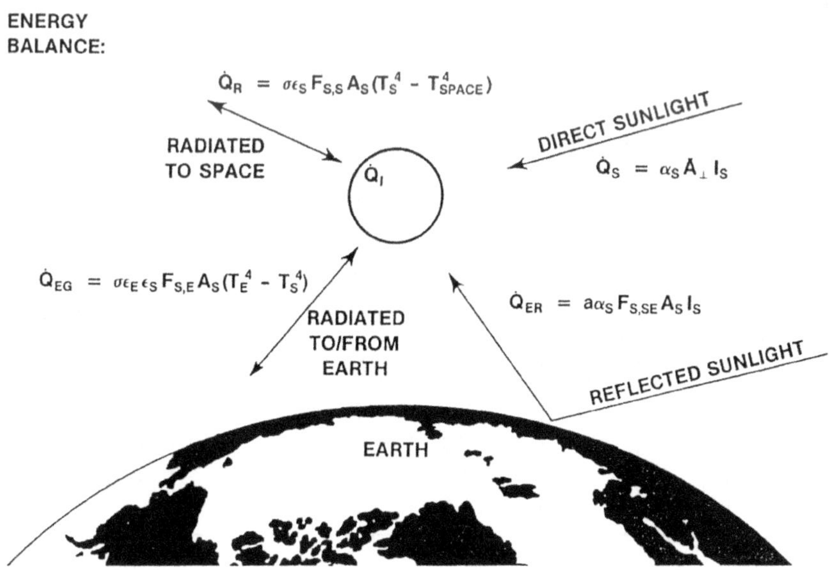

Source: Space Vehicle Design, page 392, by M. D. Griffin and J. R. French.

THERMAL CONTROL, CONTINUED

SPACECRAFT SYSTEMS DESIGN AND ENGINEERING THERMAL CONTROL, CONTINUED

where

α = Earth albedo (0.07–0.85)
α_S = spacecraft absorptivity (solar)
$A_\perp$ = orbit-averaged spacecraft projected area perpendicular to sun
A_S = spacecraft surface area
ε_S = spacecraft emissivity (IR)
ε_E = Earth emissivity (IR) $\cong$ 1
$F_{S,S}$ = view factor, spacecraft to space
$F_{S,E}$ = view factor, spacecraft to Earth
$F_{S,SE}$ = view factor, spacecraft to sunlit Earth
I_S = solar intensity, 1356 W/m^2
T_E = Earth blackbody temperature, 261 K

$$F_{S,S} + F_{S,E} = 1$$

If $\varepsilon_E = 1$, $T_{SPACE} = 0$, and $F_{S,S} + F_{S,E} = 1$, solve to yield the average satellite temperature.

$$\sigma\varepsilon_S A_S T_S^4 = \sigma\varepsilon_S F_{S,E} A_S T_E^4 + \dot{Q}_S + \dot{Q}_{ER*} + \dot{Q}_1$$

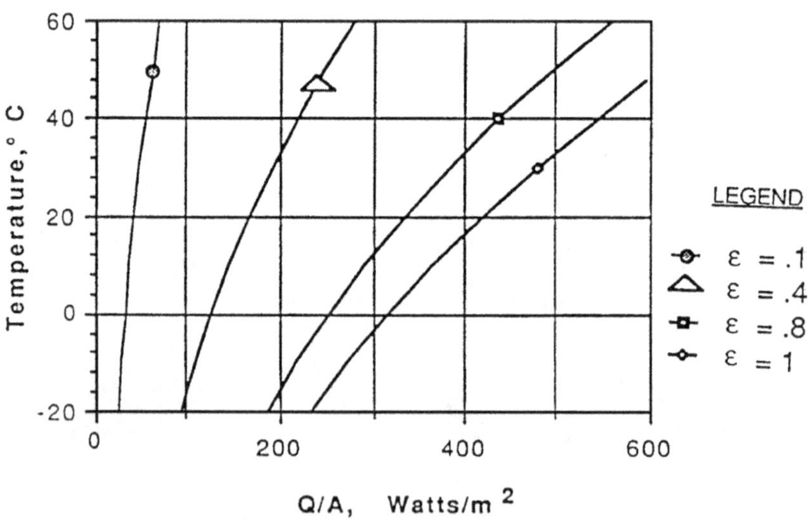

Source: "Spacecraft Systems Design Handbook" by W. J. Larson. Copyright © 1988, W. J. Larson. Reproduced with permission of the author.

FLUID SYSTEMS FOR HEAT TRANSFER

			Properties							
			Critical Point			Triple Point				
								ρ, M/L	
Fluid	Symbol	Molecular Weight	P, MPa	T, K	ρ, M/L	P, MPa	T, K	Liquid	Vapor
Helium	He	4.0026	0.2275	5.2014	17.399	0.00496	2.172	36.5343	0.2904
Nitrogen	N_2	28.0134	3.4100	126.26	11.21	0.1246	63.15	30.977	0.2396
Oxygen	O_2	31.9988	5.0422	154.481	13.63	0.149 E–03	54.359	40.620	0.3275 E–03
Argon	Ar	39.948	4.8980	150.86	13.41	0.0689	83.8	35.40	0.1015 E–02
Methane	CH_4	16.0430	4.599	190.55	10.23	0.01174	90.68	28.1511	0.0157 E–03
Hydrogen (normal)	H_2	2.0159	1.3152	33.217	14.936				
Water	H_2O	18.0153	22.1123	647.383	17.74				
Carbon dioxide	CO_2	44.0100	7.375	304.217	10.64				
Hydrazine	N_2H_2	32.050	14.690	653.16					
Ammonia	NH_3	17.0306	11.422	406.161	13.835				

COMMUNICATIONS

NOISE SOURCES

Sun	$T = 10^4$–10^{10} K Communication is effectively impossible with the sun in the antenna fov
Moon	Reflected sunlight, but much weaker; communication usually unimpeded
Earth	Characteristic temperature 254 K
Galaxy	Negligible above 1 GHz
Sky	Characteristic temperature 30 K
Atmosphere	Noise radiated by O_2 and H_2O absorption and re-emission, typically less than 50 K below 20 GHz
Weather	Heavy fog, clouds, or heavy rain can outweigh other sources (except sun), especially above 10 GHz
Johnson noise	Due to resistance or attenuation in medium (atmosphere, wire, cable, etc.); proportional to temperature, resistance
Electronics noise	Receiving equipment (antenna, amplifiers) makes a significant contribution, typically 60 K cryogenically cooled equipment may be used for extremely low signal levels; pre-amplifier customarily mounted on antenna

COMPOSITE LINK NOISE PLOT (EXCLUDING WEATHER)

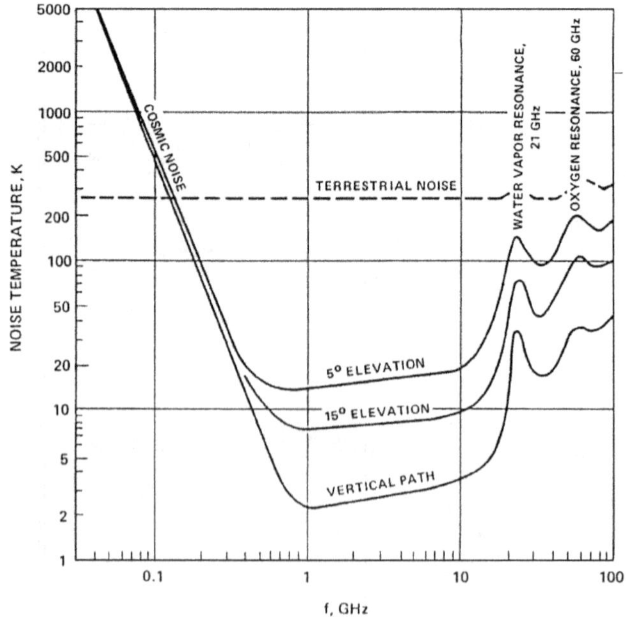

COMMUNICATIONS, CONTINUED

RECEIVER NOISE

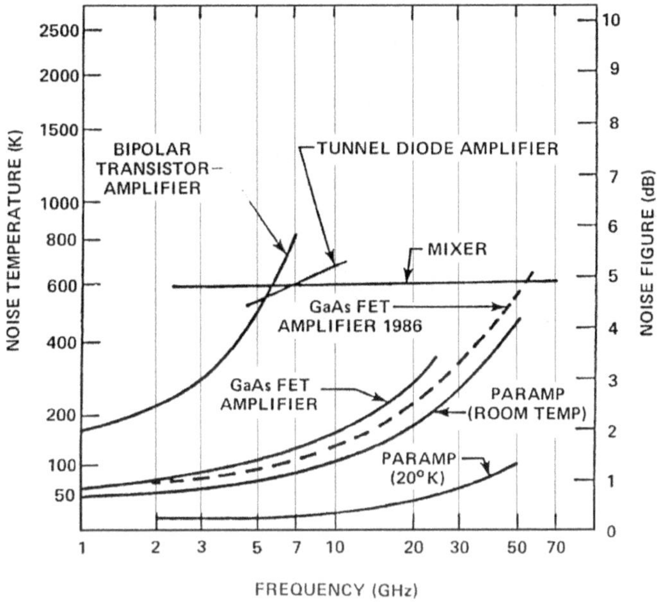

TOTAL ABSORPTION LOSS (EXCLUDING WEATHER)

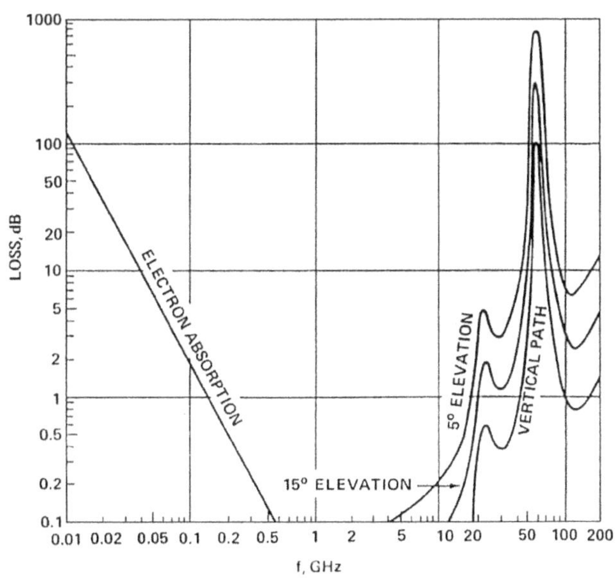

COMMUNICATIONS, CONTINUED

BEHAVIOR OF ANALOG AND DIGITAL MODULATIONS

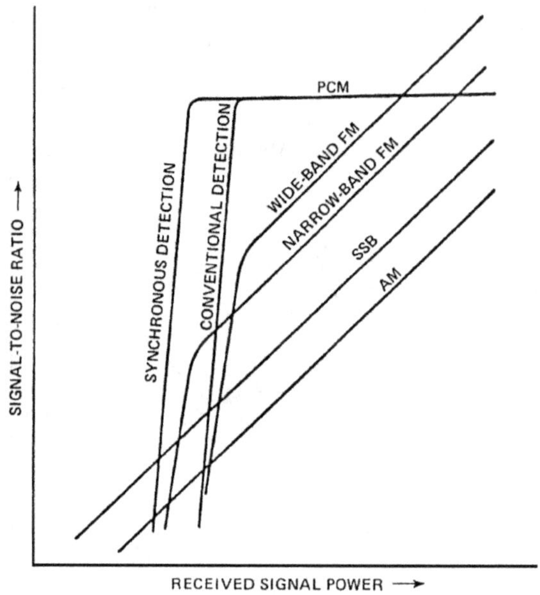

STRUCTURE DESIGN AND TEST

ACCELERATION

Booster	Axial, *g*	Lateral, *g*
Atlas–Centaur	5.5	2
Titan II	3–10	2.5
Ariane	4.5	0.2
Delta	5.8–6.9	2.5
	5.7–12	
Long March	4.1	0.6
Shuttle	−3.17 + 4.5	+ − 1.5
		+ 4.5–2

STRUCTURE DESIGN AND TEST, CONTINUED

LOW-FREQUENCY VIBRATION

Vehicle	Frequency Range	Acceleration
Atlas–Centaur	0–50 Hz	1.0 *g* axial
		0.7 *g* lateral
Ariane	5–100 Hz	1.0 *g* axial
	2–18 Hz	0.8 *g* lateral
	18–100 Hz	0.6 *g* lateral
Delta	5–6.2 Hz	0.5* amplitude axial
	6.2–100 Hz	1.0 *g* axial
	5–100 Hz	0.7 *g* lateral
Long March	5–8 Hz	3.12 mm axial
	8–100 Hz	0.8 *g* axial
	2–8 Hz	2.34 mm lateral
	8–100 Hz	0.6 *g* lateral

RANDOM VIBRATION

Vehicle	Frequency Range	Acceleration
Atlas–Centaur	20–80 Hz	+9 dB/oct
	80–200 Hz	0.03 g^2/Hz
	200–1500 Hz	–9 dB/Hz
		2.7 gms overall
Ariane	20–150 Hz	+6 dB/oct
	150–700 Hz	0.4 g^2/Hz
	700–2000 Hz	–3 dB/Hz
		7.3 gms overall
Long March	10–100 Hz	+3 dB/oct
	100–800 Hz	0.04 g^2/Hz
	800–2000 Hz	–12 dB/Hz
		6.23 gms overall
Shuttle	20–100 Hz	+6 dB/oct
	100–250 Hz	0.015 g^2/Hz sill
		0.15 g^2/Hz keel
	250–2000 Hz	–6 dB/Hz

STRUCTURE DESIGN AND TEST, CONTINUED

LOAD DEFINITIONS

Load	Specified acceleration
Limit load	Maximum expected acceleration
Yield load	Member suffers permanent deformation
Ultimate load	Structural member fails
Safety factor	Ratio of ultimate load to limit load
Margin of safety	Safety factor minus one = SF − 1

LOAD FACTORS

USE OF LOAD FACTORS

Gross factors used in preliminary design
Sometimes multiplied by an uncertainty factor
Design LL = load factor ∗ uncertainty factor
Typical uncertainty factor < 1.5
Gross load factors replaced by nodal accelerations after coupled modes analysis

LIMIT, YIELD, AND ULTIMATE

Expendable launch vehicles
- Ultimate = 1.5 × limit
- Yield = 1.0 × limit
- Sometimes yield = 1.1 × limit

For shuttle ultimate = 1.4 × limit
Pressure vessels are higher
- Typically qualification load = 1.2 × limit
- Yield = 1.2 × 1.1 × limit
- Ultimate = 1.5 × limit

STRUCTURE DESIGN AND TEST, CONTINUED

SPACECRAFT STRUCTURE DESIGN/VERIFICATION

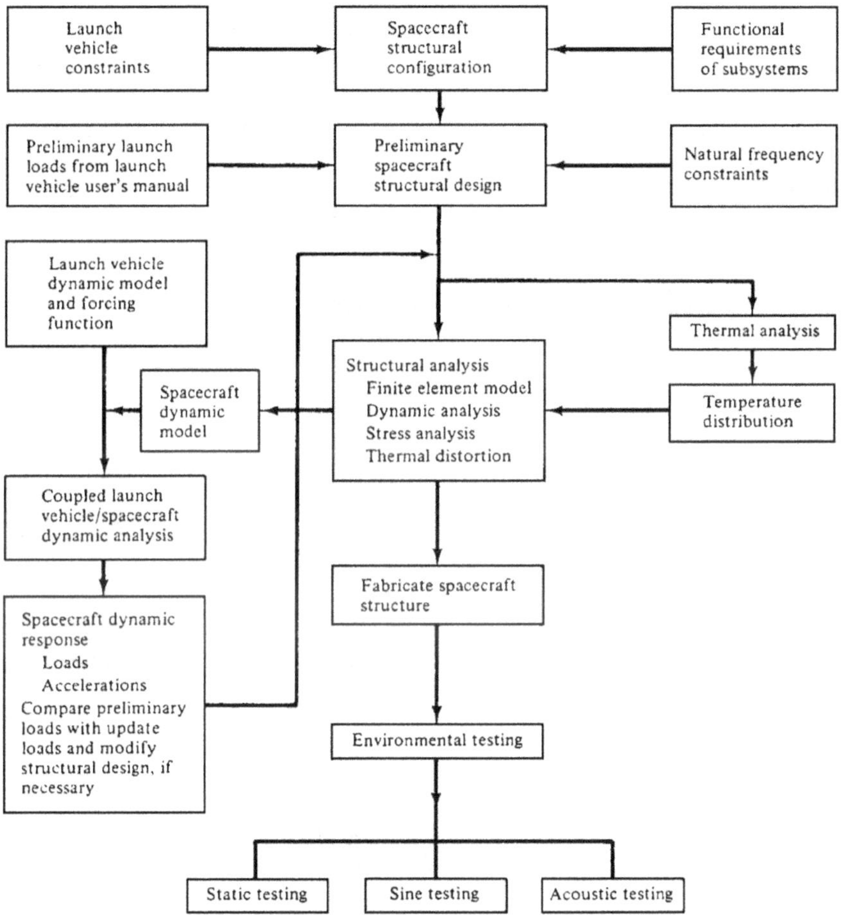

BIBLIOGRAPHY

Agrawal, B. N., *Design of Geosynchronous Spacecraft*, Prentice–Hall, Englewood Cliffs, NJ, 1986.

Battin, R. H., *An Introduction to the Mathematics and Methods of Astrodynamics*, AIAA Education Series, AIAA, Washington, DC, 1987.

Brown, C. D., *Spacecraft Mission Design*, 2nd Ed., AIAA Education Series, AIAA, Reston, VA, 1998.

Chetty, P. R. K., *Satellite Technology and its Applications*, TAB Professional and Reference Books, Blue Ridge Summit, PA, 1991.

Collicott, H. E., and Bauer, P. E. (eds.), *Spacecraft Thermal Control, Design, and Operation*, Vol. 86, Progress in Astronautics and Aeronautics, AIAA, New York, 1983.

Colwell, R. N. (ed.), *Manual of Remote Sensing*, 2nd Ed., American Society of Photogrammetry, Falls Church, VA, 1983.

Corliss, W., *Scientific Satellites*, NASA SP-133, 1969.

Garrett, H. B., and Pike, C. P., *Space Systems and Their Interactions with Earth's Space Environments*, Vol. 71, Progress in Astronautics and Aeronautics, AIAA, New York, 1980.

Griffin, M. D., and French, J. R., *Space Vehicle Design*, AIAA Education Series, AIAA, Washington, DC, 1991.

Horton, T. E. (ed.), *Spacecraft Radiative Transfer and Temperature Control*, Vol. 83, Progress in Astronautics and Aeronautics, AIAA, New York, 1982.

Hughes, P. C., *Spacecraft Attitude Dynamics*, Wiley, New York, 1986.

Isakowski, S. J., Hopkins, J. P., Jr., and Hopkins, J. B., *International Reference Guide to Space Launch Systems*, 4th Ed., AIAA, Washington, DC, 2004.

Kaplan, M. H., *Modern Spacecraft Dynamics and Control*, Wiley, New York, 1976.

Pocha, J. J., *Mission Design for Geostationary Satellites*, D. Reidel, Boston, 1987.

Shahrokhi, F., Greenberg, J. S., and Al-Saud, T. (eds.), *Space Commercialization: Launch Vehicles and Programs*, Vol. 126, Progress in Astronautics and Aeronautics, AIAA, Washington, DC, 1990.

Shahrokhi, F., Hazelrigg, G., and Bayuzick, R. (eds.), *Space Commercialization: Platforms and Processing*, Vol. 127, Progress in Astronautics and Aeronautics, AIAA, Washington, DC, 1990.

Shahrokhi, F., Jasentuliyana, N., and Tarabzouni, N. (eds.), *Space Commercialization: Satellite Technology*, Vol. 128, Progress in Astronautics and Aeronautics, AIAA, Washington, DC, 1990.

Sutton, G. P., *Rocket Propulsion Elements, An Introduction to the Engineering of Rockets*, 6th Ed., Wiley, New York, 1992.

Wertz, J. R. (ed.), *Spacecraft Attitude Determination and Control*, D. Reidel, Boston, 1978.

Wertz, J. R., and Larson, W. J., *Space Mission Analysis and Design*, Kluwer Academic, Boston, 1999.

www.ingramcontent.com/pod-product-compliance
Lightning Source LLC
Chambersburg PA
CBHW051536101025
33826CB00004B/4
9781600869129